McGRAW-HILL PUBLICATIONS IN THE
BOTANICAL SCIENCES

EDMUND W. SINNOTT, Consulting Editor

AN INTRODUCTION TO
PALEOBOTANY

McGRAW-HILL PUBLICATIONS IN
THE BOTANICAL SCIENCES

Edmund W. Sinnott, *Consulting Editor*

ARNOLD   An Introduction to Palebotany

CURTIS AND CLARK   An Introduction to Plant Physiology

EAMES   Morphology of the Angiosperms

EAMES   Morphology of Vascular Plants: Lower Groups

EAMES AND MACDANIELS   An Introduction to Plant Anatomy

HAUPT   An Introduction to Botany

HAUPT   Laboratory Manual of Elementary Botany

HAUPT   Plant Morphology

HILL   Economic Botany

HILL, OVERHOLTS, POPP, AND GROVE   Botany

JOHANSEN   Plant Microtechnique

KRAMER   Plant and Soil Water Relationships

KRAMER AND KOZLOWSKI   Physiology of Trees

LILLY AND BARNETT   Physiology of the Fungi

MAHESHWARI   An Introduction to the Embryology of the Angiosperms

MILLER   Plant Physiology

POOL   Flowers and Flowering Plants

SHARP   Fundamentals of Cytology

SINNOTT   Plant Morphogenesis

SINNOTT, DUNN, AND DOBZHANSKY   Principles of Genetics

SINNOTT AND WILSON   Botany: Principles and Problems

SMITH   Cryptogamic Botany

    Vol. I.   Algae and Fungi

    Vol. II.   Bryophytes and Pteridophytes

SMITH   The Fresh-water Algae of the United States

SWINGLE   Textbook of Systematic Botany

WEAVER AND CLEMENTS   Plant Ecology

There are also the related series of McGraw-Hill Publications in the Zoological Sciences, of which E. J. Boell is Consulting Editor, and in the Agricultural Sciences, of which R. A. Brink is Consulting Editor.

FRONTISPIECE.—Silicified log, about five feet in diameter, of *Araucarioxylon arizoni-cum* in the Petrified Forest National Monument of Arizona.

AN INTRODUCTION TO
PALEOBOTANY

By CHESTER A. ARNOLD

Professor of Botany and Geology
and Curator of Paleobotany
University of Michigan

NEW YORK AND LONDON

McGRAW-HILL BOOK COMPANY, INC.

1947

AN INTRODUCTION TO PALEOBOTANY

VIII

02310

PREFACE

The preparation of this book was motivated by a longfelt need for a concise yet fairly comprehensive textbook of paleobotany for use in American colleges and universities. Although separate courses in paleobotany are not offered in many institutions, fossil plants are frequently treated in regular courses in botany and paleontology. In these courses both student and instructor are often compelled to resort to widely scattered publications, which are not always conveniently available. Lack of ready access to sources of information has retarded instruction in paleobotany and has lessened the number of students specializing in this field. Another effect no less serious has been the frequent lack of appreciation by botanists and paleontologists of the importance of fossil plants in biological and geological science.

The two works of reference principally used by British and American students of paleobotany within recent decades have been Seward's "Fossil Plants" and Scott's "Studies in Fossil Botany," the former consisting of four volumes published at intervals between 1898 and 1917, and the latter of two volumes, the last edition of which appeared in 1920 and 1923. Both are now out of print, and although they will continue to occupy a prominent place among the great works in paleobotany, they are already in many respects obsolete. Since the publication of the last edition of Scott's "Studies," many new and important discoveries have been made, which have not only added greatly to our knowledge of fossil plants but which have altered our interpretations of some of them. Many of the newer contributions have resulted from techniques scarcely known to the writers of the first quarter of the present century. These new techniques have also brought about certain shifts of emphasis, which are evident when one compares certain portions of this book with the writings of 30 years ago.

The arrangement and scope of the subject matter is in part the result of 17 years of experience in teaching a small course in paleobotany open to advanced undergraduate and graduate students, most of whom were majors or minors in botany or biology. The approach to the subject is therefore essentially botanical. Paleobotany as a subdivision of paleontology can be treated either biologically or geologically, but the two approaches are so different that to try to combine them would result only in confusion and lack of clarity. The present arrangement, therefore, is followed partly because of the necessity of making a choice, but mostly because of the author's conviction that it is best for instructional purposes. The author is not unaware of the preoccupation with paleobotany of many geologists who might with good reason prefer a presentation following the geologic time scale. Their requirements are met to some extent by the inclusion of the chapter on The Sequence of the Plant World in Geologic Time, in which an effort is made to summarize the floras of the eras and periods. Then, in dealing with some of the plant groups, the most ancient members are described first, thereby giving some idea of the major steps in development from their first appearance down to the present.

In making selections of subject matter an author can hardly avoid being partial to his particular interests to the neglect of other material. In spite of an effort to avoid bias, the ready admission is made that this book is not free from it. In order to keep the volume within practical size limits, much important material had to be rejected arbitrarily. The work of Americans has been heavily drawn upon because there is a wealth of information on fossil plants of the Western Hemisphere that has been only casually, if at all, utilized by European textbook writers and that deserves to be emphasized in a book intended to meet the needs of American students. Wherever possible North American plants are used for description and illustration in place of the more familiar and more often figured Old World forms. This book, therefore, will not serve as a substitute for the larger and more comprehensive works on paleobotany, which advanced students will still find indispensable. For them its main value will be the presentation of new developments and new points of view.

Although American fossil plants have been given more emphasis than heretofore, the author has not neglected to stress discoveries which have come to be looked upon as landmarks in paleobotany. Brongniart's classification of Paleozoic fernlike foliage, Grand'Eury's elucidation of the Paleozoic gymnosperms, the discovery of the seed-ferns by Oliver and Scott, and the labors of Kidston and Lang on the plants of the Rhynie Chert all taken together constitute a major part of the framework of present-day paleobotany, and failure to place due emphasis on them would give a distorted and incomplete view of the whole subject.

Whenever possible new and original illustrations are provided, but when previously published ones are used an effort has been made to select those which have not been previously copied in textbooks. The reader may notice that many of the familiar textbook figures are absent. With only two or three exceptions all the original photographs were made by the author, as were also most of the line drawings and diagrams.

Some of the more difficult drawings were prepared by Eduardo Salgado, Philippine botanical artist employed at the University of Michigan. Most of the photographed specimens are in the collections of the Museum of Paleontology of the same institution.

Because most of the manuscript was written during the course of the Second World War, it was not possible to correspond as freely as was desired with foreign paleobotanists, and consequently the author has been forced to rely upon his own judgment concerning some matters where counsel from colleagues abroad would have been of value. Dr. Ralph W. Chaney very obligingly read and gave constructive criticisms of the chapter on flowering plants. For this the author is indeed grateful. For photographs and figures submitted by others, individual acknowledgments are made where they appear in the text.

Geological names, when applied to North America, conform in most instances to the accepted practice of the United States Geological Survey. For foreign countries the terminology current in each country is used.

CHESTER A. ARNOLD

ANN ARBOR, MICH.,
July, 1947

CONTENTS

CONTENTS

CHAPTER I

INTRODUCTION

Paleobotany is the study of the plant life of the geologic past. It is a part of the more comprehensive science of paleontology, although in general practice the latter is concerned mainly with animal remains, and plants receive only brief consideration. Just as paleontology is outlined by the overlap of the realms of biology and geology, paleobotany rests upon botanical and geological foundations. So inseparable is it from these basic sciences that except for taxonomic terms it has no terminology exclusively its own, but it utilizes those technical appellations that have become standardized for living plants and for the rock formations in which the plants are preserved.

Paleobotany can be approached from the viewpoint of botany, wherein emphasis is placed upon the plant, or from the geological angle in which the rock containing the fossils is the primary concern. Whatever the purpose in studying fossil plants may be, an intimate and thorough knowledge of the principles of both botany and geology is essential for a complete understanding of the subject.

Our knowledge of the plant life of the past has been assembled from the fragmentary fossil remains preserved in stratified rocks. These rocks are superimposed upon one another in series, with the older ones at the bottom and the younger ones toward the top. The layers vary greatly in thickness, composition, and lateral extent. Some lie nearly horizontal and maintain a fairly uniform thickness over large areas. Others may be thick at one place and thin at other places, or they may be localized and disappear altogether when traced for short distances. These differences in vertical and lateral extent are dependent upon the circumstances under which the rocks were formed and upon the amount of erosion to which they have been subjected since their formation. Stratified rocks are built up of sediments that accumulate in the bottoms of seas, lakes, lagoons, swamps, subsiding beaches, and flooded valleys.

The sequence of the stratified rocks is rendered visible to us chiefly through the processes of erosion. Streams everywhere cut deeply into the underlying formations and expose them to view in the walls of ravines and canyons (Figs. 1 and 3) and on the more gently sloping sides of valleys. Waves carve high and often fantastic cliffs along the coast

1

(Fig. 2), and differential erosion frequently produces striking landscape features on the surface of the land. In addition to the work of natural forces, the activities of man have uncovered the strata in otherwise unexposed places by means of drill holes, mine shafts, quarries, and railroad and highway excavations.

The total thickness of sediments that have accumulated since the beginning of decipherable geologic time amounts to several miles, but at no place on the surface of the earth is the entire sequence visible. This

Fig. 1.—View of part of the Grand Canyon of the Colorado River. The formations exposed to view range from the pre-Cambrian to the Permian. This is probably the most striking manifestation of stream erosion of sedimentary strata anywhere in the world.

is because nowhere, with the possible exception of some places in the deepest parts of the ocean, has deposition been going on continuously at any one place without interruption from the beginning. Subsidence of the land below water level, which is ordinarily necessary for the deposition of sediments, is followed after time intervals of varying lengths by elevation above the water, and then erosion may again remove part or all of the deposit. The material thus eroded is redeposited in adjacent areas. The deposition and formation of stratified rocks therefore go on intermittently within any particular area. The stratigraphic sequence therefore within any particular region will often show interruptions. These interruptions, which are caused by cycles of deposition and erosion, are called "unconformities." An unconformity may represent a time lapse of almost any length. Unconformities may be recognized by the difference in the character of the rocks above and below the break, or by means of conspicuous differences in the fossil content. Then there are other factors that may disturb the orderly sequence of rock strata. Some of these are faulting, folding, solution, and intrusions of lava.

Fig. 2.—Turnip Rock, along the shore of Lake Huron, Huron County Michigan. This rock is the result of the under-cutting of Mississippian sandstone by wave action.

The primeval rocks of the surface of the earth were igneous and crystalline, but before the beginning of decipherable time the processes of erosion had been at work for millions of years removing particles of rock from the higher places and depositing them in depressions and hollows. Thus by the time the oldest known life had come into existence, the original surface of the land had been rebuilt to a considerable extent. It is probable that at no place on the earth are the first formed rocks preserved at the surface. Even extremely ancient ones such as those which constitute the Adirondack Mountains or the Canadian Shield show evidence of transportation and deposition of the component materials similar to that which was responsible for the formation of the rocks of later ages.

The major divisions of geologic time (the eras) are based principally upon the life that characterized them as revealed by the fossil content of the rocks.   The oldest rocks that can be satisfactorily dated belong to the long era known as the "pre-Cambrian."   Evidences of life during this time consist mainly of limestones thought to be of algal origin, problematic bacteria, and marks probably representing the trails of primitive

Fig. 3.—Tertiary basalt formation along the Columbia River near Vantage Bridge, Washington.   The superimposed layers are the result of a succession of lava flows.   Lava sheets like these are frequently interbedded with layers of ash which furnish an excellent environment for the preservation of leaf compressions and petrified parts.

wormlike organisms.   Then came the Paleozoic era, with a multitude of marine invertebrates, fishes and amphibians, and spore-bearing plants such as lycopods, scouring rushes and ferns, and naked-seeded plants belonging to the Pteridospermae and the Cordaitales.   It was during the early or middle Paleozoic that land vegetation appeared upon the earth. The Mesozoic was the age of dinosaurs, and its seas were populated with giant squids and ammonites.   It was also an age of gymnosperms, as evidenced by the abundance of cycadophytes and conifers.   Toward the end of the Mesozoic the flowering plants underwent a phenomenal development to become the dominant plant group that they are today. The rapid rise of the flowering plants near the end of the Mesozoic was

GEOLOGIC TIME TABLE AS ADOPTED BY THE UNITED STATES GEOLOGICAL SURVEY

Era	Period (System)	Epoch (Series)
Cenozoic	Quaternary	Recent
		Pleistocene
	Tertiary	Pliocene
		Miocene
		Oligocene
		Eocene
Mesozoic	Cretaceous	Upper
		Lower
	Jurassic	Upper
		Middle
		Lower
	Triassic	Upper
		Middle
		Lower
Paleozoic	Carboniferous	Permian
		Pennsylvanian
		Mississippian
	Devonian	Upper
		Middle
		Lower
	Silurian	
	Ordovician	Upper
		Middle
		Lower
	Cambrian	Saratogan
		Acadian
		Waucoban

Proterozoic—pre-Cambrian

followed by the spread of land mammals, which marked the beginning of the Cenozoic.

The geological approach to the subject of paleobotany is mainly from the standpoint of the correlation of rock formations. Fossils are the markers of geologic time. For purposes of correlation, plant fossils are used either alone or as supplements to animal fossils. Rocks in widely separated localities containing similar floras and faunas are usually assumed to be of approximately the same age. The largest divisions of geologic time (the eras) are based upon the presence in the rocks of major groups of organisms, and the smaller units are demarked in turn by lesser groups. Final subdivisions may rest upon the presence of certain genera or species, or sometimes merely upon the relative abundance of a species or a group of species. Sometimes, however, there are discrepancies in the age as indicated by the fauna and by the flora. If we assume that the organisms are correctly determined, such discrepancies may be due to unexplained environmental factors under which the organisms lived. Those animals most used for correlation are marine invertebrates, which represent an aquatic environment. Plants, with the exception of algae, are mostly terrestrial. The greatest age discrepancies occur in rocks representing periods during which rapid faunal and floral changes were taking place. Sometimes both sets of organisms did not change at the same rate, with the result that an older fauna may overlap a younger flora, or vice versa. Usually, however, major floral changes have occurred first during geological time, to be followed by changes of similar magnitude in the fauna. For example, the Mesozoic flora began to develop in the Permian, and the plant groups that characterize the Cenozoic had become well established in the Upper Cretaceous. Thus in at least two instances during the geologic past, large-scale plant evolution has preceded comparable progress in animal evolution by a time interval approximately equal to one-half of a period.

From the botanical standpoint the aims and the objectives of paleobotany are not essentially different from those of the study of living plants insofar as they pertain to the interpretation of structure, morphology, distribution, phylogeny, and ecology. Paleobotany differs from other branches of botany principally in the techniques employed. Petrifactions require special equipment for sectioning, and various chemical treatments are often used for removing opaque substances or for swelling the tissues in compressions and in lignitized and bituminized remains. Incomplete preservation is usually the limiting factor in the study of fossil plants because the plants are seldom found whole with all parts present and all tissues intact. As a result of the fragmentary condition of most fossil plants, the bulk of the paleobotanical knowledge extant today has been assembled piece by piece from numerous detached and often isolated parts.

The task of classifying fossil plants is not only difficult, but is fraught with numerous chances for error. The paleobotanist must frequently work with plants belonging not only to extinct genera and species, but to extinct families or orders or even classes of the plant kingdom. In general it may be said (although many exceptions must be allowed) that the older a fossil plant is, the wider the categorical rank becomes which will include both it and its nearest living equivalents. For example, most of the Miocene conifers found in North America can be identified as the same as or very close to species that are in existence somewhere on the earth at the present time, but in the late Mesozoic it is usually possible to identify them only to genera. They seem to belong for the most part to extinct species. Then in the Jurassic modern coniferous genera are difficult to recognize with certainty, but the remains can sometimes be assigned to families that have modern representatives. In the late Paleozoic, where the conifers first appear in the fossil record, they cannot be classified with the living conifers into groups smaller than orders. Before this, all known gymnosperms belong to extinct orders and probably to extinct classes. Therefore, since the paleobotanist is often confronted with plants belonging to extinct groups, the foliage, stems, and reproductive organs may appear strange and unlike those exhibited by any living plants with which he is familiar. For a proper understanding of these ancient plants, the investigator must be equipped with a thorough knowledge of a wide range of plant forms and should be well versed in the essential characteristics of the families, orders, classes, and divisions of recent plants. This knowledge may then serve as a basis for the classification of the ancient forms. Classification and naming, however, are not the only objectives in the study of fossils. One also endeavors to learn as much as possible about the plants as organisms —how they grew, how they reproduced, and what their environmental adaptations might have been.

The facts derived from a study of fossil plants are of paramount importance for the bearing they have had on the broader subjects of phylogeny and evolution. It has long been hoped that extinct plants will ultimately reveal some of the stages through which existing groups have passed during the course of their development, but it must be freely admitted that this aspiration has been fulfilled to a very slight extent, even though paleobotanical research has been in progress for more than one hundred years. As yet we have not been able to trace the phylogenetic history of a single group of modern plants from its beginning to the present. The phylogeny of plants is much less understood than that pertaining to some groups of animals, such as the horses and elephants, for example. Two factors account for this (1) the imperfections of the fossil record, and (2) disagreement as to the meaning of various characteristics in a natural system of classification. Furthermore, different parts of plants have not always evolved to an equal extent or at the same rate, and within unrelated series of plants comparable parts have often developed along parallel lines. Whether these phenomena are the results of unknown factors in the environment or of other forces is not clear, but the fact remains that great caution must be exercised in postulating relationships on the basis of a few structural similarities. Some of the outstanding examples of parallel development revealed by fossil plants are the production of seedlike bodies by certain of the Paleozoic lycopods and the development of elaborate floral organs resembling the flowers of angiosperms in the Mesozoic cycadeoids. These examples do not mean that the lycopods are intimately related to any of the gymnosperms or flowering plants or that the cycadeoids were in line with flowering plant evolution. The independent formation of secondary wood in various plant groups is another outstanding example of parallel development. At some time during the past this tissue has appeared in some members at least of practically all the vascular plant groups. Resemblances that are the result of parallel development also extend to habit. The superficial resemblance of the stem and foliage of *Casuarina*, an angiosperm, to *Equisetum* is well known. Conversely, closely related plants may develop along noticeably different lines. The Paleozoic lepidodendrids and sigillarians are examples of lycopods which became large trees quite different from the small and inconspicuous species which make up the lycopod group at the present time.

The main contribution of paleobotany to natural science has been a more adequate picture than could be gained from living plants alone of the relative importance of the different groups constituting the plant kingdom. For example, paleobotanical studies have shown us that one of the smallest of living plant groups, the Equisetales, has a long history, and when regarded in the light of its past, is an assemblage of a number of families. Not only were the Equisetales at one time a large group, but at the height of their development there existed contemporaneously other closely allied groups that have been extinct for millions of years. The fossil records of the genus *Ginkgo* reveal a similar history. *Ginkgo biloba* is the lone survivor of an order of gymnosperms that contained a dozen or more genera during Mesozoic times. The lycopods and ferns have similar records. In addition to showing the true status of Recent plant groups in the past, paleobotanical research has brought to light the existence during the past of others which are entirely extinct and have left no near relatives, examples being the Psilophytales and the "seed-ferns," both from the Paleozoic. The influence of a knowledge of extinct plants on our concept of relationships between living types has been profound.

During the past many plant groups arose to a place of prominence, only to decline and have their place taken by others. Plant evolution, therefore, has not always been progressive, but has involved innumerable changes and constant experimentation with new types. The fossil record is replete with examples of the development of elaborate vegetative and reproductive structures in plant groups of which the recent members are considered primitive. Examples are the seedlike organs of some of the ancient lycopods, the complex strobili of *Cheirostrobus*, and the "flowers" of the Mesozoic cycadeoids. These structures all represent extreme specialization along particular lines within restricted groups. However, this was accompanied by corresponding loss of plasticity, and when the plants were confronted with far-reaching environmental changes they were unable to modify themselves sufficiently to cope with the new conditions. In plants, as with many animals, the highest state of development was often attained just prior to complete or near extinction.

A fact that has been well emphasized by paleobotanical studies is that the increase in complexity of plants seen in the Thallophyta, Bryophyta, Pteridophyta, and Spermatophyta series, and which was once the central theme of morphological thought, does not depict a phylogenetic sequence. It is still tenable that the Thallophyta were probably the starting point for all higher plant evolution, but it is more than likely that several of the higher types have evolved from this group independently of each other. There is no reason whatsoever to suppose that the so-called Pteridophyta arose as one complex from the Bryophyta, or that seed plants evolved as one type from the pteridophytes. The terms Pteridophyta and Spermatophyta do not embrace natural, or in all cases, even closely related groups, but they include several assemblages having independent origins. For example, the fossil record shows the lycopods and scouring rushes to be independent lines as far into the past as they can be traced. The same is true with respect to either of these types and the ferns. Moreover, the seed plants can be traced to a period almost as remote as the ferns. When, and from what earlier forms seed plants originated, we can only guess, although it is quite reasonable to assume that their ancestors were spore producers of some kind, maybe ancient fernlike plants, or possibly even the more primitive members of the very ancient Psilophytales. However, it is not advisable to assume that all seed plants arose from the same source. There is no proof that the conifers and cycads had a common origin, and the problem of flowering plant origin is the most enigmatic of all.

CLASSIFICATION OF PLANTS

The reason for classifying plants is to facilitate the arrangement of the different kinds into an orderly sequence so that their relationships to one another can be better understood. The older systems were based entirely upon living plants and whatever information was available concerning fossil species was utilized little or not at all. Before 1900, however, it became apparent that fossil plants revealed many facts concerning the status of some groups that are not expressed by living plants, and since that date the influence of paleobotany upon classification has become increasingly apparent.

There is no perfect and final classification of plants. Several systems are in use at the present time, and each has its stanch advocates. Those which are in use and which seem quite satisfactory today may be obsolete tomorrow, when they will be replaced by others more in accord with advancing knowledge, and they will be subject to change as long as a single fact concerning the morphology, cytology, structure, or life history of any species remains unknown.   The system followed in this book is believed to be the one that best expresses the relationships of extinct plants, as they are now understood, to the plant kingdom as a whole.

The plant kingdom can be logically and conveniently divided into two groups, or subkingdoms, which, without the necessity of using technical terms, may be termed the *nonvascular* and the *vascular* plants. The nonvascular plants, which lack highly developed food and water-conducting tissues, consist of two divisions, (1) the *Thallophyta*, which in turn is divisible into the *Algae* and the *Fungi*, and (2) the *Bryophyta*, which consists of the *Hepaticae* (liverworts) and the *Musci* (mosses). The study of fossil plants has brought little to bear upon the structure and morphology of the nonvascular plants, so this simple and conventional scheme based upon the living members is wholly acceptable from the paleobotanical standpoint.   It is upon the vascular plants that paleobotanical discoveries have had the most important bearings.

The body of the vascular plant has a highly organized food-and water-conducting structure called the "stele."   It contains the xylem and the phloem.   The classification scheme that has probably been used by more botanists than any other divides the vascular plants into two divisions, the *Pteridophyta* (spore bearers), and the *Spermatophyta,* (seed plants). This was the outgrowth of an older system of *Cryptogamia* and *Phanerogamia*.   The former embraced all nonvascular plants and those vascular plants which reproduce directly by spores.   The Phanerogamia are equivalent to the Spermatophyta.

The Divisions.—Under the system of Pteridophyta and Spermatophyta the main categories are as follows:

Division Pteridophyta—ferns and "fern allies"
    Class Lycopodineae—lycopods
    Class Equisetineae—scouring rushes
    Class Filicineae—ferns
Division Spermatophyta—seed plants
    Class Gymnospermae—gymnosperms
    Class Angiospermae—flowering plants

This system was founded entirely upon living forms, but with the increase in knowledge of fossil plants combined with a better understanding of the fundamental structure of the plant body, this scheme has been found to possess at least two basic weaknesses.   In the first place it assumes that seed production alone is sufficient to set one group apart from another regardless of certain anatomical similarities that may exist between them. However, we know from the evidence furnished by certain fossil forms that seeds have been developed in different groups during the past, and that, when considered by itself, seed formation is not a final criterion of affinity. The second weakness is that it ignores anatomical evidences, factors which we have every reason to believe are as basic in classification as reproductive structures. From a wider knowledge of plants, both living and extinct, it now appears that a more fundamental basis for classification consists of three characters as follows: (1) nature and relation of leaf and stem, (2) vascular anatomy, and (3) position of the sporangia. On the basis of these the vascular plants may be divided into four divisions as follows:

Division Psilopsida—the Psilotales (living) and the Psilophytales (extinct). The Psilotales were usually included within the Lycopodineae in the older classification, and the Psilophytales were usually ignored because they have no living members. Sometimes they were added as a separate class, the Psilophytineae. The Psilopsida often lack leaves, and when they are present a vascular supply is absent. The sporangia are terminal on the main stem or a side branch and consist essentially of an enlargement of the stem tip containing sporogenous tissue. The assumed affinity between the Psilophytales and the Psilotales is entirely on a structural basis. They have no connection in the fossil record.

Division Lycopsida—the lycopods (Lycopodineae of the older classification). Leaves small, simple, spirally arranged and provided with a vascular system. Sporangia adaxial, sessile, and borne on or in close proximity with leaves (sporophylls). No leaf gaps are present in the primary vascular cylinder, and the exarch condition prevails.

Division Sphenopsida (also called Articulatales, Articulatineae, Arthophyta, etc.)—the scouring rushes and relatives (Equisetineae of the older classification). The stems are jointed and bear whorls of leaves at the nodes. The leaves are attached by a narrow base or are coalescent, but sometimes broadened distally. The sporangia are on specialized stalks called "sporangiophores," usually peltate or recurved.

Division Pteropsida—ferns, seed-ferns, gymnosperms, and flowering plants (Filicineae and Spermatophyta of the older classification). The members are predominantly megaphyllous and leaf gaps are present in the primary vascular cylinder of all except protostelic forms and a few of the very ancient gymnosperms. The sporangia (wherever the position is clearly determinable) are abaxial on

modified or unmodified leaves.   The mesarch and endarch conditions prevail in the primary vascular system.

OUTLINE OF THE CLASSIFICATION OF VASCULAR PLANTS*

Pteropsida	Angiospermae.......	Monocotyledoneae	Spermatophyta
		Dicotyledoneae	
	Gymnospermae......	Gnetales	
		Coniferales	
		Cordaitales	
		Ginkgoales	
		Cycadeoidales	
		Cycadales	
		Pteridospermae	
	Filicineae...........	Filicales	Pteridophyta
		Marattiales	
		Ophioglossales	
		Coenopteridales	
Sphenopsida—Articulatae........		Equisetales	
		Calamitales	
		Sphenophyllales	
		Pseudoborniales	
		Hyeniales	
Lycopsida....................		Isoetales	
		Pleuromeiales	
		Lepidodendrales	
		Selaginellales	
		Lycopodiales	
Psilopsida....................		Psilotales	
		Psilophytales	

* Simplified and slightly modified from that given in A.J. Eames, "Morphology of Vascular Plants,
Lower Groups," New York, 1936.

THE NAMES OF FOSSIL PLANTS

Fossil plants are named according to the same set of rules that governs
the naming of living plants, but the names usually refer to parts rather
than complete organisms.   Since the generic name may apply to only a
leaf, a seed, or some other part, the genus is called a *form genus* or *artificial genus*, as contrasted with a *natural genus*.   It follows that the whole
plant may be known only as the sum of its separately named parts, and
paleobotanical nomenclature is therefore complex.   An excellent example
of the involvement resulting from the use of separate names for the parts
is furnished by the familiar and well-known arborescent lycopod *Lepidodendron*.   This name, which is now understood as being applicable to
the whole plant, was first given to a certain type of trunk which bore
leaves described independently as *Lepidophyllum*.   The cones that it
bore are known as *Lepidostrobus*, the spores as *Triletes*, and the rootlike
organs as *Stigmaria*.   One may be inclined to question the purpose of

retaining names for the separate organs when we know to which plants they belong, but two considerations are involved. The first is that in most instances where the complete fossil plant has been reconstructed by the gradual assembling of the parts, these parts had previously been named, and these names are retained as a matter of convenience for descriptive purposes. The second is that when these parts were first named, the designations were often given in ignorance or at least in doubt of the affinities of the parts. There are scores of generic names in existence which apply to parts that have never been found attached to anything else. Also a generic name may include parts which appear alike but which represent a variety of closely related parts, as for example *Stigmaria*, which belongs not only to *Lepidodendron* but also to *Lepidophloios*, *Bothrodendron*, and *Sigillaria*. So unless the root is found attached to the stem (which it rarely is) we do not know to which trunk genus it may belong. To many botanists the multiplicity of names handled by the paleobotanist may seem irksome or even ludicrous, but no system that is more satisfactory has ever been put into practice.

CHAPTER II

HOW PLANTS BECOME FOSSILS

When speaking of fossils, one ordinarily refers to the remains of organisms that lived long ago. Upon death the body immediately starts to disintegrate. If unimpeded, disintegration will go on until the whole body is broken up into simple chemical compounds, which leach away leaving no evidence of its former existence, but under certain circumstances the work of the agents of destruction becomes halted and parts of the body may remain in visible form. These remains we sometimes call fossils. Fossils are lifeless and are incapable of responding to those stimuli which influence normal living creatures. They are relics of the past and no longer participate directly in the affairs of the contemporary world. In a given instance a fossil may be a bone of some animal belonging to an extinct race that has lain buried for centuries in the accumulated debris within a cave, or it may be a shell dislodged by the waves from its resting place in the hard rocks along the shore. Or it may be the imprint of a leaf on a slab of fine-grained sandstone, or a fragment of charred wood in a lump of coal.

It is difficult to give a satisfactory definition of a fossil. The word is derived from the Latin verb *fodere*, which means "to dig"; so originally it referred to anything one might remove from the earth. In modern language its usage is restricted to organic remains taken from the earth, and excludes objects strictly inorganic in origin or objects fashioned by man. On the other hand, the restriction does not extend to the remains of man himself if they represent prehistoric man. Like many definitions, a certain element of arbitrariness must be included to give the word meaning, but as ordinarily understood a fossil is the remains of a whole organism, some part of an organism, or the direct evidence of the former existence of some organism preserved in the consolidated sediments of the earth. In addition, therefore, to actual parts, anything resulting from or indicating the former presence of organisms, such as tracks, trails, borings, teeth marks, coprolites, and in some cases even chemical precipitates are regarded as fossils. Chemical fossils include such things as sulphur and iron deposits that show evidence of former bacterial activity and limestones that have resulted from the deposition of calcium carbonate by algae. The actual remains of the causal organism may or may

14

not be present. Petroleum, on the other hand, has never been relegated to the fossil category although it is unquestionably of organic origin.

Since a fossil can be defined only in an arbitrary way, there are certain limitations that must be imposed in addition to those mentioned. Obviously, the body of an organism does not become a fossil the moment life departs (although the changes that take place at death are greater than those that come about later) and it may not necessarily become one after the lapse of a thousand or more years. Human bodies found in ancient tombs are not classified as fossils even though the bones of other contemporary human beings found in swamp or flood-plain deposits or windblown earth might be so regarded if they showed evidence of belonging to some prehistoric or uncivilized race. In Pleistocene or post-Pleistocene peat beds there are abundant plant and animal remains that are very ancient in terms of human history but which represent for the most part species still living. These do not differ in any essential respects from the similar remains that become annually entombed in the mud in the bottom of Recent swamps and lakes, although they may represent genera and species that no longer thrive in the immediate vicinities of these swamps and lakes at the present time. In other words, they represent a flora and fauna of the past which no longer exist at that particular place. Because these remains do not represent the contemporary flora and fauna, one may, if he so wishes, call them fossils, but the preserving medium, though compacted, is not consolidated or indurated, and the enclosed remains represent only the preliminary stages in fossilization. Remains held within comparatively recent deposits are sometimes called "subfossils," of which the classic example is the body of the frozen mammoth found several years ago in Siberia. This creature represented an extinct species even though its flesh had remained in a perfect state of preservation about 2,000 years.

Plant fossils are usually preserved in rocks composed of sediments deposited in water. Rocks of this category are the sedimentary or stratified rocks mentioned in the preceding chapter. They are distinct from igneous rocks which form when lava cools and from metamorphic rocks, in which the original structure of the components has been altered by heat or pressure. Fossils are formed from those organisms or fragments of organisms which became entombed in the accumulating sediments before they had a chance to disintegrate completely. It is very seldom that a plant is preserved all in one piece with its several parts attached and all tissues intact. More or less dismemberment invariably takes place. Leaves, seeds, and fruits become detached from the twigs on which they grew and the stems break loose from the roots. The softer tissues decay. The study of plant fossils is therefore mainly one of unconnected parts.

some of which are usually not fully preserved. Often the parts of many different kinds of plants are mixed together indiscriminately, and the main problem confronting the investigator is the proper sorting and classification of these.

Conditions Favoring Preservation of Fossil Plants

Success in interpreting fossil plants depends to a great extent upon the degree of preservation. Perfect preservation can hardly be said to exist. Only in extremely rare instances are any of the protoplasmic contents of the cells retained, and studies must be directed mainly to the grosser features of the plant. But complete tissue preservation is rare even in small parts and the extent to which the tissues are retained is governed by a number of interrelated factors, some of which will be discussed.

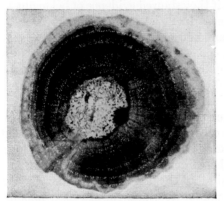

Fig. 4.—Small silicified oak (*Quercus*) twig. From the Tertiary of Oregon. Enlarged.

The chief factors governing the extent of tissue preservation in fossil plants are (1) the kinds of tissues composing the plants, and (2) the conditions to which they were subjected preceding and during fossilization. Most vascular plants and a few of the nonvascular ones are composed of a variety of tissues which are unequally resistant to destruction, and under given conditions the extent of preservation is directly proportionate to the resistance of the tissues.

Plants containing large amounts of hard tissues, such as bark, wood, sclerenchyma strands, and extensive cutinized layers are usually preserved with less destruction than flowers, thin delicate leaves, or juicy fruits. These tissues are not only more resistant to mechanical destruction but are destroyed less rapidly by microorganisms. Wood is often preserved as petrifactions (Figs. 4, 8, and 9) or lignitic compressions. Seeds, with hard layers in the integument such as are found in the cycadophytes, are often preserved as compressions (Fig. 5B), casts (Fig. 5A), or petrifactions. The cutinized exines of spores and pollen grains (Figs. 7A and B) and fragments of leaf epidermis (Fig. 7D), are sometimes found in a state of preservation showing little alteration in rocks at least 300 million years old. In fact, there is no limit to the length of time cutinized parts may retain their identity. Suberized bark layers are also extremely resistant to destructive forces. In stems and roots it is only rarely that the delicatedly constructed phloem and middle cortical tissues are preserved even though the most minute structural details may remain in the xylem or the periderm.   An excellent example of selective

(A)

(B)

Fig. 5.—(A) *Trigonocarpus triloculare.* Seed cast. From the Pottsville group near Youngstown, Ohio.   ¾ natural size. (B) *T. ampullaeforme.* Seed compression. From the Pottsville of West Virginia. Enlarged.

preservation is found in the petrified trunks of the Paleozoic arborescent lycopod *Lepidodendron*.   In this plant the trunks, which are sometimes 2 feet in diameter, are bounded by a thick hard bark layer which often remains after all other tissues have decayed.   This layer encloses delicate, thin-walled tissue, which broke down almost immediately after the death of the plant. In the center is the relatively small xylem cylinder, which in some species possesses a pith. This is often retained. In the most perfectly preserved stems all the tissues are present, but usually some of them have suffered partial or complete disintegration to an extent governed by the intensity of the forces of destruction.

The disintegration of plant tissues proceeds according to a rather definite selective order of decomposition. First to disappear are the

(A)   (B)

Fig. 6.—*Betula lacustris.* (a) Nearly complete leaf from which the parenchymatous mesophyll tissue has decayed leaving the venation system nearly intact. About natural size. (B) Portion of the same leaf shown in 6(A) enlarged to show detail of the vascular skeleton. From the Trout Creek diatomite beds. Miocene. Harney County, Oregon.

protoplasmic contents of the cells. Rarely are these preserved, although there are several instances on record of preserved nuclei, and mitotic figures have been reported within a germinated megaspore of a Carboniferous lycopod. Those tissues consisting of thin-walled cells usually disappear next. These are followed in order by wood and bast fiber cells, suberized cork cells, and cutinized spores and epidermal fragments. The indestructability of the epidermis of some plants is shown by a layer of Lower Carboniferous "paper coal" in Russia some 2 feet in thickness, which is built up almost exclusively of the foillike cuticles of certain lycopods. All of the tissues of the interior of the plants had decayed.

The second factor influencing the extent of tissue preservation is the intensity and duration of the destructive forces to which the plant is subjected before it becomes covered to a sufficient depth to prevent decay, and the extent of destruction will range from partial to complete. The

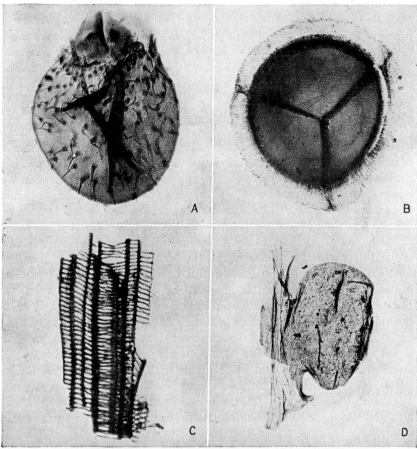

Fig. 7.—Resistant plant structures isolated from macerated bituminous coal. (*A*) Cutinized megaspore exine with slender spinelike appendages (*Lagenicula* type). × about 40. (*B*) Cutinized megaspore exine with prominent tetrad scar and an equatorial band of fringelike appendages (*Stichostelium* type). × about 40. (*C*) Group of scalariform tracheids probably belonging to a fern or lycopod. (*D*) Cutinized epidermis of a fernlike pinnule (*Sphenopteris*).

disintegrative forces are of two kinds, (1) decay and hydrolysis, and (2) the mechanical action of water, wind, scouring sand, and rolling stones.

The ideal environment for the preservation of plant material is an enclosed or protected body of water such as a small lake, embayment, or swamp in which only fine-grained sediments are accumulating and with sufficient rapidity to cause quick burial. A low oxygen content and a relatively high concentration of toxic substances also retard decay. Some protection from high winds and an absence of strong currents is essential, because intensive wave action or rolling boulders will soon reduce even the most resistant plant tissues to pulp. Plants may be transported by wind or water to a place of deposition or they may be preserved *in situ*. An outstanding example of *in situ* preservation is furnished by the plants in the Rhynie chert bed, of Middle Devonian age in Scotland, where numerous small, rushlike plants are preserved upright where they grew. In an environment of this kind mechanical disintegration is usually reduced to a minimum and comparatively little fragmentation of the plants takes place.

The depth of water into which plant material sinks is important from the standpoint of both decay and mechanical destruction. The activities of saprophytic fungi are greatly reduced at a depth of a few feet, and logs submerged in deep water may be preserved indefinitely even when not covered with silt or sand. Moreover, deep water affords good protection from wave action, and the sediments that accumulate on the bottom are usually fine textured.

Although plant tissues may escape the ravages of extensive weathering and decay-producing organisms, they may still be seriously damaged by crushing from the weight of the overlying sediments. Previous infiltration of mineral matter into the cell cavities tends to lessen the effects of crushing by affording internal support to the cells, and where the cell cavities have become completely filled, the structure may suffer little or not at all. Unlike the effects of decay and abrasion, crushing is not always greatest in those tissues made up of thin-walled cells. In stems of *Medullosa*, for example, found in coal-balls, layers of thin-walled periderm like tissue outside the stele are often quite uncrushed although the stem as a whole is considerably flattened. In thin sections of coal similar phenomena have been observed, wherein woody constituents are completely crushed, but associated tissues made up of thin-walled cells are less collapsed. The only explanation for this anomalous behavior is that the thin walls are more permeable to mineral matter in solution. The effect of mineral infiltration is to add strength to the tissues, whereas in the less permeable thicker walled cells the cavities remain empty, rendering them more susceptible to collapse.

The Plant-bearing Rocks

Most plant fossils are preserved in sedimentary rocks of fresh or brackish water origin, although on rare occasions they are found in marine beds. Fresh-water beds are formed above sea level and are of variable composition, ranging from conglomerates through sandstones to silt-stones, shales, and coals. They may be thinly laminated or massive, and cross-bedding is frequent. The laminations are the result of variations in the texture of the sediments and often reflect seasonal changes such as periodic variations in rainfall or annual floods due to melting snow. The bedding plane surfaces are often strewn with waterworn pebbles or other objects coarser than the matrix. Only the finer textured sediments ordinarily contain well-preserved plant fossils. Conglomerates are usually devoid of plant remains except for occasional fragments of macerated and carbonized wood. The finer grained sandstones may

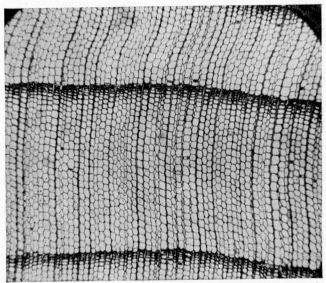

Fig. 8.—*Sequoioxylon sp.* Secondary wood showing growth rings, from the Tertiary near Sweet Home, Oregon. × about 20.

reveal excellent leaf and stem compressions, and they often contain well-preserved casts. Coarser sandstones, however, intergrade with con-glomerates, and seldom produce recognizable fossils except such objects as the bark imprints of the arborescent lycopods and casts mostly devoid of surface markings. The best casts and compressions are found in shales, or shaly or silty sandstones in which the particles are less than 0.25 mm. in diameter.

Marine deposits differ from those of continental origin generally by their finer texture and more even and regular laminations. They may consist essentially of the same material that make up fresh-water beds, such as clay, silt, volcanic dust, and organic matter, but in addition they often contain more lime and (unless formed near the shore) lack coarse sand and pebbles.   The most definite criterion of marine origin, however, is the presence of the remains of typical marine organisms such as radiolaria, foraminifera, coelenterates, bryozoa, brachiopods, or trilobites. Occasionally fresh-water invertebrates may be carried into salt water, but they are seldom present in quantity or over large areas.   Conversely, marine organisms never find their way into fresh-water deposits except where the eroded material of a previously formed and uplifted marine bed has been redeposited.   Conglomerates often contain material of this kind.   Marine limestones formed not far from the shore sometimes contain petrified wood and other plant parts, examples being the Genundewa limestone of New York and the Squaw Bay limestone of Michigan. In both of these beds wood fragments occur with the pteropod *Styliolina*, and in the latter with numerous brachiopods and cephalopods in addition.

The source of the sediments making up sandstones and shales is usually the soil and weathered rocks of uplands and plateaus through which streams cut canyons and ravines.   The eroded material either is deposited in the bed of the stream at its lowest level or is carried out beyond its mouth.   If the stream discharges its load into a small lake or bay, ultimate filling may result.   An example of such a filling and the consequent preservation of large quantities of plants is the Latah formation of Miocene age, near Spokane, Washington.   The consolidated muds contain numerous leaves, seeds, conifer twigs, and other plant parts well preserved in yellow clay.

If the body of water receiving the sediments is large, the entire bottom may not become covered and deltas will form at the mouths of the entering streams.   An example of this kind of deposit formed during the Devonian period is the great Catskill delta of eastern New York and Pennsylvania.   During the middle and latter part of the Paleozoic, a large land mass existed to the eastward of the present Atlantic shore line of Eastern North America, and as it eroded away, the sands and muds were spread over the sea bottom for a great distance to the west.   By the end of Devonian time the delta extended into western Pennsylvania, and these sediments constitute most of the Devonian rocks that are so abundantly exposed in northern and eastern Pennsylvania and adjacent New York.   This delta was a low lying terrain, which supported considerable vegetation.   The plants consisted of the fern *Archaeopteris*, primitive lycopods of the *Archaeosigillaria* type, and large forest trees such as *Aneurophyton* and *Callixylon*.

A similar deposition of sediments took place during the early Cenozoic when the Mississippian embayment extended northward up the Mississippi Valley as far as Tennessee.   As the river-borne sediments were deposited, the embayment was gradually filled, and in the sands and silts are preserved countless numbers of leaves, seed pods, logs, and other parts of the plants that grew along the swampy shores. This flora, known as the "Wilcox flora," is one of the largest fossil floras ever to be investigated.

The plant-bearing sandstones and shales of the great Carboniferous coal series of Eastern and Central North America consist of sands and muds which were washed into the swamps between periods of accumulation of vegetable matter. Beneath many of the coal seams is a layer of earthy clay known variously as the underclay, fire clay, or *Stigmaria* clay. It often contains an abundance of *Stigmaria*, the rootlike organ of several of the Paleozoic arborescent lycopods. The massive nonlaminated character of the underclay indicates that it represents the old soil in which the coal swamp plants grew, and the *Stigmariae* are the undisturbed roots of some of these trees. The underclay seldom contains any other plant fossils.

Plant remains are preserved in large quantities in the ancient volcanic ash beds that extend over large parts of Western North America (Fig. 11). In age these ash deposits range from Triassic to Pliocene, or even later, but most of them belong to the Tertiary. They cover portions of several states and in places are thousands of feet thick. They are frequently interbedded with sheets of lava of which basalt and rhyolite are the common forms throughout the Great Basin and Columbia and Snake River Plateaus (Fig. 3). Basalt solidifies in the form of large six-sided columns which stand perpendicular to the horizontal plane, and being more resistant to erosion than the ash, it produces the characteristic "rimrock" of the regions where it occurs.

In many places throughout Western North America whole forests lie buried beneath thick layers of ash. Countless trunks have become silicified, most of them after falling but some while still standing where they grew. The so-called "petrified forests" of the Yellowstone National Park contain many standing trunks, as do similar forests at other places. Probably the largest fossil stump on record stands at the summit of the Gallatin Range in the northwestern corner of the park. It is said to be a *Sequoia*, and measures 19 feet in diameter. Another stump nearly as large may be seen on a privately owned reserve near Florissant, Colorado. In the sparsely inhabited desert of the northern part of Washoe County, Nevada, huge trunks, some of them as much as 15 feet in diameter, are exposed in a hillside (Fig. 17), and judging from their perpendicular sides, the visible parts represent not the enlarged bases of the trees but the straight trunks which were broken many feet above the original ground surface.

The fine displays of upright silicified trunks on Specimen Ridge and

Amethyst Mountain along the Lamar River in the northeastern part of Yellowstone National Park consist of hardwoods in addition to coniferous types. The largest tree is *Sequoia magnifica*, believed to be closely allied to, if not identical with, the living redwood, *S. sempervirens*. Other conifers consist of two species of *Pinus* and one of *Cupressinoxylon*. The hardwoods include oaks, buckthorns, sycamores, and lauraceous genera. On the slopes of Amethyst Mountain 15 successive forests are exposed, one above the other, and each is separated from the one next above or below by a few inches or feet of ash.

Fig. 9.—Silicified log of *Pseudotsuga*. From the Tertiary volcanic ash deposit near Vantage Bridge, Washington.

In most places where erosion has uncovered the remnants of ancient forests the trees lie prostrate, having either been blown over by storms or floated in as driftwood. Sometimes many square miles of territory are strewn with these logs as in the famous Petrified Forest of Triassic age in Arizona. At this place the logs were not preserved exactly where they grew, but were transported from some unknown distance. Near Calistoga, in California north of San Franciso, hugh prostrate redwood logs are preserved in a thick bed of Pliocene ash. Living redwoods grow not far from this locality. In central Washington near Vantage Bridge, in the so-called "Ginkgo Petrified Forest," numerous logs representing ginkgos, conifers, and hardwoods have been exposed in the steep hillsides (Fig. 9). Under one log in a near-by locality a cache of fossil hickory nuts was recently unearthed, having originally been hidden there by some animal. Countless silicified logs representing for the most part several genera of hardwoods can be seen in the White River ash beds of Oligocene age in the Shirley Basin of central Wyoming. These logs, as well as several others mentioned, have never been thoroughly investigated, so we know relatively little of the kinds of trees represented by these striking manifestations of past ages. Occurring with these buried logs, or in nearby places, leaf remains are often found, and it is from these that the bulk of our knowledge of these ancient forests has been secured.

At some places, as for example at Trout Creek in southeastern Oregon, large quantities of plant remains are preserved in diatomaceous earth. During Miocene time, leaves, winged fruits, and seeds were carried from the surrounding hills by the wind in countless numbers, and as they settled to the bottom of an ancient lake were covered with diatomaceous ooze. At this place the diatomite accumulated to a depth of about 100 feet, and it was then covered by a tremendous fall of ash and a thick sheet of lava. The weight of the ash and lava compressed the diatomite, which, being of a fine-textured chalklike consistency, preserved the finest details of the venation of many of the leaves (Fig. 6).

Volcanic rocks resulting from the cooling of molten lava seldom contain recognizable organic remains, although a few exceptions of relatively little importance are known. In the lava beds of the Craters of the Moon National Monument in southern Idaho, the attention of the visitor is directed to cavities in the lava formed by the decay of tree trunks, which were partly enclosed while the lava was in a plastic state. For some unexplained reason, the trunks were not completely burned at the time. The inner surface of some of these cavities, or molds, show imprints of the outer bark surface. It has not been possible to identify any of these trees, but they were probably nonresinous species such as walnuts, oaks, or other hardwoods.

Waters discharged from hot springs sometimes provide sufficient minerals for the preservation of large quantities of plant remains. An outstanding example of preservation by this method is the Rhynie chert bed near Aberdeen, in Scotland, previously mentioned, (Fig. 29). During Middle Devonian time in this particular locality there was a small valley surrounded by pre-Cambrian mountains. The soil in this valley was saturated with water supplied by hot springs. A dense growth of rush-like plants covered the wet ground, but toxic conditions resulting from lack of aeration of the soil prevented to a great extent the decay of the vegetable debris with the consequent formation of peat. The magmatic waters which seeped into the peat were highly charged with silicates which accumulated as a result of evaporation. When the saturation point was reached, the silicates became altered into a cherty matrix that cemented the plant material together into a solidified mass. The plants

belong to various members of the Thallophyta and to an ancient group
of vascular plants known as the Psilophytales.   Many of the stems were
preserved in an upright position attached to horizontal underground rhizomes and with spore cases at their tips.

Coal seldom reveals well-preserved plant remains to the unaided eye,
but spores (Figs. 7A and B), epidermal fragments (Fig. 7D), and fragments of charcoal showing remnants of vascular elements (Fig. 7C) can
often be isolated by maceration.

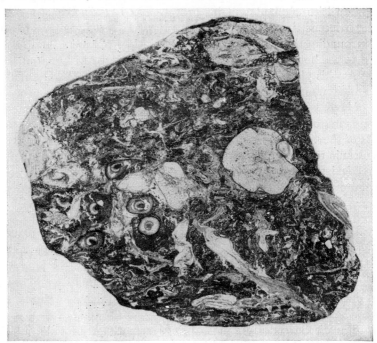

FIG. 10.—Section of a small coal-ball from the McLeansboro formation of Illinois.
The cut surface was etched for about 15 seconds in dilute hydrochloric acid to increase the
visibility of the plant structures.   Natural size.

Important sources of plant fossils in coal are *coal-balls*.   These are
irregular or subspherical masses of mineral matter, usually of calcium or
magnesium carbonate and iron pyrites, which vary in size from a fraction
of an inch to 2 or 3 feet, and in weight from an ounce to two or more tons.
Most of them, however, weigh but a few pounds.   Coal-balls are generally
of localized occurrence.   They may be congregated in definite bands or
in irregularly scattered pockets.   They have been known in Europe for
many years but were not discovered in North America until 1922.   In
North America they were originally discovered in Illinois, but they have
subsequently been found in Indiana, Iowa, Kansas, and probably else-

where, although they are unknown in the coal fields of the Appalachian region or in western Pennsylvania or Ohio. They occur mostly in rocks of middle Pennsylvanian (Conemaugh) age although a few are older. In Europe they are found mostly in the Westphalian and Yorkian, and consequently are older than those found in North America.

The conditions under which coal-balls formed appear to vary somewhat, but in Great Britain and Europe they exist only in coal seams overlain by marine limestones. It is thought that the calcium and magnesium carbonates making up the mineral content of the balls were derived from sea water, which covered the peat deposits soon after their formation but before they became sufficiently compacted to flatten the plant parts completely. However, in Illinois the coal-balls are not overlain by marine beds and the petrifying minerals must have come from some other source. These coal-balls also contain considerable pyrites and some are composed almost entirely of this mineral. It is believed that here the free carbon liberated during partial disintegration of the plant substances reduced sulphates present in the water, with the contemporaneous deposition of calcium, magnesium, and iron carbonates which formed the coal-balls.

Many coal-balls are barren of recognizable plant remains but others contain fragments of stems, roots, petioles, foliage, seeds, sporangia, and liberated spores (Fig. 10). The parts are often partially crushed and the outermost tissues have generally disappeared. Portions of stems frequently extend completely through the coal-balls and for some distance into the surrounding coal where they, too, have been altered into coal.

THE FOSSILIZATION PROCESS

As previously explained, most plant fossils have lost some of their tissues as a result of decay or abrasion before burial. The final form of the fossil, however, is determined to a considerable extent by conditions prevailing after submergence.

As soon as wind- or stream-borne plant material is brought into quiet water and becomes saturated, it begins to sink, and the individual fragments, if undisturbed, will usually repose themselves in a position nearly parallel to the horizontal or slightly sloping bottom. If mud or sand is also present, it will settle with the plants and separate them to an extent depending upon the relative abundance of the two components. Abundant sediments will cause rapid accumulation and the plants will be well separated from one another. In rocks thus formed, it is usually possible to collect better specimens than where the sediments settled so slowly that there is little or no intervening matrix between the leaves. If the water contained but little sediment, or if the plant material was brought in in large quantities, it is often impossible to recognize the outlines of the individual leaves or to separate them from one another (Fig. 11). In some of the Eocene beds along Puget Sound, leaves are present in such quantities that they tend to obscure one another, and it is extremely difficult to secure a representative collection of the flora.

The organs produced in greatest numbers by most plants are leaves, and consequently they are most abundantly represented in the fossil record. Our knowledge of Upper Cretaceous and Tertiary floras is based largely upon leaves. The surface of a leaf is often slightly curved and

Fig. 11.—Block of light gray volcanic ash bearing numerous compressions of dicotyledonous leaves.   From the Upper Eocene of Lane County, Oregon.   Reduced.

it will, unless disturbed, come to rest on the bottom with the convex surface uppermost. The weight of the accumulating sediments will flatten it somewhat, but it will seldom lie perfectly flat. The overlapping subdivisions of fronds such as those borne by pteridosperms, ferns, or cycads may be pressed flat or they may remain turned slightly from the plane of the rachis with thin films of sediments separating the overlapping edges. As the sediments increase in thickness, compaction results, and the less resistant and more compressible plant part is flattened to a mere fraction of its original thickness.   If the plant part is cylindrical in cross section and made up principally of such hard resistant tissues as compose trunks and branches of trees, the weight of the overlying sediments will produce a thick lens-shaped object (Fig. 12). The curvature becomes less with increased pressure. The first internal manifestation of change in shape of the object will be the collapse of the thin-walled cells, followed by those with thicker and harder walls. Prolonged pressure will bring about further flattening and ultimately a thin film of black shiny coallike material remains, which is only slightly thicker at the mid-region than at the margins.

Experiments have shown that fine muds on being compressed into shale may be reduced one-third in thickness, whereas the reduction in thickness of an enclosed plant part may amount to as much as 95 per cent. The compaction results in a reduction of pore space within the sediments and the displaced water is forced upward into the less compacted layers. Some organic compounds are also expelled from the cells, and these may escape in the form of marsh gas or humic substances. The latter may impart a dark color to the sediments.

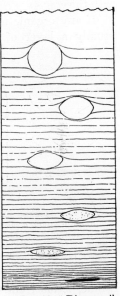

If the compressed plant part is a thin leaf of small vertical dimension, there will be less alteration in shape as a result of flattening than if the part possesses considerable thickness. In either case, being firmly encased along both sides, as well as above and below, it is impossible for the plant part to expand laterally during flattening. This means, therefore, that the cross dimension of the compression will be approximately equal to the diameter of the original specimen. Since the whole circumference surface must be forced into this area, it follows that a certain amount of buckling and warping of the compression surface invariably follows. However, since the lower side

Fig. 12.—Diagram illustrating the progressive flattening of a cylindrical or spherical object as a result of the weight of accumulating sediments.

of the specimen rests in a more stationary position on the firmer sediments of the bottom, most of the downward movement during compression must take place from above, and it is on the upper surface that most of the buckling will take place. Trunk compressions often show the results of such flattening by being split lengthwise into narrow strips that overlap slightly along the slickensided edges.

If the compressed plant part is a hollow cylinder such as a cordaitean trunk or a stem of *Calamites*, the internal cavity often becomes filled with sand or mud before flattening takes place. In such a case the central sediment-filled core will become flattened into a lens-shaped mass in cross section. It will be surrounded by a carbonaceous film resulting from the compression of the woody cylinder, but extending out into the bedding plane along either side will be a narrow band, or "wing," of carbonaceous substance which corresponds in width to the original thickness of the wood (Fig. 13E).

A very narrow border of similar origin is often present at the margins of the compressions of Carboniferous fernlike pinnules. The width of this border represents the approximate original thickness of the leaf. It is usually more distinct if the leaf happened to settle into the sediment with the convex surface upward than if turned the other way (Fig. 13A).

If any organic matter remains on the surface of a compression, it is usually a layer of structureless carbon, although sometimes the cell pattern of the cutinized epidermis is retained and may be studied by the employment of maceration or film-transfer techniques. Occasionally, especially with leathery leaves or tough fruits, the tissues are retained in a mummified condition, and by following suitable procedures these may be made to swell sufficiently so that they can be embedded and sectioned for microscopic study. Successful cuticular and epidermal studies have been made of the Paleozoic conifers and of the Upper Triassic

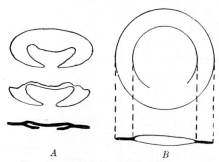

Fig. 13.—Diagrams showing the effect of compression upon a plant part. (A) Thick leaf or pinnule with enrolled margin. (After Walton.) (B) Formation of compression border as a result of the flattening of a tubular cylindrical body. (After Walton.)

gymnosperms from Greenland. The well-preserved fruits of the Caytoniales from the Yorkshire coast were embedded and sectioned. Frequently all traces of organic matter have disappeared, leaving nothing except a mere imprint in the rock. This imprint may retain the shape and surface features of the part. Sometimes such remains become covered with a mineral film of secondary origin that produces striking contrasting effects.

If a rock containing a compression is split open, one surface will usually bear the nearly intact compression and the other will show the impressed counterpart. The well-known Mazon Creek nodules, from the middle Pennsylvanian of Illinois, are outstanding examples of compressions and counterparts (Fig. 14). These smooth oval masses of fine-grained rock, varying in length from an inch to a foot, will often, when split, reveal a perfectly preserved pinnule, a portion of a frond, a cone, or even a seed. One-half of the split nodule contains the main part of the compression or the positive, and the other bears the negative which is mainly an imprint of the positive.

Thick bulky plant parts such as trunks and large seeds or hard dry fruits are often preserved in the compressed condition, while retaining a considerable portion of their original thickness. Compressed wood is one of the main constituents of lignites. The flattening has forced the air

FIG. 14.—*Pecopteris Miltoni*. Compression of terminal portion of a pinna exposed on the surfaces of a split nodule. From the Pennsylvanian of Will County, Illinois. Natural size.

and water out of the cell cavities and intercellular spaces, but it is often possible by the judicious use of caustic alkalies to produce enough swelling of the cell walls to enable them to be studied. Many of the Cretaceous plants found at Kreischerville on Staten Island, at Cliffwood, New Jersey, and at Gay Head on Martha's Vineyard Island are preserved as compressed lignites. Lignitized wood usually has a dark brown color.

When moist, it has a firm cheesy consistency and can be crushed in the hand, but after drying it becomes hard and fractures readily.   The fossil fruits from the Eocene at Brandon, Vermont, are also lignitized.

A second type of plant fossil that is closely related to the compression is the *cast*.   A cast results from the filling of a cavity formed by the decay of some or all of the tissues of a plant part.   It may be formed within the pith cavity of a trunk before the woody cylinder has broken down, as in *Cordaites* and *Calamites*.   In these forms the pith casts will show surface

FIG. 15.—Cast of stump of *Eospermatopteris* exposed by blasting away the enclosing sandstone.   Hamilton group, Middle Devonian.   Riverside Quarry, Gilboa, New York. (*Photo by courtesy of New York State Museum.*)

markings that are the counterparts of the inner surface of the wood. Casts of whole trunks, or of stumps or seeds, are frequently formed when the plant part decays before any great amount of compaction of the sediments has taken place.   The removal of the plant tissues leaves a cavity, or mold, which becomes subsequently filled with mud or sand and which, upon hardening, becomes a cast.   Besides clastic material, a cast may be formed of crystalline substances.   Calcite casts of *Lepidodendron* cones occur in the Coal Measures of Great Britian.   Other minerals that may produce casts are iron pyrites, sphalerite, chalcedony, agate, and opal.

Excellent examples of sandstone casts are the stumps of *Eospermatopteris* in the Middle Devonian of Gilboa, New York (Fig. 15).   When unearthed during quarrying operations for the construction of the Gilboa dam, more than fifty of these stumps were found situated where they grew. They were standing in a thin bed of shale that represents the old soil, and radiating outward in all directions from the enlarged stump bases were the carbonized remains of the slender roots. The sand composing the casts is the same as that surrounding them. Only the merest traces of any of the original organic matter remain and not in sufficient quantity to show much of the internal anatomy.

Stumps of *Lepidodendron* similarly preserved may be seen on display in Victoria Park in Glasgow. Casts of stigmarian roots are familiar objects in coal mines and quarries and are the most common Carbonifer-

FIG. 16.—Cast of *Stigmaria*, the rootlike organ of a Paleozoic arborescent lycopod. From the glacial drift near Jackson, Michigan. ⅔ natural size.

ous fossils (Fig. 16). *Trigonocarpus*, the seed of one of the Carboniferous pteridosperms, is often partly preserved as a cast, the portion inside the stony layer becoming filled and forming a cast of the inner portion of the seed. Many excellent *Trigonocarpus* casts have been found in the Pottsville sandstones of eastern Ohio (Fig. 5*A*).

In *Calamites* the pith sometimes attained a diameter of 6 inches. As the plant, grew, the pith tissue broke down leaving a central hollow which upon burial became filled with mud or sand. The resulting cast usually shows the position of the joints of the stem and the longitudinal course of the primary vascular bundles outside the pith.

Although casts seldom show any of the original tissue structure, they are, nevertheless, valuable fossils because they often give a good idea of the external form of parts of many extinct plants.

The third type of plant fossil, the *petrifaction* (Figs. 4, 8, and 9), is the rarest but in many respects the most important. In a petrifaction, the original cell structure is retained by means of some mineral that has infiltrated the tissues. Petrifactions permit morphological studies of plant parts, which could not be made from casts or compressions. From petrifactions the structure of many reproductive organs and other elaborately constructed parts have been worked out in detail, and the information thus obtained has been of great value in determining the taxonomic position of many fossil types in the plant kingdom.

About twenty mineral substances are known to cause petrifaction, but those most frequently encountered are silica, calcium carbonate, magnesium carbonate, and iron sulphide, either pyrite or marcasite. Of these, silica is the most prevalent and has preserved the largest quantities of plant material although iron sulphide compounds are very widely spread. Small fragments and splinters of wood impregnated with iron sulphide are often scattered throughout shales, sandstones, and coal seams.

The processes causing petrifaction of plant tissue are not well understood although they evidently involve infiltration followed by precipitation due probably to interaction between soluble mineral salts and certain compounds released during partial disintegration of the cell walls. The explanation often given in textbooks that petrifaction is a "molecule-by-molecule replacement of the plant substance by mineral matter" is erroneous and misleading. This definition assumes that the original plant substances have been removed, which does not necessarily happen. In many petrifactions, in fact in most of them, the presence of some of the original compounds of the cell walls can be shown by chemical tests. The mineral matrix can often be removed with suitable acids without destroying the tissue structure, and if replacement had occurred, the whole mass would be expected to dissolve in any reagent that reacted with the matrix. The preparation of thin sections of coal-ball petrifactions by the peel method is dependent upon the presence of nonmineralized substances in the cell walls. The hydrochloric acid that is applied to the prepared surface dissolves the mineral matter from the cell cavities and leaves the cell walls standing in relief. Then when a thin solution of cellulose nitrate or similar material is spread over the etched surface and allowed to dry, a thin layer of embedded plant tissue can be removed with it.

When petrifaction begins, the buried plant material absorbs soluble mineral substances from the surrounding water. These may be in the form of carbonates, sulphates, silicates, phosphates, or others. The various products of decomposition released by bacterial action or by hydrolysis include free carbon, carboxylic acids, humic acids, and hydrogen sulphide, and some of these behave as reducing agents that precipitate insoluble compounds within the plant tissues. Thus soluble silicates are reduced to silica (SiO_2) and iron compounds react with hydrogen sulphide to produce pyrite or marcasite (FeS_2), both being common petrifying minerals. Just how basic compounds such as calcium and magnesium carbonates are precipitated within the cells is less known, but apparently they form along with coal-balls or other calcareous masses in coal seams and shales. Calcification and pyritization of plant tissues often take place concurrently, with either one or the other of the minerals predominating or with deposition in approximately equal quantities. Silica usually exists alone; it seldom combines with other petrifying minerals.

While mineral deposition is going on within plant tissues, water is being expelled as a result of compaction of the sediments. This causes the sediments and the entombed plant parts to solidify, after which the petrifaction process is complete. Although the deposition of mineral substances within plant tissues is dependent upon the release of certain substances from the cell walls, there is no evidence of direct replacement of any of the unaltered organic constituents of the plant by the petrifying minerals. In fact, all evidence is decidedly against it. In case of silification, a direct reaction would produce organic silicates, and these are not what we find in silicified wood. The minerals react only with the decomposition products, not to replace them but to produce insoluble compounds usually of simpler molecular structure.

Chemical analyses of silicified woods show that some of the original constituents of the wood are often present. Cellulose and lignin can be detected, and the lignin content is in higher proportion than in non-fossilized woods because it decomposes more slowly than the cellulose.

One of the marvels of paleobotany is the perfection with which minute histological details in very ancient woods are often preserved. In woods from the Devonian, pit structure, crassules, and the layers of the cell wall are sometimes as distinctly visible as in Recent woods. Age alone is not a factor in preservation, and one can tell little or nothing of the age of a specimen from the extent to which details are visible.

The color of petrified plant tissues ranges from nearly pure white and various shades of gray to red, yellow, amber, brown, and jet black. White usually results from weathering, and such woods are often very fragile unless secondary infiltration has cemented the cells together. The brown, yellow, and red tints may be due to the presence of small quantities of other minerals, especially iron, although brown is often the result of retention of humic compounds. Woods that appear black to the unaided eye are often light brown in thin section.

Silicified plants occur in rocks of all ages from the Silurian to the Recent.   The silica is believed to have originated mainly from two sources, volcanic ash and the waters from hot springs.   Ash has provided the silicates for the petrifaction of most of the logs in the Mesozoic and Cenozoic deposits of the western part of North America.   The upright trunks exposed on Amethyst Mountain and Specimen Ridge in Yellowstone National Park were preserved by the heavy layer of ash that was thrown from near-by volcanoes during early and middle Cenozoic times.

Fig. 17.—Silicified redwood stump 47 feet in circumference.   From the early Middle Tertiary near Leadville, Washoe County, Nevada.

In these instances burial was rapid and decay organisms had no opportunity to function over any length of time.   The weathering of the ash released large quantities of soluble silicates, which saturated the tissues. The same process went on at countless other places throughout the Great Plains, Rocky Mountains, Great Basin, and coastal regions of the western part of the continent and the number of examples that could be cited is unlimited (Fig. 17).

Silicification resulting from deposition of minerals from hot springs is not as frequent as from the weathering of volcanic ash although the phenomenon is by no means a rare one.   Probably the best known example of this kind of petrifaction is the flora in the Middle Devonian Rhynie chert in Scotland (Fig. 26). Other instances of silicification independent of volcanic activity are the *Callixylon* logs in the Upper Devonian black shales in the east central United States and the *Lepidodendron* trunks of lower Pennsylvanian age in Colorado.

Several forms of silica have been identified in silicified woods, but only the cryptocrystalline and amorphous varieties commonly retain the tissues in their original form. The formation of crystals, even very small ones, is destructive to cell arrangement. Crystals may, however, form in cavities such as knotholes, stem hollows, or large cracks in otherwise well-preserved logs. Amethyst Mountain received its name from the amethyst crystals in the silicified logs scattered over the slopes. Among the cryptocrystalline types agate is frequently present, and splendid examples are found in the Eden Valley petrified forest, of Eocene age, in western Wyoming. Chalcedony characterizes the logs in the Petrified Forest National Monument of Arizona, and jasper is frequent. In northern Nevada cones and small limbs have been preserved with fire opal.

Calcified plants are found mostly in the Paleozoic. Although calcification is less common than silicification, it has been an important process, and the plants preserved by it have added greatly to the main body of paleobotanical knowledge. The most prolific sources of calcified plants are coal-balls. Fragments of calcified wood are occasionally found in marine or brackish water limestones, especially those thin limestone layers often interleaved in shale formations. An example is the *Callixylon* wood scattered in small pieces throughout the Genundewa limestone of Upper Devonian age in central New York. This limestone is nodular or concretionary at places and is made up mostly of the shells of the minute pterapod *Styliolina fissurella*, which apparently supplied most of the calcium carbonate.

As previously stated, iron pyrites (pyrite or marcasite) is widely distributed in shale and sandstone as a petrifying mineral, and is often a constituent of coal-balls. It is an opaque substance which renders almost impossible the study of thin sections with transmitted light. Reflected light can sometimes be employed with a certain degree of satisfaction when only medium or low magnifications are used.

Pyrites apparently forms under conditions of extreme stagnation, and the mineral often composes thin layers in dark shales containing abundant iron and organic matter. Pieces of secondary wood and fragments of small stems completely infiltrated with pyrites are scattered sparingly among the lenses of the Tully pyrite and the underlying Ludlowville shale of upper Middle Devonian (Hamilton) age in western

New York. Although iron pyrites is probably the most prevalent of all petrifying minerals, plant material so preserved is often of limited use because of difficulties in microscopic study.

The Relative Value of Compressions and Petrifactions

Although petrifactions have been of immense value in paleobotanical researches in revealing the internal make-up of extinct plants, we are mainly dependent upon compressions and casts for external form and habit. Compressions and petrifactions are seldom found together, with the result that correlation of external and internal morphology often presents serious difficulties. This problem is especially well brought out in connection with the study of Paleozoic seeds, where the same type of seed may bear more than one generic name, depending upon how it is preserved. Likewise, many of the well-known Paleozoic fernlike leaf types may belong to fructifications with which we are well acquainted, but in the absence of attachment they cannot be brought together and must therefore be treated separately. The Calamopityaceae provides an example of a group of plants of which the stems are preserved in great detail but of which we know little of the habit or external appearance. Numerous other examples could be cited. A notable exception in which it has been possible to bring about a correlation of external and internal structure is the plants of the Rhynie chert. Broken surfaces of the chert have revealed the outer contour of the silicified stems, and these, combined with serial anatomical sections, served as a basis for the well-known reconstructions of the Rhynie plants.

Although the paleobotanical investigator may be fully aware of the limitations imposed by compressions or petrifactions when found alone, he seldom has any choice in the matter, and is confronted with the problem of the interpretation of the material as it exists. To a great extent the study of the two types of fossils has been pursued along different lines, with compressions engaging the attention of those primarily interested in stratigraphic correlation, and petrifactions being sought after by those concerned more with comparative morphology and evolutionary considerations. The monumental paleobotanical works of Zeiller in France, Kidston in Great Britian, and White in the United States have dealt mainly with compressions in an effort to clarify stratigraphic problems, whereas the contributions by Williamson and Scott in England and Renault in France have resulted in setting forth in great detail the affinities of numerous Carboniferous plants as interpreted from petrifactions.

As a result of the partial divorcement of the study of compressions and petrifactions, there has been a tendency on the part of some investigators to express scorn for those types of fossils not pertinent to their particular fields of interest. This attitude is most frequently displayed by anatomists and morphologists who feel that affinities are better expressed by internal anatomy than by external form and that foliar characteristics as displayed in compressions do not provide reliable clues to relationships. This biased viewpoint insofar as it endeavors to depreciate the scientific value of compressions is unjustified and ignores the important fact that in most instances neither type of preservation by itself reveals all the characteristics of the plant. A paleobotanist cannot afford to disregard compressions any more than one can overlook habit and leaf form in living plants.

OBJECTS SOMETIMES MISTAKEN FOR PLANT FOSSILS

Animal remains, inorganic objects such as concretions, and cultural objects are sometimes mistaken for plant fossils. The resemblance to a plant part may be so close as to deceive at least temporarily even the professional paleobotanist. Remains of graptolites and the "pinnules" of some crinoids, as *Botryocrinus*, have been misinterpreted as Devonian or pre-Devonian plants. The organism named *Psilophyton Hedei*, from the Silurian of the island of Gotland, and once figured as the oldest example of a land plant, is now thought to be an animal closely related to the graptolites. Other examples of a similar nature could be cited. A cavity in a piece of rock from Greenland, long supposed to be a palm fruit, is now thought to be the footprint of some ancient animal.

Mineral matter often crystallizes along the bedding planes of shales to produce dendritic growths simulating plants. One such example from the Silurian was named *Eopteris*, and is sometimes cited as the oldest example of a fern. Concretionary objects are collected by amateurs who think they have discovered fossil nuts, eggs, oranges, or pears. Some large oval sandstone concretions in the Upper Devonian of West Virginia were described as tree trunks, and closely banded volcanic or metamorphic rock is often mistaken for petrified wood. Ice marks and fracture surfaces are frequently compared to seaweeds and other plants.

Cultural objects mistaken for fossil plants sometimes have ludicrous consequences. An oval body with four longitudinal grooves, and described as a cucurbitaceous fruit, was found upon careful examination to be a fire lighter made of baked clay. Several years ago a paleobotanist of international fame described an object externally resembling an ear of maize from an unknown locality in Peru. It was thought to be a fossil and was named *Zea antiqua*, although it did not differ in any important respect from recent Peruvian varieties. This specimen was cited in several papers as evidence of the antiquity of the maize plant, but finally, when cut, it was found to be nothing other than a baked clay object fashioned by a clever craftsman.

It is seldom impossible to distinguish between organic remains and other objects if they are carefully examined.   A plant fossil will usually show at least traces of carbonized fibrous tissues or cutinized epidermal cells, although seaweeds and other thallophytes may present special problems.   The best safeguard against going astray is experience bolstered by a critical attitude.

References

ADAMS, S.F.: A replacement of wood by dolomite, *Jour. Geology*, **28**, 1920.

ARNOLD, C.A.: The petrifaction of wood, *The Mineralogist*, **9**, 1941.

FILICIANO, J.M.: The relation of concretions to coal seams, *Jour. Geology*, **32**, 1924.

RAISTRICK, A., and C.E. MARSHALL: "The Nature and Origin of Coal and Coal Seams," London, 1939.

SEWARD, A.C.: "Fossil Plants," Vol. 1, Cambridge, 1898.

ST. JOHN, R.N.: Replacement vs. impregnation in petrified wood, *Econ. Geology*, **22**, 1927.

STOPES, M.C., and D.M.S. WATSON: On the present distribution and origin of the calcareous concretions in coal seams known as "coal balls," *Royal Soc. London Philos. Trans.*, **B 200**, 1908.

THIESSEN, R., and G.C. SPRUNK: Coal Paleobotany, *U. S. Bur. Mines Tech. Paper*, 631, 1941.

TWENHOFEL, W.H.: "Principles of sedimentation," New York, 1939.

WALTON, J.: On the factors that influence the external form of fossil plants, with descriptions of the foliage of some species of the Paleozoic equisetalean genus *Annularia* Sternb., *Royal Soc. London Philos. Trans.*, **B 226**, 1936.

CHAPTER III

THE NONVASCULAR PLANTS

Organic remains in the rocks furnish conclusive evidence that plants belonging to the division Thallophyta were in existence ages before the initial appearance of vascular plants. In fact, approximately three-fourths of known geologic time had elapsed before the latter group came into existence. What the first plants were like and when they came into being are integral parts of the problem of the origin of life itself, whose beginnings are blotted out by the imperfections of the fossil record. The first living organisms were probably not distinct as either plants or animals, but were rather naked bits of the simplest kind of protoplasm, which took on an amoeboid or plasmodial form. This protoplasm must, however, have possessed both the remarkable property of sustaining itself and the ability to derive the energy necessary for its metabolic processes from sources outside. These sources may have been light, heat given off from the earth, or chemical reactions between inorganic substances in the immediate environment.

Although it is possible that life had been in existence for a billion years or more before any creatures evolved which were sufficiently indestructible to leave recognizable remains, it is more likely that this initial period was of relatively short duration because the first protoplasm was probably a virile substance which caused the initial stages of organic evolution to follow one another in rapid sequence. It does not seem likely that the original protoplasm could have remained in a state of almost complete quiescence for a very long interval and retain its dynamic properties. Geologically speaking, the interval between the origin of life and the appearance of cellular plants was probably not long.

BACTERIA

Evidence that bacteria or organisms similar to them were in existence during pre-Cambrian times is both presumptive and direct. Bacteria perform such an essential role as the agents of disintegration of organic and inorganic substances that it is impossible to comprehend how the balance of nature could ever have been maintained without their aid. The conclusion is that bacteria have been in existence nearly as long as other organisms and that during this time they have functioned similarly to the way they do at present.

Direct evidence that bacteria existed during the latter part of the long pre-Cambrian era is supplied by minute dark objects found in the upper Huronian rocks of the Upper Peninsula of Michigan. These supposed bacteria consist of filamentous chains, which are sometimes branches of elongated cells resembling the living *Chlamydothrix*, an iron-depositing organism. The iron ores in which these remains were found are believed to have been deposited by organisms resembling those that make similar deposits today. Associated with these probable bacteria are other slender branching filaments that may represent blue-green algae. Some of these threads even show evidence of a surrounding gelatinous sheath suggesting that which encases the threads or cell colonies of the blue-greens at the present time. In view of the many similarities between living bacteria and blue-green algae, it is interesting to find remains resembling these organisms intimately associated in the ancient pre-Cambrian rocks. Slightly later in the pre-Cambrian, in the Keween-awan limestones of Montana, minute spherical bodies resembling *Micrococcus* have been found. These also are associated with remains believed to be blue-green algae.

During the latter part of the Paleozoic beginning with the Devonian, and throughout the eras that followed, there is no lack of evidence that bacteria were everywhere present. Probably the most extensive studies ever made of ancient bacteria were by the French paleobotanists C.E. Bertrand and Bernard Renault, who examined especially prepared thin sections of decayed silicified and calcified plant tissue and coprolites of reptiles and fishes. In material ranging in age from Devonian to Jurassic they found a great variety of spherulitic and rodlike objects, which they described as species of *Micrococcus* and *Bacillus*. *Micrococcus Zeilleri* is the name given certain small round bodies either single or grouped which they found on the surfaces of the thin cuticular membranes which make up the so-called "paper coal" found at Toula in Russia. Renault believed that these objects were the remains of the organisms responsible for the disintegration of the cellular tissues of the plants that compose the coal seam. He had no proof of this, however, and other investigators have been inclined to discount his theory. *Bacillus Permicus* was discovered in sections of a coprolite from the Permian of France. It consists of cylindrical rods 12 microns or more long and about 1.4 microns broad. The rods are rounded at the ends, and are single or are joined into chains of two or three cells.

These are but two examples of supposed fossil bacterial organisms that have received specific names. Many others could be cited, but to do so would serve no useful purpose because of our meager knowledge of them. No satisfactory test has been devised for distinguishing between the remains of bacterial cells and small particles of flocculent material or minute crystals. External resemblance to living bacteria is almost the only clue, but since cultural and physiological characteristics are so essential in the study of these organisms, bacteriologists are inclined to attach little importance to the minute lifeless objects found in the rocks. Although some supposed fossil bacteria have been shown to be inorganic or the decomposition products of other organisms, one can hardly doubt that they are probably present in quantity in some sediments, and attention to the development of new techniques for their study would certainly yield unexpected results.

FUNGI

Throughout the long geologic past fungi have played the same role in nature as at present, that of acting as scavengers, and thereby preventing an endless accumulation of dead vegetable matter. Others have acted as parasites which preyed upon living plants. Many blanks in the fossil record are the direct result of the destructive activities of fungi. It was mainly when deposition took place rapidly or where plant material accumulated under conditions unfavorable for fungal growth that the destructive activities of fungi were evaded.

The fungi as a group are quite old. Well preserved mycelium and spores have been found within the tissues of the very ancient vascular plants (Fig. 18). Throughout the Devonian and Carboniferous there is ample evidence of fungous activity within plant tissues although the actual remains of the organisms are seldom present.

Fungous spores and mycelia resembling those of Recent Phycomycetes occur in the tissues of Carboniferous plants. Some of these have been described under such names as *Urophlyctites*, which has been found parasitic upon *Alethopteris*, and *Zygosporites*, which resembles a mucor. Under the comprehensive name of *Palaeomyces* a number of fungal types have been described from the silicified stems and rhizomes in the Middle Devonian Rhynie chert in Scotland (Fig. 18A). These remains consist of branched, nonseptate hyphae and large, thick-walled resting spores. Some of the larger spores had been secondarily invaded by other fungi, which produced smaller spores within. These fungi bear some resemblance to the Peronosporales and Saprolegniales, although their actual affinities are unknown.

The foliage of Carboniferous and later plants sometimes bears small spots resembling the hard fruiting bodies of the Pyrenomycetes. *Hysier-ites Cordaitis* occurs on the long strap-shaped leaves of Cordaites. Other pyrenomycetous fruiting bodies have been found on the leaves of Eocene and Miocene angiosperms.

The Discomycetes are very meagerly represented in the fossil series although such names as *Pezizites, Aspergillites,* and others implying resemblances to discomycetous genera have been assigned to Cenozoic and more recent remains.

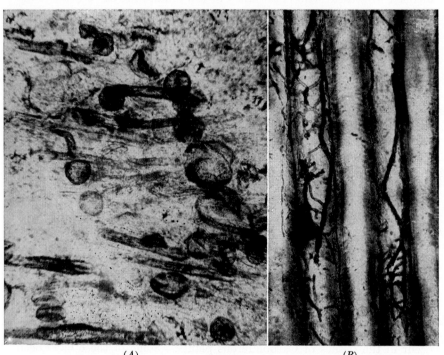

(A)   (B)

Fig. 18.—(*A*) *Palaeomyces Asteroxyli.* Hyphae and vesicles in the disintegrated cortical tissues of *Asteroxylon Mackiei.* Rhynie chert. Aberdeenshire, Scotland. × 200. (*B*) Non-septate fungous hyphae in the secondary wood tracheids of *Callixylon Newberryi.* New Albany shale, Upper Devonian. Indiana. × 200.

A number of leaf spots resembling the aecial, uredinial, and telial sori of the Pucciniaceae have been observed on foliage of Upper Cretaceous and Cenozoic plants. These have received such names as *Puccinites, Aecidites,* and *Teleutosporites.* The latter has been reported parasitic upon a Carboniferous *Lepidodendron,* although the identity is too uncertain to constitute sound evidence of a Carboniferous rust. Sporophores of the bracket fungi occur occasionally in the Cenozoic, and a fossil "earth star," *Geasterites florissantensis,* came from the Tertiary of central Colorado.

Among the Fungi Imperfecti *Pestalozzites*, so named because of its resemblance to *Pestalozzia*, occurs on Miocene palm leaves. Numerous other examples of imperfects might be cited, although their importance is relatively slight.

The fossil record has thrown no light on the problem of the evolution or origin of the fungi. Morphological studies of recent fungi indicate that the class is polyphyletic, different groups having developed from various algal types through the loss of chlorophyll and the resulting inability to sustain themselves independent of previously synthesized organic material. Therefore, it is probably useless to look for a common starting point in the time series for all fungi. Different ones originated at different times. Some have probably become extinct, and others are very recent.

Algae

The algae rival the bacteria in their claims to antiquity, and the influence they have exerted upon the course of the life history of the earth since pre-Cambrian times has been inestimable. Algae rank second in importance to the vascular plants as vegetable organisms in the geological series. Geologically they are significant for the roles they have played in limestone and petroleum formation. Limestones believed to have been formed by algae occur in the oldest rock formations, and the deposition of calcium carbonate as a result of the activities of these organisms has been going on continuously in the warmer seas since the earliest decipherable geologic times. This deposition has been going on not only in the ocean but in fresh-water lakes and streams where suitable conditions exist. The so-called "water biscuits" found in some of our inland lakes and the rough limy coating often surrounding pebbles in stream beds are examples of deposition of calcium carbonate by Recent algae.

Although there is substantial evidence that algae have been in existence since pre-Cambrian times, the fossil record shows us rather little concerning the evolution of this group of thallophytes through the vast stretches of the Paleozoic, Mesozoic, and Cenozoic eras. They appear to have undergone relatively few changes in form and structure or in their mode of living. This comparative stability is the result of the aquatic environment in which abrupt changes were lacking as compared with the environment of land plants where temperature, rainfall, topography, and soil conditions are guiding factors.

Fossil algae are often difficult to recognize with certainty because in many instances no traces of the original cell structure remains. There are numerous examples on record of forms originally supposed to be algal fossils that were later shown to be worm borings, scratches, tracks, or inorganic formations caused by chemical action or by deformation brought about by earth movements. Not only are algae sometimes difficult to recognize, but their proper classification, once they have been identified as such is equally problematic. Modern algae are classified mainly on the basis of the color pigments in the cells and on the reproductive organs, features scarcely discernable in the fossil condition.

It is a common practice among authors on paleontologic subjects to include under the category of fossil algae not only those objects which show actual remains of the organisms, but also structures believed to have resulted from their activities. The latter include many limestone deposits known as *stromatolites*. Stromatolites may be nodular, columnar, spheroid, or biscuit-shaped masses made up of superimposed or concentric laminae, or they may consist of irregularly expanded sheets or lenses. They may be solid, spongy, or vesicular. Microscopically they show nothing except mineral crystals, their algal origin being inferred from their resemblance to limestone masses formed at present. Whether stromatolites are the result of algal activity or of inorganic agencies has been the subject of much dispute, but if they are organic the algae that produced them outrank all other pre-Silurian vegetable forms in abundance, variety, and importance as rock builders. Of the noncalcareous remains often interpreted as algae are numerous impressions and markings on rocks, which resemble branched seaweeds or fresh-water forms. The interpretation of such objects is invariably open to question although many have been described under generic and specific names.

Before discussing the general subject of calcareous algae and algal limestones, a few remarks will be made concerning the geologic distribution of some of the classes of algae. The blue-greens (Myxophyceae) are believed to be the oldest. Certain very extensive limestone beds of the pre-Cambrian are believed to be the result of their activity. They were active limestone formers throughout the Paleozoic, although by this time other groups of algae are believed to have come into existence. The oldest noncalcareous alga showing cell structure is *Gloeocapsomorpha* probably a blue-green, from the Ordovician of Esthonia. It has cells 0.01 to 0.08 mm. long, which are held together in groups by a transparent matrix. The resemblance to the living genus *Gloeocapsa* is obvious. Somewhat older is *Morania*, from the Middle Cambrian, which resembles *Nostoc*, and *Marpolia*, which has some likeness to the recent *Schizothrix*.

The Chlorophyceae are believed to have existed in large numbers during the Paleozoic. The oldest fossil believed to represent this class is *Oldhamia* from the Lower Cambrian. It exists as an impression of an articulated stem bearing whorls of closely set branchlike structures.

Because of its resemblance to the recent *Halimeda,* it is provisionally classed with the Codiaceae.

The algal family most copiously represented in the fossil series is the Dasycladaceae, one of the families of the Siphonocladales. This family embraces 10 Recent genera and a greater number of extinct ones. Its members are responsible for calcareous deposits ranging from the Ordovician to the present except that none are known from the Devonian or Lower Carboniferous. The plants of this family consist of upright elongated coenocytic axes bearing radiating lateral appendages, which may be simple or branched. In many forms the tips of the appendages are terminally enlarged so that they nearly touch and form a globular or cylindrical structure which may in some forms exceed a centimeter in diameter.

Rhabdoporella, Cyclocrinus, and *Prismocorallina* are early forms attributed to the Dasycladaceae that date back to the Ordovician. *Rhabdoporella* is a small cylindrical unbranched form, which may be the ancestral type for the other members of the family. *Cyclocrinus* produces large delicate heads about 8 mm. in diameter, with a thinly calcified surface consisting of compactly fitting enlarged tips of the appendages that arose from the swollen axis tips. In *Prismocorallina* the branches were irregularly placed, not forming a definite surface layer. The branches were twice divided to form two terminal branchlets.

The Charales have a long record that extends into the Devonian. Genera in which the spherical oögonia are surrounded by five spiral cells occur frequently in the Cretaceous and Cenozoic, although *Gyrogonites,* which is one of the best known genera, dates from the late Triassic. *Palaeochara* is a late Carboniferous genus with six spiral cells. *Palaeonitella,* from the Middle Devonian Rhynie chert, is a doubtful member of this order, but other possible Devonian ones are *Trochiliscus* and *Sycidium.*

Although the red algae (Rhodophyceae) are among the most important of the rock builders in the Recent tropical seas, some investigators believe that they played a lesser role during the early Paleozoic. Nevertheless, certain very abundant Paleozoic algae are assigned to this class. One of the best known is *Solenopora* (Fig. 19), which ranges from the Ordovician to the Jurassic. *Solenopora* produced biscuit or headlike limestone masses with a lobate surface which, when sectioned vertically, are seen to consist of faintly visible radiating lines (the algal filaments) and wavy concentric bands which probably represent seasonal accretions. A Recent member of the Rhodophyceae which extends back into the Triassic is *Lithothamnium.* The thallus of this plant forms a hard stony mass on the surface of a submerged rock, and the exposed surface is irregular or warty.   In section the thallus mass shows two layers, a lower one consisting of cells radiating outward from a central point, and an upper one of cells arranged in horizontal and vertical rows.   The genus *Lithophyllum* is believed to have been in existence since the Cretaceous, but the possibility that it may have a longer history is suggested by the Big Horn dolomite of Ordovician age in Wyoming.   In structure this dolomite resembles those deposits made by modern species of *Lithophyllum* although no cells are preserved in either the fossil or the Recent deposits.

The brown algae (Phaeophyceae), being noncalcareous, have not entered as strongly into the fossil record as the three classes just men-

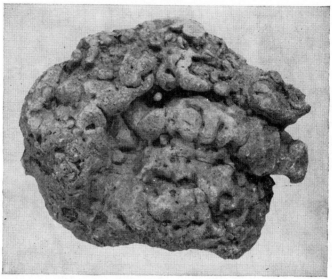

Fig. 19.—*Solenopora compacta.*   Upper Ordovician.   ½ natural size.

tioned.   Various imprints resembling *Fucus* and other genera of seaweeds have been found in rocks of all ages from the early Paleozoic to the present but their value as authentic occurrences of these plants is seldom great. Most of those objects which are indiscriminately called "fucoids" and which are found especially in the Devonian are inorganic.   The genus *Prototaxites*, once supposed to be a brown alga, is provisionally referred to a higher group known as the Nematophytales, and is described on a later page.

The recent discovery of well-preserved zygospores of desmids in the Middle Devonian Onondaga chert of central New York furnishes satisfactory evidence of the antiquity of this group of organisms.   One type with four spines resembles the living genus *Arthrodesmus*.   Other globose forms with slender furcate spines resemble *Cosmarium, Staurastrium,* and *Xanthidium.* Desmids have seldom been reported elsewhere although they have been mentioned from the Devonian, Jurassic, and Cretaceous.

The fossil record of the diatoms is confined mostly to the Cretaceous and Cenozoic, although well-preserved remains have been found in the Jurassic. They probably existed earlier, but because of the destructibility of the siliceous valves they are readily destroyed either from crushing or from the solvent action of alkaline waters.

Diatomaceous earth beds, which are deposits made up of the siliceous valves of diatoms, form in both fresh and salt water. In some lakes at the present time, notably Klamath Lake in southwestern Oregon, a soupy diatomaceous ooze is accumulating on the bottom. In the Cenozoic deposits of the Great Basin, diatomaceous earth beds are interlayered with basalt and volcanic ash. The ash apparently supplied an abundance of soluble silicates, which the diatoms were able to utilize. The Trout Creek diatomite bed of middle Miocene age in southeastern Oregon contains large quantities of leaf impressions and other plant remains.

Ranking next to the calcareous algae in importance in geologic history are those which have played a leading role in coal and petroleum formation. Certain types of carbonaceous deposits such as boghead and torbanite, and some oil shales, are thought to have received their oil content from algae similar to the recent *Botryococcus Brownii.* Boghead and torbanite resemble cannel coal except for their composition. Many years ago it became known that these oil-bearing coals are made up principally of "yellow bodies," which are merely colonies of unicellular organisms held together by a rubbery or gelatinous matrix. A hollow type of "yellow body" resembling *Volvox* was named *Reinschia,* and a similar but solid one was called *Pila.* The morphology of *Reinschia* and *Pila* has long been a matter of disagreement. They were originally thought to be algae, but some investigators believed they were spore cases. The question now seems definitely settled because studies rather recently made in peat deposits have shown the presence there of organisms identical with both *Botryococcus Brownii* and the "yellow bodies." In all of them the colony consists of a framework formed of material secreted from the cell cavities. Each colony has a dome-in-dome structure and is globular when small. The rubbery matrix that binds the cells together is highly resistant to decay, and yields paraffin upon distillation. *B. Brownii* is one of the outstanding examples of an oil-producing alga. It has probably been in existence since the lower Carboniferous.

Much of the petroleum in the pre-Carboniferous shales of Eastern North America is believed to have been derived from organisms of a different kind. Two of these, *Sporocarpon* (Fig. 20*B*) and *Foerstia* (Fig. 20*A*) are small thalloid bodies a few centimeters long. The form

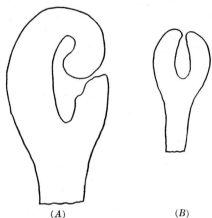

known as *Sporocarpon furcatum* is slightly unequally two-lobed and somewhat resembles the pincher of a lobster. *Foerstia ohioensis* is similar but the longer of the two lobes sometimes curves inward like a shepherd's crosier. The bodies of both consist of several layers of heavily cutinized cells, and within the tissue of the body cutinized spores are borne in tetrads.

Sporocarpon and *Foerstia* are found in the Upper Devonian black shales of Kentucky and Ohio. Although their affinities are unknown, they are similar in appearance to certain modern brown algae that inhabit the zone between the tides, but being cutinized were able to withstand longer exposure to dry air. They resemble *Nematothallus* to some extent but differ in the cellular structure of the tissues.

(A)   (B)

Fig. 20.—(*A*) *Foerstia ohioensis.* Drawn from photograph by White. × about 10. (*B*) *Sporocarpon furcatum.* (*After Dawson.*) × about 10.

Tasmanites (*Sporangites*) is another problematic organism in the Upper Devonian black shales. It is found with *Sporocarpon* but is more widespread and abundant, and at some places it imparts a yellowish tint to the fractured surface of the shale. *Tasmanites* is unicellular and resembles the exine of a rounded spore without a triradiate ridge. The surfaces of some forms bear minute warts, but others are smooth.

Pachytheca (Fig. 21) is a small spherical body, supposedly an alga, from the Upper Silurian and

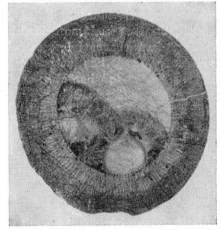

Fig. 21.—*Pachytheca sp.* Downtonian, early Devonian. Staffordshire, England. × 18.

Lower Devonian. It consists of a central mass of intertwined tubes surrounded by a layer of outwardly radiating ones. The latter contain small cellular filaments that project slightly beyond into what was probably a gelatinous sheath around the outside. *Pachytheca* may belong to the blue-green algae.

Parka, from the Lower Devonian of Great Britian, is a flat thalluslike body covered by a surface layer of roundish hexagonal cells. The shield-shaped structure, which is a centimeter or less in diameter, was probably borne on an upright stipe. Numerous spores without triradiate ridges were produced in cavities within the thallus.

THE NEMATOPHYTALES

This name was recently proposed for certain Silurian and Devonian plants of unknown relationship in which the plant body consists of a

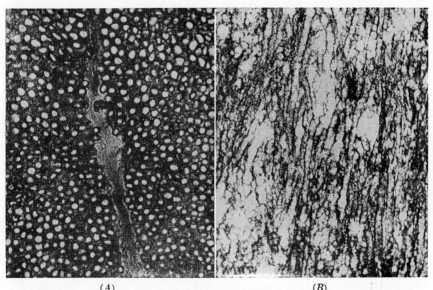

| (A) | (B) |

FIG. 22.—(A) *Prototaxites Logani.* Transverse section. Upper Devonian. Scaumenac Bay, Quebec. ✕ about 100. (B) *Prototaxites Logani.* Longitudinal section. Upper Devonian. Scaumenac Bay, Quebec. ✕ about 100.

system of interlaced tubes. *Nematothallus* is a thin, flat, leaflike compression from the Downtonian (early Devonian) of England. The expanded plant body, of which the original size and shape are unknown, is composed of a system of tubes of two sizes. The smaller of these averages about 2.5 microns in diameter, and the larger range from 10 to 50 microns. No transverse walls are present, but the large tubes are reinforced on the inside with closely spaced annular bands. Their outer surface is cutinized and inside the tubes are spores measuring 12.5 to 40 microns, which appear to have been produced in tetrads.

The name *Prototaxites* was proposed by Sir William Dawson in 1859 for some large silicified trunklike organs from the Devonian rocks of the Gaspé Peninsula. The specimens were rather large, some being as much as 3 feet in diameter, with a smooth or irregularly ribbed and nodose surface. Internally the trunks reveal a series of concentric layers resembling the annual rings of temperate zone trees, and radiating outwardly are narrow strips of tissue simulating medullary rays.

Microscopic examination of the tissue reveals a system of nonseptate interlacing tubes, which in general follow a vertical course. In *Prototaxites Logani*, the species described by Dawson, the tubes are of two sizes, large ones ranging from 13 to 35 microns in diameter and smaller ones 5 to 6 microns across (Fig. 22). The smaller tubes fill the interstices between the larger ones, and because of partial decay often give the appearance of a structureless matrix. In other species of *Prototaxites* (*P. Ortoni* from Ohio) the tubes are all of the same diameter or (as in *P. Storriei* from Wales) they may intergrade from large to small. No spores or fructifications of any kind have been assigned with certainty to any species of *Prototaxites*.

When Dawson proposed the name *Prototaxites* he was impressed by the presence within the large tubes of structures resembling the spiral bands found within the tracheids of the Recent *Taxus*. Later investigations, however, showed the supposed wall sculpturing was more likely the result of partial disintegration of the walls, and that the tubes do not resemble the tracheids of either ancient or modern conifers. They bear more resemblance to the tubes in the stemlike stalks of some of the giant seaweeds, and consequently the names *Nematophycus* and *Nematophyton* were proposed by Carruthers and Penhallow, respectively, as substitutes for *Prototaxites*. Although these substitute names have been widely used, neither is valid because under the rules of nomenclature adopted by botanists a name cannot be rejected merely because of a mistake in judgment in selecting it.

Although the affinities of *Prototaxites* are only conjectural, it is often given a tentative place among the brown algae. On a structural basis it may be compared with *Macrocystis*, *Lessonia*, and *Laminaria*. Its inclusion by Professor Lang in the Nematophytales is justified on the basis of the similarity between the tubular construction of it and *Nematothallus*, regardless of the fact that no spores or cutinized structures are known for the former. It seems not impossible that *Nematothallus* may ultimately prove to be the leaflike appendage of *Prototaxites*. In the Lower Devonian of Germany impressions of fan-shaped appendages of the *Psygmophyllum* type were once assigned to trunks believed to represent *Prototaxites*, and the plant was described as *Prototaxites psygmophyl-*

loides. Dr. Høeg, who has recently reexamined the material, believes, however, that the trunk of *P. psygmophylloides* is parenchymatous and not tubular, and that the designation *Prototaxites* is not appropriate for it. He has proposed the name *Germanophyton.*

Because of our sketchy knowledge of the Nematophytales, it is impossible to say to what extent they represent a natural plant group or whether our concept of them as a group distinct from the algae is justified. Students of phylogeny have long searched for evidence of connecting links between the typical seaweeds and the lowest land plants. Such links, it has been postulated, must have been semiaquatic, and inhabited the littoral zone, where they were at times submerged but at other times exposed to the air for long intervals. When the water level receded permanently, those farthest from it became stranded. Some were able to adapt themselves to a land habitat, and others perished. One of the structures that developed as a result of life out of water was the cuticle that is present in *Nematothallus* and other forms such as *Parka* and *Sporocarpon.* That the Nematophytales gave rise to permanent land plants is improbable, but they may represent remnants of a group of early Paleozoic plants that was transitional between vascular plants and some lower aquatic form.

Sporocarpon, Foerstia, and *Parka* are on an approximate evolutionary level with the Nematophytales in that they have cutinized surface tissues and firm-walled spores. Being cellular instead of tubular, they probably belong to a distinct though related group.

ALGAL LIMESTONES

Much of our knowledge of ancient algae has been gained from studies of limestones believed to have been formed wholly or in part from the activities of these organisms. In many instances (as in stromatolites) no traces of the cellular structure are retained, and in such cases, the algal origin is inferred from the general resemblance of the limestones to those in process of formation at the present time.

The most important rock builders among living algae belong to the Myxophyceae, the Corallinaceae of the Rhodophyceae, the Dasycladaceae of the Chlorophyceae, and the Charales. These organisms are capable of thriving under a wide range of conditions, and some, such as those responsible for the deposition of the delicately tinted tufas and sinsters in Yellowstone National Park, live in water of high temperature. Calcareous algae range from the arctic to the tropics, and sometimes exist at great depths. In the warmer seas they are active in the building of coral reefs. Studies made at the great reef of Funafuti in the South Seas showed that the two most active organisms are *Lithothamnium* and

Halimeda. The foraminifera ranked third and the corals fourth in the amount of lime deposited. These algae, or nullipores, grow among the corals, and the deposited lime serves to cement the tubular coral growths together into a compact and resistant mass.

The deposition of calcium carbonate by algae results from the fact that they obtain carbon dioxide for photosynthesis from calcium bicarbonate in the water, leaving calcium carbonate as a residue. The presence, therefore, of fossil algal reefs in sedimentary deposits is generally

Fig. 23.—*Cryptozoon minnesotensis.* Limestone head showing concentric layers. Ordovician. Cannon Falls, Minnesota. Slightly reduced.

interpreted as an indication of clear water in which the sunlight was able to penetrate to the bottom. A tropical climate is not necessarily assumed, although higher temperatures are known to favor the development of limestone partly because of increased algal growth, and partly because of the lesser solubility of the precipitated calcium carbonate in warm water. However, in cold climates calcareous algae flourish in streams and lakes supplied by hot springs.

Most limestones of algal origin consist of small heads or large masses cemented together laterally. These heads may vary from a few inches to several feet in extent. The individual head usually has in the center or toward the base a pebble, shell, or other object which served as the original substratum to which the algal filaments were attached. The calcium carbonate may be deposited as a crust surrounding the surface of the algal filaments, or it may settle out as a precipitate to form a lime mud after death of the plants. In the former case the filaments may be preserved, but seldom are in the latter. Since the limy substance increases by accretion, definite layering often results. In many limestones these layers are nearly concentric around the central nucleus, but in others they may be thicker toward the top, or they may be irregularly formed.

Algal limestones exhibit such a variety of forms that classification is difficult. In general, however, they may be grouped under (1) simple or pisolitic forms and (2) large colonial forms. In the simple forms, the limestone consists principally of small rounded or angular masses, usually under an inch in diameter, which are cemented together by lime mud. When sectioned, these masses show irregular concentric layering around a central nucleus. These simple forms are the most common type, and are formed in quite shallow lakes, estuaries, and seas. They are produced by a variety of organisms.

The colonial forms produce large "biscuits," or heads with a rough irregular surface. Sometimes they are made up of irregular concentric layers, but often they are composed of branching fingerlike processes. Frequently they are more or less coalescent, producing large compound colonies. *Solenopora*, is a well-known example of this type (Fig. 19).

In addition to the simple and colonial forms, a third type of algal limestone may be mentioned that consists of broken fragments of branching forms cemented together within a matrix. This is rarer than the nodular or large colonial forms and occurs principally in the late Cretaceous and Cenozoic.

It has long been customary when describing algal limestones to assign them to genera and species, but since many show no traces of the causal organism, the descriptions must be based upon external form, internal structure, etc. The genera to which they are assigned are form genera, and may include many different biological types. Thus, *Pycnostroma* is the form genus for many simple pisolitic limestones, and *Cryptozoon* (Fig. 23) embraces large spherical types made up of concentric layers. In some limestones it is possible to recognize traces of the organism, examples being *Solenopora*, *Lithothamnium*, and *Chlorellopsis*.

Of the very large number of algal limestones that have been described, only a selected few will be mentioned here. Among the most important are those described by the late Doctor Walcott from the Big Belt Mountains of Montana. These are of Algonkian (pre-Cambrian) age, and are of great scientific interest because they represent some of the oldest known rocks thought to be of organic origin. They cover an area of more than 6,000 square miles and are of great vertical extent. Actual remains of algae in these rocks are scanty, and for several years considerable doubt existed in the minds of certain authors whether these irregular masses resulted from the activities of algae or whether they were formed by inorganic agencies. However, it is now generally believed that they were precipitated by algae, probably of the blue-green group.

The most abundant form genus of the Big Belt Mountain area is *Newlandia,* which exists as irregular subspherical or frondlike growths constructed of concentric or subequidistant thin layers that are connected

Fig. 24.—*Gouldina magna.* Portion of a large colony showing wavy growth layers. Weber shale, lower Pennsylvanian. Park County, Colorado. Slightly reduced.

by irregular broken partitions. Four so-called "species" of *Newlandia* have been described and figured. *Collenia,* another genus, consists of more or less irregular dome-shaped bodies a foot or less in diameter, which grew with the arched surface uppermost. They appear to have increased in size by the addition of irregular external layers. *Camasia* is a compactly layered growth with numerous tubelike openings that give a spongoid appearance in cross section. These openings are smallest at the base. Other form genera such as *Kinneyia, Greysonia,* and *Copperia* are inorganic, and the names serve only for distinguishing certain types of rock formations in the localities where they occur.

The *Cryptozoon* reefs of Upper Cambrian age near Saratoga Springs, New York, are also examples of very characteristic formations of disputed origin. These reefs consist of flattened spherical masses, 2 feet or less in diameter, made up of concentric layers and so constructed that they resemble heads of cabbage. Whether these heads are of vegetable or animal origin, or are merely masses formed of inorganic colloidal material, has been a subject for considerable speculation. Recent studies strongly support the theory of algal origin.

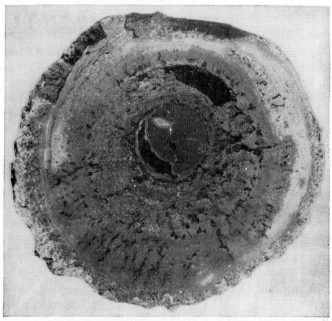

Fig. 25.—*Chlorellopsis coloniata.* Thick calcareous crust surrounding a silicified twig. Green River formation, Eocene. Sweetwater County, Wyoming. Slightly reduced.

Several forms of algal limestones have recently been described from the Pennsylvania rocks of central Colorado. They consist of simple forms (*Pycnostroma*) and large colonies that are described under such names as *Shermanophycus, Gouldina, Stylophycus, Artophycus, Calyptophycus,* and others believed to belong to the Myxophyceae. The outer surface of *Shermanophycus* is irregularly pitted and resembles a cauliflower. Internally, the colony consists of a thin basal mass of irregular concentric structure surrounded by fingerlike branching outgrowths of arched layers. *Gouldina* differs from *Shermanophycus* in that the central portion is relatively smaller and less distinct from the wavy outer portion. The filaments appear to have grown perpendicular to the surface in *Gouldina* but parallel to it in *Shermanophycus*. A cross section of *Gouldina* shows a series of concentric or eccentric layers that become more undulating farthei out from the center (Fig. 24). *Calyptophycus* is an interesting form in that it produces wartlike crusts on logs and other objects. It sometimes forms a thick sheath around casts of *Calamites* and other stems.

Another notable occurrence of algal limestones is in the middle Eocene Green River formation of northwestern Colorado and adjacent parts of Wyoming and Utah. Some of the largest reefs extend over several square miles and apparently formed on the bottoms of clear, warm, shallow lakes. Others known as the fringing type were formed near the shores. The algal colonies consist of aggregates of dome-shaped or roughly spherical masses that when sectioned are seen to be made of irregular concentric laminae typical of algal limestones in general. The organism mainly responsible for the formation of these Green River reefs is *Chlorellopsis coloniata*, and at one locality in Sweetwater County, Wyoming, it has formed thick lime deposits around the silicified branches of trees (Fig. 25).

Chlorellopsis coloniata is a spherical organism surrounded by a thin shell of interlocking grains of calcium carbonate. The spheres range from 110 to 140 microns in diameter, and usually occur in clustered colonies although rarely in chains. The genus is named from the living genus *Chlorella*.

Bryophyta

Fossil remains of the Bryophyta are rare, a fact which may seem strange in view of their abundance at the present time but which is not difficult to explain when the very fragile nature of the plant body is considered.

Until 1925 very little was known concerning liverworts in the Paleozoic and Mesozoic. The names *Palaeohepatica* and *Marchantites* had been applied to thalloid objects bearing some superficial resemblance to liverworts, but in most instances their affinities were admittedly uncertain. To date the only Paleozoic liverworts of which we possess any amount of knowledge are from the Middle and Upper Coal Measures of Great Britain. These are all referred to the form genus *Hepaticites*, which has been instituted for fossil plants which are obviously hepatics but cannot be assigned to true genera.

No reproductive structures have been found attached to any of the Paleozoic species of *Hepaticites*. The vegetative features, however, suggest affinity with the anacrogenous Jungermanniales. All species are dorsiventral. *H. Willsi* is a ribbonlike, dichotomously branched thallus with a growing point at the apical notch of each lobe. *H. Kidstoni* is more elaborately constructed. It has a central axis bearing four rows of leaflike lobes, two large and two small. The large lobes are rounded, broad objects attached in two opposite rows. The smaller lobes are on one side. Whether the latter are on the upper or the lower surface is unknown because the orientation of the material is undetermined.

Two species of Carboniferous mosses, *Muscites polytrichaceus* and *M. Bertrandi*, came from the late Carboniferous of France. All the material is fragmentary, and it is impossible to compare these forms with modern genera.

Our knowledge of Mesozoic Bryophyta is meager. Two species of *Hepaticites* have been described from the lowermost Jurassic of east Greenland and one from the late Triassic of England. *Naiadita*, a fossil of long disputed affinities and also from the late Triassic of England, is now definitely referred to the Bryophyta, and is probably the most completely known fossil representative of this group of plants.

Naiadita, with one species, *N. lanceolata*, is a small plant 1 to 3 cm. high, with a slender sparsely branched axis. The linear to rounded leaves are arranged in a loose spiral and are attached by a broad base. The shoots produce terminal gemmae cups, which bear gemmae resembling those of *Marchantia*.

Many of the shoots of *Naiadita* bear archegonia. These are lateral structures produced along the sides of the stem. There is no evidence that they are axillary, although they may replace leaves. The archegonia are sessile, but the spore capsules are borne on basal stalks about 3 mm. long. A "perianth" of four lobes develops around the base of the archegonium.

The mature sporophyte is a globular body less than 1 mm. in diameter that contains several hundred lenticular spores 80 to 100 microns in diameter.

The entire plant was attached by slender rhizoids.

Naiadita was originally supposed to be a monocotyledon allied to *Naias*. Later it was identified as a water moss, and then as a lycopod. Its Bryophytic affinities are now established beyond all doubt, although no recent family combines all its features. It is thought to be closest to the Riellaceae of the Sphaerocarpales, and in some respects resembles the genus *Riella*.

Fossil mosses in the Mesozoic rocks are almost nonexistent.[1] Probably the only reported occurrence is a fragment of *Sphagnum* leaf from the Upper Cretaceous of Disko Island, western Greenland.

[1] See, however, STEERE, W.C.: Cenozoic and Mesozoic bryophytes of North America, *Am. Midland Naturalist*, **36**, 1946, which appeared after this chapter was in press.

In the Cenozoic a number of liverworts and mosses are known which, like the flowering plants, are referable to modern genera. Most of the fossil mosses, however, come from Pleistocene deposits, more than 300 species having been reported from Europe and a much smaller number from North America. Probably the largest Pleistocene moss flora in North America is from the Aftonian interglacial deposits of Iowa where 33 species, have been found, four of which were described as new.

The fossil record of the Bryophyta, meager though it is, does show that the group is an ancient one which has been in existence since the Carboniferous at least. Furthermore, there is evidence that the two groups, the Hepaticae and the Musci, have existed as distinct lines since that time. The fossil Bryophyta throw no light on the question of vascular plant origin, and there is no reason to assume that they ever served as intermediate links in the evolution of vascular plants from algal ancestors.

References

BASCHNAGEL, R.A.: Some microfossils from the Onondaga chert of central New York, *Buffalo Soc. Nat. Sci. Bull.*, **17**, 1941.

BLACKBURN, K.B., and B.N. TEMPERLEY: I. A reinvestigation of the alga *Botrycoccus Brownii* Kützing, II. The boghead controversy and the morphology of the boghead algae, *Royal Soc. Edinburgh Trans.*, **58**, 1935.

BRADLEY, W.H.: Algal reefs and öolites in the Green River formation, *U. S. Geol. Survey Prof. Paper* 154 G, 1929.

DIXON, H.N.: Muscineae, *Foss. Cat.*, II. Plantae, Pars 13, Berlin, 1927.

FENTON, C.L., and M.A. FENTON: Algal reefs or bioherms in the Belt series of Montana, *Geol. Soc. America Bull.* **44**, 1933.

————: Pre-Cambrian and early Paleozoic algae, *Amer. Midland Naturalist*, **30** (1), 1943.

GOLDRING, W.: Algal barrier reefs in the Lower Ozarkian of New York, *New York State Mus. Bull.* 315, Albany, 1938.

GRUNER, J.W.: Discovery of life in the Archean, *Jour. Geology*, **33**, 1925.

HARRIS, T.M.: Naiadita, a fossil bryophyte with reproductive organs, *Ann. Bryol.*, **12.** 1939.

HOWE, M.A.: Plants that form reefs and islands, *Sci. Monthly*, **36**, 1933.

JOHNSON, J.H.: Algal limestones: their appearance and superficial characteristics, *Mines Mag.*, Oct., 1937.

————: Algae as rock builders, with notes on some algal limestones from Colorado, *Colorado Univ. Studies*, **23**, 1936.

————: Lime secreting algae and algal limestones from the Pennsylvanian of central Colorado, *Geol. Soc. America Bull.*, 51, 1940.

————: Geological importance of calcareous algae with annotated bibliography, *Colorado School of Mines Quart.*, **38** (1), 1943.

KIDSTON, R., and W.H. LANG: Old Red Sandstone plants showing structure, from the Rhynie Chert bed, Aberdeenshire, pt. V, Thallophyta, *Royal Soc. Edinburgh Trans.*, **52**, 1921.

—, ———: On the presence of tetrads of resistant spores in the tissue of Sporocarpon furcatum Dawson from the Upper Devonian of America, *Royal Soc. Edinburgh Trans.*, **53**, 1924.

LANG, W.H.: On the plant remains from the Downtonian of England and Wales, *Royal Soc. London Philos. Trans.*, **B 227**, 1937.

PIA, J.: "Pflanzen als Gesteinsbildner," Jena, 1926.

STEERE, W.C.: Pleistocene mosses from the Aftonian interglacial deposits of Iowa, *Michigan Acad. Sci. Papers*, **27**, 1941 (1942).

WALCOTT, C.D.: Pre-Cambrian Algonkian algal flora, *Smithsonian Misc. Coll.*, **64**, 1914.

WALTON, J.: Carboniferous Bryophyta, I. Hepaticae, *Annals of Botany*, **39**, 1925; II. Hepaticae and Musci, *ibid.*, **42**, 1928.

WHITE, D., and T. STADNICHENKO: Some mother plants of petroleum in the Devonian black shales, *Econ. Geology*, **18**, 1923.

CHAPTER IV

THE EARLY VASCULAR PLANTS

One of the decisive events in the evolutionary history of the plant kingdom was the appearance of the vascular system. This development probably took place soon after plants commenced sustained life upon land either during the early part of the Paleozoic era or possibly as long ago as the latter part of the pre-Cambrian. The problem of when vascular plants first emerged from the great flux of aquatic or semi-aquatic vegetation of the primeval world is increased not solely by the fragile construction of these ancient organisms but also by the extensive metamorphism to which the rocks of these early times were subjected and which obliterated almost the last traces of any kind of life ever contained in them. Metamorphism has most severely effected the pre-Cambrian rocks. Those of the Cambrian, Ordovician, and Silurian are less metamorphosed, but being principally in marine formations, the organic remains they contain are mostly invertebrates and algae. Only recently have we possessed explicit information on any pre-Devonian vascular plants, and nowhere have they been found in quantity. However, the ultimate discovery of land plants in the Silurian had long been anticipated, because the degree of development exhibited by some early Devonian types could be explained only by assuming that they were the products of a long evolutionary sequence.

The time and place of the origin of vascular plants, and the manner in which they developed, are some of the basic problems of evolutionary science upon which paleontological investigations have thrown no direct light. Theorists have speculated at great length upon these matters but perforce without the support of much evidence from the fossil record, which does not begin until land plants began to be permanently preserved. A theory that has found wide acceptance and is not contradicted by any facts revealed by the fossil record is that the vascular system developed among certain highly plastic marine thallophytes which grew along the shores probably at about the upper limits of the tide levels. At places where the shore line receded because of land uplift, these plants became stranded in temporary pools, which were isolated from the main body of water. As a result of this gradual change from an aquatic to a more terrestrial environment most of the stranded species became extinct, but a few of the more adaptable ones gradually became transformed to permanent life on land. The first bodily modification was probably the transformation of holdfast organs either into rhizomelike structures bearing rhizoids or into roots, and this was accompanied or closely followed by the development of cutinized surface layers that retarded water loss. Then, with the increase of the size of the plant body certain tissues became specialized for the conduction of food and water. This last development not only made further increase in size possible, but increased the life span of the plant by rendering it more capable of coping with environmental fluctuations. The spore-producing organs were elevated to the highest extremities where the widest possible dispersal was assured, but since water was still necessary for fertilization the spores germinated and produced gametophytes close to the surface of the ground. In this way alternation of generations with independent gametophytes and sporophytes was established at an early date.

The Nematophytales described in the preceding chapter possessed a cuticle but no vascular system. Too little is known, however, of this vaguely defined plant assemblage to enable us to look upon it with certainty as an intermediate stage in the evolution of the land habit although it is suggestive of such, and it may ultimately show something of the course followed during this early subaerial transmigration. Like most theories that have to be constructed from a meager array of facts, the postulation that land plants developed in the manner just described probably creates more problems than it solves. It leaves us completely in the dark as to whether all vascular plants are the ultimate descendants of a single transmigration that occurred only once, or whether several such movements have occurred at different places and at well-separated intervals extending even to within comparatively recent times. The diversity of form displayed among vascular plants may indicate the latter, and this may be the basic cause of many of the difficulties in the plotting of phylogenetic sequences.

THE PSILOPHYTALES

Our knowledge of the Psilophytales, the simplest and possibly the oldest of the known vascular plant groups, began in 1858 when Sir William Dawson published his first description of *Psilophyton princeps* from the Gaspé sandstone. The group was neither formally recognized nor named, however, until Kidston and Lang published the first of a series of articles in 1917 describing the plants from the Rhynie chert. In this paper *Rhynia Gwynne-Vaughani* was described and its evident relationship to *Psilophyton* was stressed.   Dawson's account of *Psilophyton* was founded upon detached fragments in which the internal structure was not well preserved, and the anatomical evidences of primitiveness so admirably revealed by the Rhynie plants were lacking.   During the long interval between Dawson's latest contributions on the subject and the discovery of the Rhynie plants, little was added to our knowledge of the morphology or relationships of *Psilophyton*, and some incredulity was expressed concerning the genuineness of Dawson's interpretation of

Fig. 26.—Section of Rhynie chert showing numerous stems of *Rhynia* cut transversely and obliquely.   × 2½.

it.   But the genus took on added significance after the Rhynie discoveries, and Dawson's restoration (Fig. 32) is now believed to be essentially correct.   *Psilophyton* is the type genus of the order.

The Psilophytales range from the Middle Silurian to the Upper Devonian but they are most prevalent in the Lower and Middle Devonian.   Known instances of their occurrence in the Upper Devonian are few and probably not entirely free from doubt.   Most of the psilophytes that have been reported from the Upper Devonian were either wrongly determined or the rocks were erroneously dated.

Characteristics

Although considerable diversification exists among the Psilophytales, the more typical ones exhibit the following characteristics:

1. The plant body shows a relatively low degree of organ differentiation and is simply and sparsely branched. Most members have rhizomes bearing rhizoids and aerial stems, but lack roots.

2. The stems are naked or spiny, with stomata on the surface, or with small simple leaves.

3. The vascular system, where known, is a simple strand that is either circular or lobed in cross section, with the phloem conforming to the outline of the xylem. Occasionally some of the lobes may become locally separated from the main strand but this is not a regular feature.

4. The spore-bearing organs are terminal, either on the main shoots or their branches or on short lateral shoots, forming a racemelike inflorescence. They are not associated with foliar organs. The sporangium is essentially an enlargement of the stem tip within which a single spore cavity had formed. The walls are several cells thick, and the spores are borne in tetrads. All known members are homosporous.

Classification

Since the Psilophytales as a group is somewhat arbitrarily defined, any proposed classification of its members is tentative, and subject to future modification or even complete replacement. If more were to be found out about some of the less well-preserved plants now placed in the Psilophytales, they would probably be transferred to other groups, and additional information on almost any of the genera might necessitate a rearrangement of generic and family limits within the order. Kräusel and Hirmer have both proposed classifications of the Psilophytales. Kräusel has set up nine families including more than 20 genera. Some of the families are well founded whereas others contain only one or two very inadequately known genera. Those families outlined in the following paragraphs are believed to be the most firmly established and to conform best to the Psilophytales as they are here defined. Members of two additional families proposed by Kräusel, the Drepanophycaceae and the Protopteridaceae, are discussed in this book under the lycopods and ferns, respectively.

Rhyniaceae.—The plant body consists of a cylindrical or lobed rhizome, which bears rhizoids and aerial shoots. It is rootless and leafless. The aerial shoots are simple or equally or unequally dichotomously forked, with stomata in the epidermis. The sporangia are terminal on the main shoots and borne singly or grouped into synangia. The family contains the genera *Rhynia, Horneophyton, Sporogonites, Cooksonia, Yarravia* and *Hicklingia,*.

Zosterophyllaceae.—The plant body consists of a profusely and irregularly branched rhizome bearing leafless aerial shoots. The sporangia are terminal on short side branches in racemose clusters at the tips of the aerial shoots. *Zosterophyllum* and *Bucheria* constitute the family.

Psilophytaceae.—The plant body is similar to that of the Rhyniaceae. The aerial shoots, especially in the larger portions, are ornamented with small, spinelike emergences probably glandular in nature. The sporangia are small, oval, and terminal on small branch ramifications. The essential genera are *Psilophyton, Dawsonites,* and *Loganella.*

Asteroxylaceae.—The plant body consists of a cylindrical branched rhizome of which some of the smaller subdivisions act as absorptive organs, and others of which become aerial shoots. The aerial shoots are covered in part with small simple leaves but the smaller twigs are spiny or naked. The xylem strand of the aerial shoot is stellate in cross section. *Asteroxylon* and possibly *Schizopodium* constitute the family.

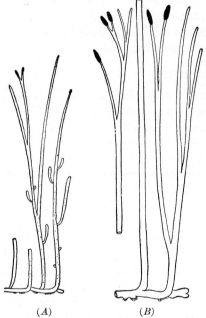

Pseudosporochnaceae.—The plants are comparatively large (2 m. or more high) and upright. The swollen base merges with a straight trunk that splits at the top into a digitate crown of slender branches. The smallest branches are dichotomously forked and some bear oval sporangia at the tips; others are prolonged as filiform segments probably serving as leaves. *Pseudosporochnus* is the only genus.

THE RHYNIACEAE

Rhynia.—This genus with its two species was discovered in the

(A)   (B)

FIG. 27.—Restorations of *Rhynia.* (A) *R. Gwynne-Vaughani.* (B) *R. major.* (*After Kidston and Lang.*) About ½ natural size.

Middle Devonian Rhynie chert in 1913, and subsequently described by Kidston and Lang in a series of papers published in the *Transactions of the Royal Society of Edinburgh.*

Rhynia is a simple plant of gregarious habit with a body consisting of a cylindrical subterranean rhizome that bore upright, leafless, simply branched, aerial shoots (Fig. 27). It has no roots or foliar structures of any kind. Instead of roots, the rhizome bore rhizoids which arose from the underside and which functioned to absorb water and minerals from the soil. The smooth aerial stems taper gradually from the base to the tips. Some of them are simple and others are forked once or twice into branches of equal size. The tips of the branches either are pointed or terminate in upright oval sporangia, which are slightly broader than the stems. The stomata are scattered over the surface of the aerial shoots among the epidermal cells.

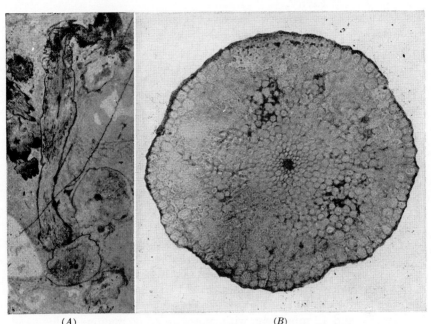

(A)   (B)

Fig. 28.—*Rhynia Gwynne-Vaughani.* (A) Transverse section of rhizome with a portion of the aerial stem attached. × 9. (B) Cross section of stem. × about 30.

The simplicity of the plant is exhibited by the internal anatomy as well as by the habit. The vascular system consists of a slender cylindrical column of annular tracheids surrounded by a layer of phloem four or five cells in radial extent (Fig. 28). The phloem is a simple tissue made up of elongated elements with oblique end walls. These evidently functioned as sieve tubes although no sieve plates have been observed. Whether they are absent or obscured by poor preservation is unknown. Immediately outside the phloem lies the broad cortex with no intervening pericycle or endodermis. In the largest stems the central tracheids of the xylem mass are the smallest although usually they are all very much alike in size.   Sometimes the vascular strand stops before it reaches the tips of the smallest branches.

The sporangia are oval or slightly cylindrical spore sacs somewhat pointed distally but at the base rather broadly attached to the tips of the aerial stems.   The sporangial wall is several cells thick and is rather distinctly differentiated into two layers, an outer one and an inner one, the latter probably having functioned as a tapetum.   The spores, which were heavily cutinized and borne in tetrads, occupied the large central cavity.

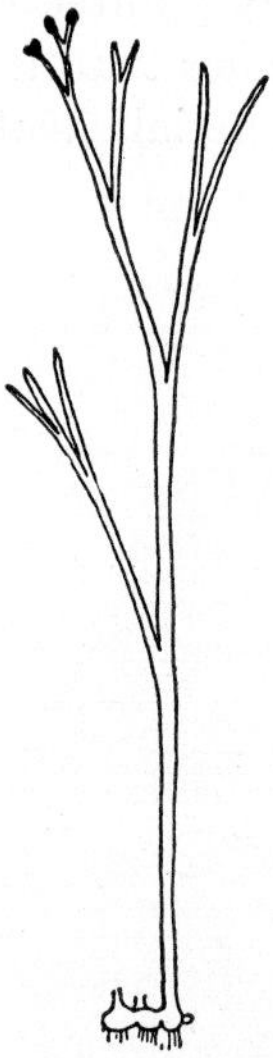

The two species, *Rhynia Gwynne-Vaughani* (Fig. 27*A*) and *R. major* (Fig. 27*B*), are much alike, but differ in several minor respects.   *R. Gwynne-Vaughani* is the smaller of the two and in some respects the more complex.   The aerial stems, which are about 2 mm. in diameter, were stubbed with small hemispherical outgrowths which sometimes developed into adventitious branches.   These branches possessed vascular strands of their own, which, however, were not connected with the vascular supply of the main stem. These side shoots were narrowly constricted at the point of attachment and became readily detached. The plant thus appears to have had a means of vegetative reproduction.

Fig. 29.—Restoration of *Horneophyton Lignieri.* (*After Kidston and Lang.*)   About ½ natural size.

The sporangia of *Rhynia Gwynne-Vaughani* measure from 1 to 1.5 mm. in diameter and are about 3 mm. long.   The spores are about 40 microns in diameter.

Rhynia major is a larger plant with aerial stems about 5 mm. in diameter and 40 to 50 cm. high.   The sporangia are as much as 4 mm. in diameter and 12 mm. long, with spores about 65 microns in diameter. Aside from over-all dimensions, the main difference between the two species of *Rhynia* is the presence in *R. Gwynne-Vaughani* of lateral propagative shoots.

Horneophyton.[1]—This genus with one species, *Horneophyton Lignieri* (Fig. 29), was found with *Rhynia* in the Rhynie chert.   It closely resembles *Rhynia* and sometimes occurs with it, but frequently it is the sole constituent of the chert.   The aerial stems range from 1 to 2 mm. in diameter and are dichotomously branched after the manner of *Rhynia*, but the rhizome is different.   The rhizome of *Horneophyton* is a lobed

[1] *Horneophyton* was originally named *Hornea* by Kidston and Lang, but the later discovery that the latter name had been used in 1877 for a member of the Sapindaceae necessitated the substitution of the new name.

tuberous body lacking a vascular system but bearing nonseptate rhizoids. In the upper part the vascular strand of the vertical shoot ends as an expanded vertical cup. Saprophytic or mycorhizic fungous hyphae occupy the spaces between the parenchymatous cells of the rhizome.

The sporangium of *Horneophyton* is a distinctive organ constructed similarly to that of *Rhynia* but it has a columella of sterile tissue projecting into the spore cavity from the base (Fig. 30). The spores, which are about 60 microns in diameter, are borne in tetrads (Fig. 30*B*). The diameter of the sporangium is only slightly greater than that of the

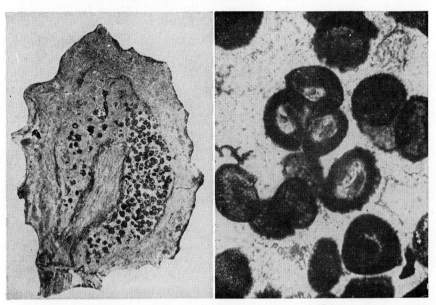

(A)   (B)

FIG. 30.—*Horneophyton Lignieri.* (A) Longitudinal section of sporangium showing the central columella and the overarched spore mass. × 25. (B) Spores, some still joined in tetrads. × 325.

stem which bears it. The sporangia are simple where they terminate a simple branch, but where they surmount branches in the act of dichotomy, the sporangium is a double one. The structure shows that the sporangium is essentially a stem tip modified for spore production. The wall surrounding the spore cavity is several cells thick, as in *Rhynia*, and lining the inside of the spore cavity is a well defined tapetum that extends over the columella. The central columella is continuous with the phloem of the main stem, and since the tapetum is a nourishing tissue for the developing spores, its continuity with the phloem indicates that the latter tissue served for the conduction of food as in Recent plants.

Other Members of the Rhyniaceae.—*Sporogonites exuberans,* from the Lower Devonian of Norway, is a small plant consisting of a slender unadorned stalk bearing a terminal spore case that varies from 2 to 4 mm. in diameter and 6 to 9 mm. in length. The basal half of the capsule is made up of sterile tissue, and above this and occupying the interior of the upper half is a dome-shaped spore mass. The spores are 20 to 25 microns in diameter and were formed in tetrads. There is no suggestion of any special mechanism for dehiscence. No vascular strand is visible in any of the stems, but this is believed to be due to faulty preservation.

Sporogonites has been assigned to the Psilophytales because of its general resemblance to the aboveground part of *Horneophyton*. It might also be interpreted as the sporophytic structure of a large moss, but since no mosses have been recognized with certainty in the Devonian, the chances of *Sporogonites* belonging to the Bryophyta are relatively slight. Its correct assignment, of course, depends upon the presence or absence of a vascular system. A second species, *S. Chapmani,* was found in the Upper Silurian of Australia.

Cooksonia, with two species, *C. Pertoni* and *C. hemispherica,* is an imperfectly known genus from the Downtonian (Lower Devonian) of Wales. It resembles *Rhynia* in having straight, upright, leafless, dichotomously branched stems. The distinguishing feature is the sporangium, which is slightly broader than long with a rounded upper surface. Very little is known of its internal anatomy.

Yarravia, with two species, *Y. oblonga* and *Y. subsphaerica,* is probably the oldest known member of the Psilophytales. It is from the *Monograptus* beds of Australia, which are thought by most geologists to belong to the Middle Silurian. It is a remarkable plant, in some respects differing from other members of the Psilophytales. The plant has a synangial fructification that is terminal on a straight naked stalk. The radially symmetrical synangium consists of five or six linear-oval sporangia fused laterally at the base but with the tips free. The two species are distinguished by the shape and size of the synangium. That of *Y. oblonga* measures about 5 mm. in diameter by 9 mm. in height, whereas that of *Y. subsphaerica* is about 1 cm. in each dimension. The arrangement of the sporangia in *Yarravia* is suggestive of the medullosan fructification *Aulacotheca,* but since the latter occurs in the Middle Carboniferous there is no reason to suspect an intimate relationship.

Hicklingia is a psilophyte from the middle Old Red Sandstone of Caithness in Scotland and is apparently related to *Rhynia,* although it is not sufficiently known to permit exact comparison. The plant produced a bushy growth of shoots from an indistinct rhizomatous base. The stems are naked, 1 to 2 mm. in diameter, and as much as 15 cm. long.

Branching is by repeated dichotomies with occasional transitions to a more or less monopodial habit. Some of the branchlets bear oval, coil-shaped sporangia at their tips.

Comparisons of the Rhyniaceae.—The presence of a vascular system in all members of the Rhyniaceae in which the tissues are preserved shows that these plants belong to the vascular cryptogams or Pteridophyta, and constitute the most primitive known representatives of this group. Of living plants, the Rhyniaceae are closest to the Psilotales, an order consisting of the two tropical and subtropical genera *Psilotum* and *Tmesipteris*. Roots are lacking in both of these genera, but the underground rhizomes are provided with nonseptate rhizoids much like those of *Rhynia*. *Psilotum* has no true leaves and the green, upright, aerial stems, always of small diameter, are dichotomously branched. *Tmesipteris* bears appendages somewhat resembling leaves, but they are flattened laterally and merge with the stem by strong decurrent bases. In both genera the sporangia are borne along the sides of the stems, not directly upon them but upon short lateral extensions originally regarded as sporophylls but now believed to be short branches. Since the Psilotales have no known fossil representatives it is unlikely that there is a direct connection between them and the Devonian Rhyniaceae regardless of the rather striking morphological similarities. Either the two groups represent independent lines of development or there is a long interval in their history yet to be discovered.

The rhizome of *Horneophyton* bears some resemblance to the protocorm of certain species of *Lycopodium* and *Phylloglossum*. This protocorm is a tuberous, parenchymatous body that develops before the root and is connected to the soil by rhizoids. In some species of *Lycopodium* the protocorm becomes fairly large and during the first season of growth functions as the main plant body.

The simplicity of the plant body of the Rhyniaceae receives special emphasis if it is compared with certain thallophytes and bryophytes. The low degree of organ differentiation is reminiscent of the generalized plant body of many Thallophyta, and except for the absence of a vascular system some of the higher algae have a body scarcely less complex. Some algae have rhizomelike basal portions which support cylindrical dichotomously forked upright stems which bear tetraspores at the tips.

Analogies between the Rhyniaceae and certain bryophytes are suggested by the simple leafless stems and terminal spore cases. In drawing comparisons between these two groups we are limited to the sporophyte stages because of our complete ignorance of the gametophytes of any of the Psilophytales. The dome-shaped columella in the spore case of *Horneophyton* resembles the similarly situated object in the capsule of *Sphagnum*. In *Sphagnum* the capsule, however, is nearly sessile on the gametophyte up to the moment when the sporophyte is mature, and it has a specialized means of dehiscence. No annulus or other dehiscence mechanism is known in any of the Rhyniaceae.

Some morphologists have laid stress on the similarity between the tuberous rhizome of *Horneophyton* and the foot at the base of the sporophyte in some species of the liverwort *Anthoceros*. They have endeavored to homologize the whole sporophyte of the Rhyniaceae with *Anthoceros* by interpreting the naked aerial stem portion between the rhizome and the spore case of the former as sterilized tissue derived from the lower part of a capsule of the *Anthoceros* type. In *Anthoceros* the sporophyte is an elongated body in which the sporogenous tissue extends from the base just above the foot to the tip. The sporogenous tissue is not continuous but is interrupted at intervals by masses of sterile cells. It has been suggested, but not proved, that the ancestors of the Rhyniaceae were plants much like *Anthoceros*. From the swollen basal portion comparable to the foot, which at the beginning of its growth was in close proximity to a gametophyte body of some kind, an elongated portion arose which consisted mostly of sporogenous tissue. Gradually the lower part above the foot became sterile, and this elongated still more to become the aerial stem. This contained a vascular strand (a structure not present in *Anthoceros*) which was able to carry nutrients to the sporogenous tissue which was elevated at some distance above the ground. The *Anthoceros* sporophyte, like that of the Rhyniaceae, bears chlorophyll and has stomata in the epidermis.

It is not likely that *Anthoceros* and the Rhyniaceae constitute a direct phylogenetic series even though they are similar in some ways. *Anthoceros*, as far as we know, is a Recent genus, and there is no evidence of its existence during the Paleozoic. But comparisons between these ancient vascular plants and modern liverworts are enlightening if properly interpreted. They show what some of the developmental trends might have been during that period of the history of the earth when vascular plants were in their youth. Also something is revealed of the general architectural plan according to which plants were built before all the organs took shape and of what the ancestral forms of the first vascular plants might have been.

THE ZOSTEROPHYLLACEAE

Zosterophyllum.—Of the three species that have been assigned to this genus (two from the Lower Devonian and one from the Upper Silurian) *Z. myretonianum*, from the lower Old Red Sandstone of Scotland, is the best known. In fact it is probably the most complete

Lower Devonian plant and only some of the details of the internal anatomy are lacking. The plant is small. The basal part consists of a profusely branched rhizome mass that bears sparsely branched upright aerial stems. The rhizomes are about 2 mm. in width and are nearly constant in diameter. In the center is a vascular strand about 0.25 mm. in diameter that contains annular tracheids. The branching of the rhizome is peculiar. Some of the branches continued to grow away from main central axes but others grew backward in the opposite direction, thus producing a tangled and complicated system (Fig. 31a). The plants apparently grew in dense tufts. The upright stems taper slightly and are covered by a heavy cuticle. No stomata have been

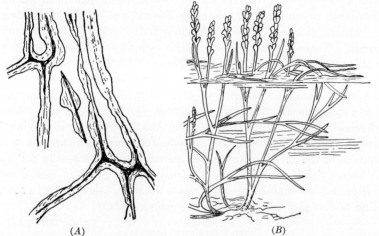

(A) (B)

Fig. 31.—(A) *Zosterophyllum myretonianum.* Obliquely and backwardly directed rhizome-like branches containing a vascular strand. (*After Kidston and Lang.*)
(B) *Zosterophyllum rhenanum.* Reconstruction. (*After Kräusel and Weyland.*)

observed but in the absence of leaves it is assumed that they were present in the stem epidermis. The sporangia were borne in a loose spikelike cluster on the upper part of some of the aerial shoots. Each sporangium is reniform and shortly stalked. The spores, which are unknown in this species, escaped through a slit along the top.

Zosterophyllum australianum, from the Walhalla (Upper Silurian) series of Australia, is similar in most respects to the preceding species, although the only parts known are the upright fertile stalks. The sporangia are 3 to 8 mm. broad and are spirally disposed around the upper part of the stalk. Dehiscence was by a terminal slit. The spores are about 75 microns in diameter. *Z. rhenanum* is from the Lower Devonian of Germany, and resembles *Z. myretonianum* except that it appears to have been partly aquatic (Fig. 31b). The slender flattened stems of the lower parts probably grew under water, and from these slender stalks arose branches which bore sporangia above the surface.

Since it occurs principally in the late Silurian and early Devonian, *Zosterophyllum* is one of the oldest members of the Psilophytales. It may, however, extend into the Upper Devonian because fragments of a plant known as *Psilophyton alcicorne*, from Perry, Maine, exhibit branching typical of *Zosterophyllum*. The genus agrees with other members of the Psilophytales in the relative simplicity of the plant body. The distinguishing feature is the position of the sporangia on short lateral shoots.

Bucheria.—This genus, with one species, *Bucheria ovata*, was discovered a few years ago in the Lower Devonian of Beartooth Butte, in northwestern Wyoming. It resembles *Zosterophyllum* in having slender stalks which bear closely set rounded bodies which are apparently sporangia. These lateral bodies number as many as 16, and are in two rows on opposite sides of the stem. Those of both rows are directed to one side, and are sessile and somewhat pointed distally. Most of them show a longitudinal median slit, which is probably a line of dehiscence.

Zosterophyllum and *Bucheria* have been placed in the family Zosterophyllaceae which is distinguished from the Rhyniaceae in having the sporangia borne in spikelike or racemelike clusters near the upper extremities of the aerial shoots. The sporangia are essentially psilophytic in being strictly cauline and terminal, although they are situated upon short lateral shoots instead of upon the main stems. The presence of a line of dehiscence is also a distinguishing feature.

The Psilophytaceae

Psilophyton.—This genus, which represents one of the most widely distributed of Devonian types, has been reported from several countries including Belgium, Canada, France, Great Britain, Norway, and the United States.

The remains of *Psilophyton* are usually fragmentary, and great care is necessary in identifying them. The name has long served as a ready catchall for indeterminate plant fragments consisting mostly of straight or crooked stems which may be either branched or unbranched, but which cannot with good reason be assigned to any other plant group. Because of the loose application of the name, *Psilophyton* has been greatly overemphasized as a constituent of the Devonian, especially in Eastern North America, where many supposed examples really belong *Aneurophyton*, *Drepanophycus*, *Protolepidodendron*, *Zosterophyllum*, or other genera. In North America the genus has been authentically identified at only a few places in Eastern Canada, eastern New York, and Wyoming. At other places the identifications have not been confirmed.

Psilophyton princeps (Fig. 32) is the best known species. It is a small plant, probably not more than a meter high, which grew in dense clumps along streams or in wet marshy soil of shallow bogs and swales. In the Gaspé sandstones the stems often lie in one direction on the slabs indicating inundation by overflowing streams. There is no evidence that the plant had roots, but, like *Rhynia*, a rhizome supported the aerial shoots. It also appears that small hairlike rhizoids aided in absorption from the soil. All parts of the plant are small, and neither the rhizomes nor the aerial shoots often exceeded 1 cm. in diameter.

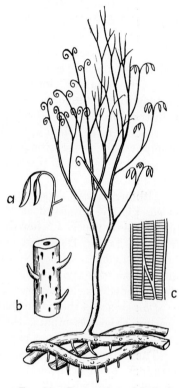

The aerial shoots of *Psilophyton princeps* are considerably more branched than those of *Rhynia*, and although the dichotomous plan is retained throughout, one branch often developed as a central leader and became larger and longer than the other, which terminated as a side shoot. Some of the branchlets are fertile, bearing at their tips small oval sporangia. The vegetative branchlets are often circinately coiled at the tips and resemble unfolding fern fronds. Covering most of the aboveground parts of the plant are short, rigid, pointed spines, which Dawson, in his original description, referred to as leaves. These spines are numerous on the larger stems but are sparsely

Fig. 32.—Restoration of *Psilophyton princeps: a,* paired sporangia; *b,* short section of stem, natural size, showing spines; and *c,* scalariform tracheids of the aerial stem. (*After Dawson.*)

scattered or sometimes absent from the smaller twigs. Little is known of the vascular system other than that it contains spiral and scalariform tracheids.

The sporangia are borne singly on the tips of shortly bifurcated branchlets and have the appearance of being produced in pairs. They vary somewhat in size, but usually range from 1 to 1.5 mm. in width and 4 to 5 mm. in length. They are generally ovoid, and round at the distal end, but sometimes pointed. The wall appears to consist of two tissue layers, although the exact structure is not distinctly visible. The spores range from 60 to 100 microns in diameter, have smooth walls, and were formed in tetrads. Since there is no segregation of spores on the basis of size, the plant apparently is homosporous.

Maceration techniques have revealed something of the structure of the epidermis and spines, both of which have been found well preserved in material from Eastern Canada and Scotland. The epidermal cells are isodiametric or slightly elongated and in the center of each there is a slightly projecting papilla. The stomata are scattered among the ordinary epidermal cells of the stem surface and consist of two guard cells surrounding an elongated pore. It is therefore apparent that the photosynthetic tissue was situated in the outermost tissue of the cortex as in the Rhyniaceae. The spines appear to have been of the nature of emergences rather than leaves because there is no evidence of a vascular supply within or of stomata on the surface. On exceptionally well-preserved material the tips appear to be slightly enlarged and to contain a dark substance which indicates that the spines functioned as glands.

Although about 20 species of *Psilophyton* have been named, none are as completely preserved as *P. princeps*, and many probably represent other generic types. A few that may be mentioned are *P. Goldschmidttii*, from the Lower Devonian of Norway, and *P. wyomingense* from the Lower Devonian of Beartooth Butte in Wyoming. These, however, differ from the type species only with respect to minor details, and well-founded specific differences are generally lacking.

Dawsonites is the name that has been applied to smooth sporangium-bearing twigs resembling the fertile branchlets of *Psilophyton*. It is strictly a form genus not essentially different from *Psilophyton*.

The name *Loganella* has been given certain profusely branched compressions found in the early Devonian at Campbellton, New Brunswick. The stems are without spines and the sporangia are unknown, but the combination of dichotomous and sympodial branching resembles that of *Psilophyton* more than *Rhynia*.

THE ASTEROXYLACEAE

Asteroxylon.—This genus has two species, *Asteroxylon Mackiei*, which was discovered in association with *Rhynia* and *Horneophyton* in the Rhynie chert, and *A. elberfeldense* from the Middle Devonian near Elberfeld, Germany.

Asteroxylon Mackiei, the first species discovered, is a larger plant than either *Rhynia* or *Horneophyton*, and it is preserved in a more fragmentary condition. It has not been possible to reconstruct the plant with the same degree of certainty as *Rhynia* or *Horneophyton* were restored because none of the stems were found in the upright position attached to the underground parts. Nevertheless, abundant and well-preserved material was found in the chert, from which it was possible to draw inferences concerning the habit even though actual proof was sometimes lacking.

The plant had a leafless dichotomously branched rhizome without roots or absorptive hairs, but certain of the smaller branches penetrated into the peaty substrate and thereby probably functioned as roots. The aerial stems were leafy and were connected with the rhizomes by a transi-

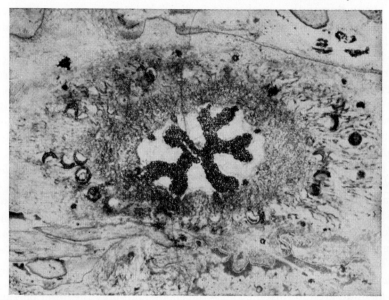

Fig. 33.—*Asteroxylon Mackiei.* Cross section of the aerial stem showing the stellate xylem strand, the foliar traces, and the partially disintegrated tissues of the cortex. Resting spores of the endophytic fungi are present in the outer cortex. × 18.

tion zone within which there were stomata in the epidermis and small scale leaves on the surface. The main stems, which are believed to have stood erect, bore lateral branches of smaller diameter. The latter branched by dichotomy, and dichotomous branching is occasionally observed in the main stems. In diameter the stems range from 1 mm. at the outer extremities to 1 cm. toward the base of the main axis. The leaves are simple structures 5 mm. or less in length and are provided with stomata. At the base of the stem they become smaller and intergrade with the scale leaves of the transition zone. In habit the plant has been compared with recent species of *Lycopodium.*

The vascular system of the rhizome and its rootlike branches consists of a cylindrical rod of spirally thickened tracheids surrounded by phloem, with no apparent distinction between protoxylem and metaxylem. In the aerial stem the xylem becomes stellate with four or more lobes (Fig. 33). In this part of the plant the protoxylem is situated near the outer extremities of the lobes but remains slightly immersed. The metaxylem is made up of spiral tracheids without conjunctive parenchyma. The phloem surrounds the xylem as in the rhizome and fills most of the space in the embayments between the xylem arms. Small leaf traces depart from the tips of the lobes, which terminate at the bases of the leaves without entering them.

The cortex of the aerial shoot is divisible into a narrow outer layer of tangentially extended cells and an inner layer that is sometimes uniform but often differentiated into three zones. The outer layer of the inner cortex consists of compact tissue, but the middle one is trabecular, consisting of vertically radiating plates of cells separated by large intercellular spaces. The innermost cortex is compact and as a rule is not clearly demarked from the outermost tissues of the stele, although rarely there is a suggestion of an ill-defined endodermis. The cortical tissues also contain nonseptate fungous hyphae and resting spores that belong to the form genus *Palaeomyces* (Fig. 18*A*).

No fructifications have been found attached to the stems of *Asteroxylon Mackiei*, but certain associated sporangia are believed to belong to them. These consist of pear-shaped sporangia about 1 mm. long attached to slender, leafless, branched axes. There is a line of specialized dehiscence cells at the enlarged apex. The spores are all of one kind and are about 64 microns in diameter.

The other species of *Asteroxylon*, *A. elberfeldense*, is larger than *A. Mackiei* and probably grew to a height of a meter. Like *A. Mackiei*, it has an underground rhizome and aerial shoots, but only the lower portions of the latter are leafy. The higher branches are spiny (like *Psilophyton*) and the uppermost ones are smooth. The shape of the xylem strand in the lower part of the upright stem is similar to that of *A. Mackiei* but it has a small pith which is a distinguishing feature. The protostelic condition prevails in the higher parts. The sporangia are borne at the ends of dichotomously forked twigs. They are not sufficiently preserved to show details of their structure.

Asteroxylon is an important genus from the phylogenetic standpoint in that it shows certain developmental trends not so clearly revealed among the simpler members of the Psilophytales. It is more complex than *Rhynia* or *Horneophyton*, but at the same time it is simpler than most other so-called "vascular cryptogams" both living and fossil. Anatomically and in habit it is suggestive of a *Lycopodium*, although the organization of the body into leafless rhizomes and leafy aerial stems is paralleled somewhat in the Recent Psilotales. The outstanding psilophytalean characteristic of *Asteroxylon* is the rootless underground rhizome supporting the rather simply branched aerial shoots; and the fructifications, if correctly assigned, furnish additional evidence of psilophytalean affinities. Points of advancement over the simpler psilophytes are the presence of leaves on the aboveground parts of the plant and a slightly greater degree of complexity of the vascular system.

THE PSEUDOSPOROCHNACEAE

Pseudosporochnus.—This genus, represented by one or more species, has been found in the Middle Devonian of Bohemia, New York, Norway, and Scotland. The Bohemian material is best preserved and has been most thoroughly investigated. The plant is different from other members of the Psilophytales in that it became a small tree probably 2 m. or more in height. A swollen base supported an upright trunk, which at the top split into a digitate crown of branches, all about the same size. These branches then forked freely to produce a bushy crown. The smallest branches are dichotomously forked, and some of them bear terminal, oval sporangia. The vegetative tips are filiform and probably functioned as leaves. Nothing is known of the internal anatomy of the plant.

MISCELLANEOUS PSILOPHYTES

There are many plant forms from the Devonian rocks that are evidently psilophytic but not sufficiently preserved to permit exact placement within the group. Most of them have the status of form genera. A few that deserve mention are *Thursophyton, Hostimella, Haliserites, Sciadophyton,* and *Taeniocrada.*

The name *Thursophyton* refers to small branched stems bearing simple, spirally or irregularly dispersed leaves The leaves are laterally expanded at the point of attachment and some of them may be spinose. The genus is provisional and the leafy stems probably belong for the most part to *Asteroxylon* or to primitive lycopods.

Hostimella is a form genus to which smooth, dichotomously or sympodially branched stems may be assigned. Some of the fragments referred to this genus belong to *Psilophyton, Asteroxylon,* and *Protopteridium.*

Haliserites is the name given to certain naked, dichotomously forked twigs from the Lower Devonian. Some of them are 1 cm. in diameter and show a central strand of scalariform tracheids Fructifications are unknown. It is probably not a readily distinguishable type.

Sciadophyton, which Kräusel has made the type of a separate family, the Sciadophytaceae, is a small plant consisting of a rosette of slender shoots attached to a central structure that anchored the plant to the soil. The shoots are rather lax, and some are forked and bear terminal enlargements, which are probably sporangia. *Sciadophyton* has been found in the early Devonian of Germany and the Gaspé coast in Canada.

Taeniocrada consists of flat, ribbonlike axes, sometimes as much as a centimeter broad, which are much branched and possess a central vascular strand. Toward the extremities some of the branches become cylindrical and bear terminal sporangia. The plant was originally described as an alga, but the presence of the vascular strand definitely removes it from this category and places it within the Psilophytales. The smooth stems resemble those of the Rhyniaceae, but the rather profuse branching and the small size of the fertile tips have more in common with the *Psilophyton* habit of growth. Its exact relations are uncertain.

Morphological Considerations

Although the Psilophytales are morphologically the most primitive of known vascular plants, their claim to antiquity is challenged by certain very ancient lycopodiaceous types. The Psilophytales are represented in the Silurian by *Cooksonia*, *Yarravia*, and *Zosterophyllum*. The contemporaneous or older *Baragwanathia* has a more lycopodiaceous aspect. In the early Devonian the Psilophytales are accompanied by *Drepanophycus* and *Protolepidodendron* both of which bear sporangia on the upper surface of sporophylls having the form of spinelike organs with a vascular strand. However, the affinities of many of these very early plants are conjectural, and it is believed that plants such as *Baragwanathia* and *Drepanophycus*, in which the sporangia are attached either to or close to leaflike structures instead of being terminal on the main stem or its branches, should be excluded from the Psilophytales. In delimiting the Psilophytales it seems advisable to follow the principle established by Kidston and Lang and to place emphasis on the terminal position of the sporangium, while at the same time not ignoring the other characteristics of the group, which are outlined at the beginning of the present chapter. However, if this principle is too rigidly applied, difficulties may arise with respect to certain other Devonian plants like *Protopteridium*, which has terminal sporangia but is more fernlike in other respects. Then *Calamophyton* and *Hyenia*, plants obviously early members of the sphenopsid complex, bear their sporangia upon short lateral stalks which may be the prototypes of the equisetalean sporangiophore.

These examples of plants that appear more or less intermediate between the Psilophytales and other groups serve all the more to stress the generalized character of the Psilophytales. They show evidence of a certain amount of divergence within the group while at the same time pointing back to the more primitive members from which they all probably developed at an earlier date.

The evolutionary significance of the Psilophytales is mainly that they demonstrate some of the basic characters possessed by the oldest vascular stock. The outstanding one is probably the general lack of organ differentiation within the plant body. Then, too, the psilophytic sporangium is clearly a part of the stem. This is strongly revealed in the Rhyniaceae. In *Zosterophyllum*, on the other hand, there is a tendency toward the formation of specialized sporangial branches although the terminal position is none the less retained. Even in *Asteroxylon*, where leaves are present, the leaves do not bear any intimate relation to the fructifications.

The fact has been stressed by authors that the Rhynie flora, upon which our concept of the Psilophytales is largely based, developed under a rather adverse environment, and that the relative simplicity of the plants may be an adaptation to special conditions. Such plants as *Rhynia* and *Horneophyton* are obviously xerophytes since their transpiration surfaces are relatively small. The atmosphere surrounding them contained the sulphurous vapors discharged from fumaroles, and the soil in which they grew was saturated with the acid waters from hot springs. However, it is very doubtful whether the environment can account for all the primitive features of the plants. It might explain the reduced transpiration surfaces but not the close relation between the sporangium and stem or the lack of roots. These latter features are primitive ones which have been retained from their still more ancient ancestral stock.

OTHER ANCIENT VASCULAR PLANTS

In addition to psilophytes, lycopods, and sphenopsids, the early land floras contained representatives of other groups, which, though often well preserved, are difficult to classify. A few of these will be briefly discussed.

Schizopodium.—This genus has two species, *S. Davidi*, from the Middle Devonian of Australia, and *S. Mummii*, from the Hamilton beds of New York. The only parts known are the stems. The stems are small, 3 to 15 mm. in diameter, and within each is a lobate stele 1 to 7 mm. in diameter. The xylem strand resembles that of *Asteroxylon*, but at the outer extremity of each lobe is a small zone of radially aligned tracheids, which resemble secondary xylem. However, no cambium can be identified in the otherwise rather well-preserved stem outside the xylem, and the supposition is that this tissue resulted from radial divisions of cells in the procambium and is therefore primary in origin. The protoxylem is situated just back of these masses of radially aligned cells and the central part of the strand consists of irregularly disposed tracheids such as commonly make up metaxylem.   In some stems one or more of the lobes may become detached from the main strand.   The protoxylem consists of annular and scalariform tracheids, but the metaxylem and the radially aligned tissue to the outside contain tracheids with bordered pits with oblique apertures.   These pits are distributed over the tangential as well as the radial walls.   The phloem forms a thin zone that follows the outline of the xylem, and in this respect differs from *Asteroxylon* where it fills the bays between the lobes.   The cortex consists of three rather distinct zones.   The epidermis is not preserved.   No leaves or leaf traces have been observed in connection with any of the stems.   Branching is dichotomous.

The shape of the xylem strand in *Schizopodium* suggests a relationship with *Asteroxylon* but it differs from this genus in the pitted tracheids, the absence of leaves, and in having tissue resembling secondary wood. In some respects the fern genus *Cladoxylon* is approached except that the stele is much less divided.

Barinophyton.—This strange plant of unknown affinities is rare but widely distributed.   It occurs in the Middle and Upper Devonian and has been found in Eastern Canada, western New York, Maine, Germany, New South Wales, and Norway.

The best known species of *Barinophyton* is *B. citrulliforme* from the late Upper Devonian of New York.   Only the fertile shoots are known. These are upright leafless stalks 20 cm. or more high and 6 to 7 mm. in diameter, which bear near the top alternately arranged, slightly flattened fertile branches with two rows of large, oval, appressed sporangia presumably on the lower surface.   These fertile branches are about 6 cm. long, and depart from the main axis at an angle of about 50 deg., but they appear to droop slightly, probably from the weight of the large sporangia.   The sporangia, which measure about 5 mm. in width by 7 mm. in length, are flattened between fleshy bracts of about the same size.   The large smooth spores are 0.3 mm. in diameter.   Judging from their size, they are probably megaspores.

The leafless stem of *Barinophyton* is suggestive of a psilophyte but the general habit and mode of branching is more indicative of a fern. It is possible that the plant produced separate fertile and vegetative shoots which are quite different in appearance.   This seems especially probable from the size of the fructifications, which appear too large to be nourished entirely by the leafless stems on which they are borne.

The name *Pectinophyton* (suggested by the comblike arrangement of the sporangia) has been used by European authors for material believed to be generically identical with *Barinophyton*.

Palaeopitys.—The name *Palaeopitys Milleri* was given to a fragment of poorly preserved secondary wood discovered many years ago by the Scottish geologist, Hugh Miller, in the Old Red Sandstone near Cromarty in Northeastern Scotland. Miller believed this wood fragment to belong to a gymnosperm. Further search in the same locality has yielded additional material, one specimen of which is a decorticated stem with primary and secondary wood. The secondary wood is a layer about 1 cm. thick around a primary strand about 1.5 mm. in diameter. The secondary wood contains small compactly placed tracheids and small rays. On the radial and tangential walls of the tracheids are as many as four rows of slightly oval, transversely elongated bordered pits. The primary xylem consists of a cylindrical core of tracheids with protoxylem masses near the periphery. The affinities of *P. Milleri* are unknown. The tangential pitting is a feature shared with *Aneurophyton* and *Schizopodium*. Other anatomical features suggest an ancient and primitive gymnosperm or pteridosperm, but there is no positive evidence that it is either.

Aneurophyton and Eospermatopteris.—These two genera are described together because it is now believed that the plants to which the names were originally applied are identical, with *Aneurophyton* having priority. *Aneurophyton* was proposed by Kräusel and Weyland in 1923 for material from the Middle Devonian at Elberfeld, Germany, only a short time before Miss Goldring described *Eospermatopteris* from rocks of similar age in eastern New York.

Aneurophyton was probably a large plant. The upper part of the straight trunk was considerably branched, and the branches seem to have possessed some of the characteristics of fronds, although the vascular strand in the smaller subdivisions is the same as that in the larger branches. The branchlets are alternately or suboppositely arranged, thus producing a monopodial rather than a dichotomous system (Fig. 34). They bear small, bifurcated foliar structures with slightly recurved tips, (Fig. 35*B*). The leaves are only a few millimeters long, are pointed, and lack a vascular supply. They seem to be arranged in three rows. The sporangia are small oval spore cases produced in loose clusters on the slightly recurved branch tips resembling modified foliar appendages, (Fig. 35*A*). Little is known of the spores, and the sporangia have no specialized means of dehiscence.

The petrified stem of *Aneurophyton* was first recognized in Germany, but it has recently been found that small pyritized stems from the Hamilton beds of western New York, which Dawson had many years ago named *Dadoxylon Hallii*, also belong to *Aneurophyton*. The primary xylem is a small three-angled tracheid mass without a pith and the protoxylem is probably situated somewhere within each of the three lobes

FIG. 34.—*Aneurophyton sp.* Large branch bearing numerous slender branchlets. Upper
Devonian. Eastern New York. Natural size.

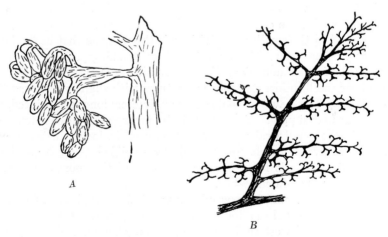

FIG. 35.—*Aneurophyton germanicum.* (*A*) Sporangia. (*After Kräusel and Weyland*)
Enlarged. (*B*) Vegetative branch. (*After Kräusel and Weyland.*) About natural size

(Fig. 36). The primary xylem is surrounded by secondary xylem made up of compactly placed tracheids and small, uniseriate rays. The tracheids bear multiseriate bordered pits on the radial and tangential walls, a type of distribution not common in woody stems but present in a few other Devonian plants. The structure of the phloem is unknown. The cortex contains small scattered sclerenchyma strands and just beneath the epidermis these strands are partly coalescent forming a regular layer. The largest stems observed with the cortex preserved are about 1 cm. in diameter but wood fragments from larger specimens have been found. As stated, the larger stems and smaller branches are similar in structure.

Fig. 36.—*Aneurophyton Hallii*. Transverse section of pyritized stem showing the three-angled primary xylem and the extensively developed secondary xylem. Tully pyrite, upper Middle Devonian. Erie County, New York. × 6⅔.

The name *Eospermatopteris* was given to casts of stumps and compressions of branch systems and fructifications found in the Hamilton group of Middle Devonian age near the village of Gilboa in Schoharie County, New York. The stump casts from this locality had been known for many years; in fact, Sir William Dawson first described them under the genus *Psaronius* believing that they belonged to tree ferns. When extensive quarrying operations were started in the vicinity of Gilboa about 1920, nearly 50 specimens of various size were uncovered, and these, in combination with other vegetative fragments, served as the basis for the well-known restoration of *Eospermatopteris* that has been reproduced in several recent textbooks. The restoration shows a tree 20 feet or more high with a swollen base and a crown of large drooping fernlike fronds surmounting the straight unbranched trunk. Some of the fronds are shown bearing pairs of small oval bodies at the tips of the smooth bifurcated ramifications which were interpreted at the time of their discovery as seeds.   These bodies, which vary somewhat in size, average about 3 by 5.5 mm. in breadth and length, and are believed to have been borne within cupules or envelopes, which probably opened at the tip in a manner similar to the cupules of some of the pteridosperms. In fact, it was the close similarity between these objects and the cupulate seeds of the Lyginopteridaceae that caused them to be interpreted originally as seeds.   The restored plant is often cited as the oldest seed plant.

The stump casts are large onion-shaped objects as much as 3 feet in diameter through the largest portion (Fig. 15), although most of them are smaller.   The lower part is rounded but the upper portion merges with the trunk with varying degrees of tapering so that the lower end of the trunk is seldom more than one-half the diameter of the largest part of the stump.   All the trunks were broken off 5½ feet or less above the stumps so that no extended portions were found on any of the specimens. In all observed instances the stumps rested in a thin bed of dark shale, which separated the massive sandstone layers.   Attached to the stumps at the broadest place were the carbonized remains of slender roots, which radiated away in all directions.   All these observations furnish ample evidence that the dark shale represents the original soil in which the trees were rooted and that the stumps were preserved *in situ*.   The only remnants of preserved tissues in the stumps were strips of carbonized fibrous strands forming a reticulate pattern over the surface.   Two pattern types were recognized.   In *Eospermatopteris textilis* the arrangement is a distinct network, but in *E. erianus* the strands extend in a more or less parallel direction.   It is now believed, however, that these two patterns are merely variations of one type.   With the exception of these fibrous strands all traces of the original tissue have disappeared, and the space formerly occupied has become secondarily filled with sand.

Our interpretation of *Eospermatopteris* has become considerably modified as a result of later studies and there remains no doubt that *Eospermatopteris* and *Aneurophyton* are names for plants of the same kind. The difference originally proposed, and unfortunately repeated in recent literature, that *Aneurophyton* reproduced by cryptogamic methods and *Eospermatopteris* by seeds, is no longer valid.   When the supposed seeds of *Eospermatopteris* were subjected to maceration, they were found to contain spores, which shows that they were simply seed-shaped spore cases.   Furthermore, these once supposed seeds were never found attached to vegetative organs of the type attributed to *Eospermatopteris*. The assumption, therefore, that the plant was seed-bearing is without the evidence necessary for proof. Moreover the identity of *Eospermatopteris* and *Aneurophyton* is undoubted because of the great similarity of vegetative parts assigned to them. The strong resemblance between the branch systems of the two plants was noted immediately after they were described, and more recently such remains have been found widely distributed throughout the Middle and Upper Devonian strata of eastern New York. Also in Germany, where *Aneurophyton* was originally found, fragments of casts showing the *Eospermatopteris* pattern were found associated with *Aneurophyton* twigs and foliage. Then an additional bit of data supporting the assumed identity of the two plants under consideration is supplied by the occurrence in eastern New York of typical *Aneurophyton* fructifications attached to vegetative remains identical with those supposedly representing *Eospermatopteris*. The only conclusion possible is that the two names designate identical plants neither of which produced seeds. *Eospermatopteris*, however, may be retained as an organ genus for bulbous stumps which probably belong to *Aneurophyton* and which show the characteristic reticulate surface pattern.

The position of *Aneurophyton* in the plant kingdom is not clear although it is one of the most completely preserved of Devonian plants. According to all available evidence it reproduced by cryptogamic methods but it is not definitely known whether homospory or heterospory prevailed. The ramification of the branches is suggestive of a frond although the similarity in anatomical structure of the larger and smaller axes and the lack of an abrupt transition between stem and leaf indicates the generalized plant body of the Psilophytales. Moreover, the sporangia are indicative of psilophytic origin. On the other hand, the vascular system with its solid, three-angled primary xylem core surrounded by compact secondary wood may indicate pteridospermous affinities. It somewhat recalls *Stenomyelon* of the calamopityean complex. *Aneurophyton* cannot with readiness be classified with the ferns because of the structure of the vascular system. It shows considerable evidence of psilophytic ancestory but otherwise seems to be in line with the primitive pteridospermous stock, which at this time apparently had not advanced beyond the fern level as far as reproductive methods are concerned.

References

DAWSON, J.W.: "Geological History of Plants," London, 1888.
DORF, E.: A new occurrence of the oldest known vegetation from Beartooth Butte, Wyoming, *Bot. Gazette*, **95**, 1933.
EAMES, A.J.: "Morphology of vascular plants," New York, 1936.
GOLDRING, W.: The Upper Devonian forest of seed ferns in eastern New York, *New York State Mus. Bull.* 251, 1924.

HALLE, T.G.: Notes on the Devonian genus Sporogonites, *Svensk bot. tidskr.*, **30**, 1936.
HØEG, O.A.: The Downtonian and Devonian flora of Spitzbergen, *Skr. Norges Svalb. Ish.*, No. 83, Oslo, 1942.
KIDSTON, R., and W.H. LANG: Old Red Sandstone plants showing structure; from the Rhynie chert bed, Aberdeenshire, pts. I–IV, *Royal Soc. Edinburgh, Trans.*, **51–52**, 1917–21.
KRÄUSEL, R., and H. WEYLAND: Beiträge zur Kenntnis der Devonflora, I. *Senckenbergiana*, **5**, 1923; II. *Senckenberg. naturf. Gesell. Abh.*, **40**, 1926; III. *ibid*, **41**, 1929.
———— and H. WEYLAND: Die Flora des deutschen Mitteldevons, *Palaeontographica*, **78** (B), 1933.
———— and H. WEYLAND: Neue Pflanzenfunde in rheinischen Unterdevon, *Palaeontographica.*, **80** (B), 1935.
LANG, W.H.: On the spines, sporangia and spores of *Psilophyton princeps* Dawson, shown in specimens from Gaspé, *Phil. Royal Soc. London Philos. Trans.*, **B 219**, 1931.
———— and I.C. COOKSON: Some fossil plants of early Devonian type from the Walhalla series, Victoria, Australia, *Royal Soc. London Philos Trans.*, **B 219**, 1930.
————, and I.C. COOKSON: On a flora including vascular plants, associated with *Monograptus*, in rocks of Silurian age, from Victoria, Australia, *Royal Soc. London Philos. Trans.*, **B 224**, 1935.
SCOTT, D.H.: "Studies in Fossil Botany," 3d ed., Vol. 1, London, 1920; Vol. 2, 1923.
VERDOORN, F.: "Manual of Pteridology," The Hague, 1938.

CHAPTER V

THE ANCIENT LYCOPODS

The lycopods (Lycopsida) are spore-bearing, vascular plants represented in the recent flora by the four genera *Lycopodium, Selaginella, Phylloglossum,* and *Isoetes.* All of these except the monotypic *Phylloglossum* are known in the fossil state. *Selaginella* is the largest genus with almost 600 species, and it, along with *Lycopodium,* is represented in the rocks by a series of fossil forms extending as far back as the Carboniferous. The somewhat enigmatic genus *Isoetes,* placed by some authors in the eusporangiate ferns, has been reported from the Cretaceous and Tertiary, and may be related to the early Mesozoic *Pleuromeia.*

All living lycopods are small plants but some of the extinct members were large trees. The group is characterized throughout by its simple foliage, which forms a dense covering on the branches and smaller stems, and in all except some of the very ancient forms the leaf tapers gradually from the base to a point at the apex. The vein of the leaf may be single or double, but it has never been observed to branch within the lamina of the leaf. The spores are borne in sporangia situated on the adaxial surfaces of the sporophylls, which in most species are organized into strobili. Some genera are homosporous but others are heterosporous. Anatomically the plants are primitive. The xylem cylinder may be a simple rod without a pith and nearly smooth on the surface, or it may be fluted to various depths, or dissected into several distinct strands. Secondary wood and pith are present in some extinct species. The protoxylem is exarch, and there are no leaf gaps in the primary vascular cylinder. The pitting in the tracheids is prevailingly scalariform.

Because of the great age of the lycopods their development constitutes an important chapter in the geologic history of the plant kingdom. Although they are distinct from all other plant groups as far into the past as they can be traced, it is probable that they evolved out of the psilophytic complex during the early Paleozoic. They developed rapidly during the Devonian and early Carboniferous, and reached their developmental climax during the middle of the latter period. They suffered a sharp decline during the Permian, and appear in greatly diminished numbers in the rocks of the Mesozoic.

The present chapter is concerned mainly with the lycopods of the Paleozoic, and for convenience in presentation the early primitive forms occurring mainly in the Devonian are treated first, to be followed by a discussion of the more highly developed genera of the Carboniferous.

THE OLDEST LYCOPODS

Most of the pre-Carboniferous lycopods are preserved as impressions and compressions of twigs and stems bearing on their surfaces the slightly enlarged bases of the leaves (Fig. 37). Often, however, they lack sufficient detail for satisfactory identification. The stems are frequently branched but sporangia are rarely present. Many of these obscurely preserved lycopodiaceous stems have been described and figured, and for want of suitable generic names have been assigned to the most

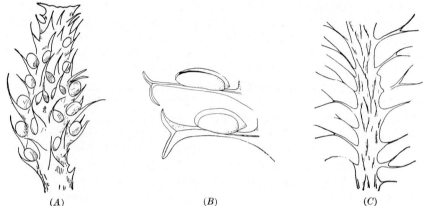

         (A)   (B)   (C)
Fig. 37.—(A) *Drepanophycus spinaeformis.* (*After Kräusel and Weyland.*) About natural size. (B) *Protolepidodendron scharyanum.* Bifurcate sporophylls bearing sporangia. (*After Kräusel and Weyland.*) × 5. (C) *Protolepidodendron wahnbachense.* (*After Kräusel and Weyland.*) × 1½.

familiar Carboniferous genera, to which they generally bear only a vague and superficial resemblance. Some of the better preserved forms have, however, received special generic names.

The oldest plant showing evidence of lycopodiaceous affinities is *Baragwanathia longifolia*, discovered a few years ago in the lower Ludlow beds of Silurian age in Australia. The plant has dichotomously branched stems that range from one to 6.5 cm. in diameter. The lax, spirally arranged leaves are almost 1 mm. broad and 4 cm. long, and each has a single vein. The reniform sporangia, which are axillary and restricted to certain portions of the stem, are almost 2 mm. in diameter and contain spores measuring 50 microns. Whether the sporangia are attached to the leaves or to the stem immediately above the leaves is not clear. The xylem strand is deeply fluted lengthwise and consists of annular tracheids. The habit of the plant is not well revealed by the fragmentary nature of the remains, but there is some evidence that there was a horizontal rhizome which bore upright branches.

In general appearance *Baragwanathia* must have closely resembled some of the herbaceous lycopods such as *Lycopodium*. The leaves are longer than in a typical *Lycopodium*, and they are supplied with vascular tissue such as is not found among the true psilophytes. The position of the sporangia is lycopodiaceous and not psilophytic. The outstanding difference between *Baragwanathia* and the true lycopods is that the xylem consists mostly or entirely of annular tracheids, a fact which is not strange when the great age of the plant is taken into consideration.

Drepanophycus (Fig. 37*A*), from the Lower and Middle Devonian of Canada, Germany, and Norway, was originally supposed to be a robust form of *Psilophyton* until it was found that it bore lateral sporangia, and that the spines scattered over the stem possess vascular strands and are therefore probably foliar structures. These stout falcate spines are mostly irregularly arranged, and are 1 or 2 cm. long. The stems are 2 to 5 cm. in diameter. It is believed that vertical shoots, many of them simply forked, arose from horizontal branched rhizomes. *Drepanophycus* is identical with the genus *Arthrostigma* (of Dawson), but the former name is now used on grounds of priority.

Drepanophycus and *Baragwanathia* are similar plants differing mainly in that the latter has longer, more slender, and more lax leaves. The sporangia appear to be similarly situated in both, although the exact relation to leaf and stem is not absolutely clear in either. Little is known of the internal structure of *Drepanophycus*.

Gilboaphyton is a genus based upon spiny stem compressions from the Middle Devonian at Gilboa, New York. The internal structure and fructifications are unknown. The slender dichotomously branched stems are similar to those of *Drepanophycus*, the chief distinction being the more regular arrangement of the spines. The stems average about 1 cm. in diameter and the spines, which are about 5 mm. long, taper from a broad base to a sharp point. The spines are alternately aligned in vertical series, and the tips are slightly incurved.

Protolepidodendron is a Devonian lycopod with dichotomously branched stems bearing small, closely set, linear leaves. In most species the leaves are bifurcated at the tip. The base of the leaf is enlarged to form a low, elongated, spindle-shaped or obovate cushion. These leaf bases usually appear on the compressed stem surfaces in vertical alignment, and those in adjacent rows alternate with each other. When the leaves are preserved they are usually attached slightly above the midportion of the cushion and frequently near the top. No abscission layer was formed before the leaves were shed, and consequently no scar such as characterizes a true *Lepidodendron* was formed.   The place of leaf attachment is usually preserved only as a small spot without any special markings or outline.

The oldest species of *Protolepidodendron* is *P. wahnbachense*, (Fig. 37C), from the Lower Devonian of Germany.   With the possible exception of *Baragwanathia* it is the oldest known lycopod.   *P. scharyanum* (Fig. 37B), from the Middle Devonian of Bohemia, is the most completely known of the genus.   It had a horizontal rhizome from which there arose dichotomously branched leafy stems.   No strobili were formed, but oval sporangia were borne on the upper surfaces of some of the leaves.   The xylem strand was a solid, three-angled core of scalariform tracheids.

Archaeosigillaria is the name for early lycopods supposed to bear certain characteristics of *Sigillaria*, although affinity with this genus is not necessarily implied.   However, it is frequently difficult to distinguish between *Archaeosigillaria* and *Protolepidodendron*, and the two names are sometimes interchanged.   *Archaeosigillaria* was first applied by Kidston to a specimen from the Upper Devonian of New York, which had previously been named *Sigillaria Vanuxemi*, and to some similar material from the lower Carboniferous of Great Britain.   The smaller stems bore spirally arranged, deltoid or acuminate leaves on fusiform bases, but on older stems the leaf bases had become hexagonal, resembling the condition in certain species of *Sigillaria*.   The surface of each leaf base bears a single print that represents the position of the vascular strand.   Nothing is known of the internal anatomy or of the fructifications.

Several years ago the compressed trunk of a large lycopod was discovered in the Upper Devonian Hatch shale in a creek bed near the village of Naples in western New York.   The plant was a small tree of which a portion of the straight unbranched trunk about 5 mm. long was preserved.   The specimen tapers very gradually from a diameter of 38 cm. at the swollen base to 12 cm. at the top where it was broken off. Nothing is known of the terminal portion or of its method of branching. The trunk is covered for its entire length with oval or spindle-shaped leaf cushions.   At the base these cushions are in vertical alignment, but at a distance of a meter or more above, the arrangement gradually changes over to a steep spiral.   The trunk therefore exhibits a unique combination of sigillarioid and lepidodendroid characters.   The leaf scar is a structure of considerable interest.   It is a small obovate or oval print rounded below and slightly cordate above, and it is situated just above the center of the cushion.   Above the notch at the upper edge is a small ligule pit which is similar to that of many of the arborescent lycopods of the Carboniferous.   The surface of the scar itself bears three small prints. The central one is the leaf-trace print, and the lateral ones mark the position of small strands of aerating tissue that accompanied the leaf trace through the outer part of the cortex. The leaves, a few of which remained in attachment, are not over 3 cm. long. They are simple-pointed, curved structures fastened to the cushions by a slightly broadened basal portion.

 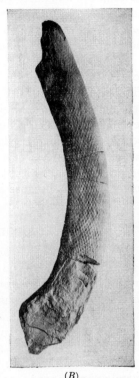

(A)   (B)

Fig. 38.—(A) *Colpodexylon Deatsii.* An ancient lycopod related to *Protolepidodendron.* Upper Devonian of eastern New York. Photo by Deats. Reduced. (B) "*Knorria*" *chemungensis.* Stem cast of an ancient lycopod showing obscurely preserved leaf bases. Originally described by James Hall in Part IV of "Geology of New York," 1843. Upper Devonian near Elmira, New York. About ⅓ natural size.

Dr. David White, who described this interesting plant, assigned it to the genus *Archaeosigillaria*, but the surface features of the trunk show rather conclusively that it is generically distinct from the material upon which Kidston founded the genus. It has been referred by some authors to *Protolepidodendron* but it differs from the more typical representatives of that genus. Probably a new genus should be established for it, but until the specimen is reexamined and the description is revised in the light of our most recent knowledge of the ancient lycopods, it seems best to retain for it the name *Archaeosigillaria primaeva* as originally proposed.

Lycopodiaceous stem impressions bearing alternately arranged, circular or somewhat rectangular leaf scars are frequently referred to *Cyclostigma*. These scars are flush with the stem surface, or slightly raised, but are never on definite cushions. The only other diagnostic markings are fine raised lines that form a network on the surface between the scars. *Cyclostigma* resembles *Bothrodendron*, a relative of *Lepidodendron*, but differs in the absence of a ligule above the leaf scar. The best known species is *C. ursinum* from the Upper Devonian of Bear Island.

Colpodexylon, a new lycopod genus with two species, was found recently in southwestern New York. One of these, *Colpodexylon Deatsii* (Fig. 38*A*), is from the Upper Devonian Delaware River flagstones, and the other, *C. trifurcatum*, is from the Bellvale sandstone of lower Middle Devonian age. The genus is characterized by a stem with a solid but deeply lobed primary xylem core and three-forked leaves that failed to absciss from low leaf cushions arranged in a close spiral simulating the appearance of a whorled condition. The three-forked leaf is unique among lycopods because in other members it is either simple or forked by a simple dichotomy.

LYCOPODS OF THE LATE PALEOZOIC

Any comprehensive account of the lycopods of the late Paleozoic deals principally with the Lepidodendrales, of which the most important genera are *Lepidodendron* and *Sigillaria*. The Lepidodendrales are arborescent, heterosporous, ligulate lycopods in which the genera are distinguished mainly by characteristics displayed on the surface of the trunk. Members of the order Lycopodiales were also in existence during the late Paleozoic as is indicated by the less conspicuous *Lycopodites* and *Selaginellites*. These are herbaceous types so named because of their general resemblance to modern species of *Lycopodium* and *Selaginella*.

Lepidodendron and Related Genera

Lepidodendron is one of the largest and best known of Paleozoic genera, and its remains are among the most abundant plant fossils found in the shales and sandstones of the Carboniferous coal-bearing formations. Judging from its prevalence, it was probably the dominant tree in many of the coal swamps. Because of the resistant character of the outer cortex the trunk surfaces are often well preserved, and the characteristic appearance renders the remains easily recognizable.

The large size attained by many of the trees is adequately revealed by trunks that have been uncovered in coal mines and quarries. In

England one trunk was reported to be 114 feet long to the lowest branches, and the branched part continued for at least another 20 feet. Petrified trunks 18 inches in diameter are known, and trunk compressions and casts a foot across are not rare. The upper part of the trunk branched by an unequal dichotomy to produce a large crown of leafy twigs. At the first forking of the trunk the larger of the two branches continued upward as the main axis, and the smaller one became a side branch. The ultimate branches, which are often but a fraction of an inch in diameter, bore simple linear or awl-shaped leaves. These leaves vary from 1 to 50 cm. or more in length and taper gradually to a sharp point. Each leaf is traversed for its full length by a single vein, and the stomata are situated in two bands on the lower surface. The leaves are arranged according to a steep spiral with a complex phyllotaxy.

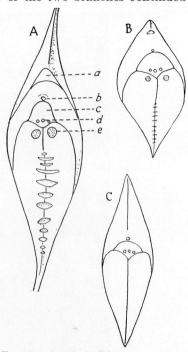

FIG. 39.—Lepidodendrid leaf cushions. (*After Zeiller.*)

(*A*) *L. aculeatum:* a, vestigial sporangial scar; *b*, ligule pit; *c*, leaf scar; *d*, bundle scar; *e*, parichnos. (*B*) *L. obovatum.* (*C*) *L. Veltheimi.*

The base of the trunk split into two large rootlike organs that immediately divided again, forming four arms extending outward into the soil at approximately right angles to each other. These large organs grew to a great length and branched repeatedly. They bore small slender lateral absorptive appendages arranged in an alternating pattern. The cones of *Lepidodendron* (*Lepidostrobus*) are elongated organs produced either at the tips of the slender twigs or laterally on the larger ones.

The leaves were borne on the summit of low, nearly contiguous pyramidal cushions of rhombic outline, which covered the stems (Fig. 39). Upon abscission, a flat rhomboidal scar truncating the apex of the cushion remained. These cushions were persistent on the trunk surface throughout the life of the tree, and nearly 100 species of *Lepidodendron* have been described on the basis of the variations in shape, size, and surface markings of the cushions as shown by compressions. Separating the cushions are narrow grooves that become wider on the older expanded portions of the trunk.

The vertical dimension of the low pyramidal leaf cushion is greater than the cross diameter, a feature distinguishing *Lepidodendron* (Fig. 40) from *Lepidophloios* (Fig. 41) in which the cross diameter is larger. The outline of the cushion is slightly asymmetrical due to the slightly oblique alignment on the trunk. The uppermost and lowermost angles of the cushion are attenuated, the lower usually more so than the upper. The

Fig. 40.—*Lepidodendron clypeatum.* From the Pennsylvanian ("Sub-conglomerate") near Pittston, Pennsylvania. Natural size.

uppermost angle extends slightly beyond the lower tip of the cushion next above. The flat leaf scar is usually situated slightly above the middle of the cushion, and it is either rhomboidal or slightly elongated transversely. The scar surface bears three prints. The central one, usually a round dot, marks the position of the vascular strand supplying the leaf, and the two lateral ones are strands of parenchyma (the parichni), which accompany the leaf trace through the outer part of the cortex and into the leaf. On the sloping surface of the cushion just below the scar and on opposite sides of the median ridge are two small but prominent prints which represent the ends of internal branches of the parichni which came to the surface below the leaf. The median ridge separating the left and right faces of the inferior part of the cushion is often marked by a series of cross striations or notches. On the cushion surface just above the scar is a small round print that represents the position of the ligule. *Lepidodendron* possesses this in common with *Selaginella* and other heterosporous lycopods. The two lateral angles of the leaf scar are usually connected to the corresponding lateral angles of the cushion by a pair of downward curved lines that separate the upper and the lower faces of the cushion. A median line may extend from the upper angle of the cushion in the direction of the scar, but in most forms it stops above a small, flat, triangular area above the ligule pit. This triangular area varies in size. Sometimes it covers most of the cushion surface above the ligule, but in other instances it is very small or absent altogether. It has been interpreted as a vestigial sporangial scar.

FIG. 41.—*Lepidophloios VanIngeni*. Des Moines series. Henry County, Missouri. Natural size.

Some departures from the form of cushion just described characterize other genera. In *Lepidophloios*, mentioned above, the transverse diameter of the cushion exceeds the vertical one, and the leaf scar is below the middle line. Anatomically, *Lepidodendron* and *Lepidophloios* are indistinguishable, and the latter is considered by some as merely a subgenus of the former. There is, however, a difference in the manner in which the cones of the two are borne, so the generic distinction is probably justified. In *Bothrodendron* the leaf scars are flush with the stem surface, no raised cushions being present.

An obvious limitation in the study of fossil plants is the impossibility of observing the changes that take place in the plant body during growth. It is generally supposed that the persistent *Lepidodendron* leaf cushion enlarged as the trunk grew, but the extent of enlargement is not known. In some so-called "species" there are separating furrows between adjacent rows of scars, although generally the cushions are in such close proximity, even in the largest trunks, that separation alone could not have accommodated the growth of the stem.

Careful observations of well-preserved imprints of the trunk surfaces of the Carboniferous lepidophytes show rather conclusively that increase in size of the leaf cushions was relatively slight. Otherwise they would have become noticably broader in proportion to the vertical dimension in passing from the uppermost parts of the plant toward the base. Consequently it is evident that some species had small cushions and others large ones. *Lepidodendron ophiurus, L. lycopodioides,* and *L. simile* are examples of small-cushioned forms with short leaves generally not exceeding a centimeter or two in length. Leaves are rarely found attached to large-leaved forms such as *L. aculeatum* and *L. obovatum,* but one such specimen of an undetermined species from the McLeansboro formation of Illinois has cushions about 3 cm. long to which are attached large leaves about a centimeter broad and at least 50 cm. long. The stem compression is about 10 cm. wide. Specimens of this kind furnish rather conclusive evidence that the large leaf cushion was not a structure that developed after the fall of the leaves but rather that it had reached its approximate ultimate size by the time the leaf was fully formed.

Although *Lepidodendron* leaves are usually regarded as small organs (Scott gives 6 or 7 inches as the maximum), it is quite likely that many of the undetermined linear compressions that are frequently observed but otherwise ignored are large lycopodiaceous leaves. The foliage-bearing twigs of the small-leaved species are more readily preserved, and these have given a somewhat erroneous idea of leaf size among the arborescent Paleozoic lycopods.

It is possible that the very long leaves of the type mentioned above grew by means of meristematic tissues located at the basal part just above the point of attachment, but there is no actual evidence in support of this possibility.

Sandstone and shale slabs bearing the cushion patterns of *Lepidodendron* and other fossil lycopods are often gathered by amateur collectors who mistake them for fish, lizards, or snakes.

The names *Bergeria, Aspidiaria,* and *Knorria* have been applied to lepidodendroid stems in various states of preservation. When the epidermis is lost, the characteristic appearance of the stem is altered. Such stems have been assigned to *Bergeria.* If decay has gone deeper, the *Aspidiaria* condition is produced. If all the tissue surrounding the xylem core has been removed, leaving only a shallowly fluted column, the specimen is known as *Knorria.* These, of course, are not true genera, or even good form genera, because they may all be represented by the same species, and are useful only for descriptive purposes, or, as Scott has pertinently remarked, they "are of botanical interest only insofar as they illustrate the difficulties of the subject."

Ulodendron is a stem having lepidodendroid leaf scars and two rows of large roundish depressions alternately arranged in vertical series on opposite sides of the stem. In the depression is a small stump that probably represents the stalk or axis of a lateral appendage, such as a branch or cone. *Bothrodendron punctatum* is often found in the *Ulodendron* condition.

Halonia differs from *Ulodendron* in having raised instead of depressed scars. These, however, often show a spiral rather than a vertical sequence. The *Halonia* type characterizes some species of *Lepidophloios*.

Leaves and cones are sometimes found attached to the smaller branches. The leaves of *Lepidodendron* were probably retained on the tree for several seasons and there is no evidence of periodic shedding. The characteristic appearance of the trunk is due to the persistence of the leaf cushions. Although a thick layer of secondary periderm was formed within the outer cortex in most species, it was not exfoliated as in most modern trees but persisted as a permanent tissue.

Anatomy.—As a result of the abundance of *Lepidodendron* and *Lepidophloios* in coal-balls and other petrifactions, the anatomy of these two genera is well known. In only a few instances, however, have anatomically preserved stems been found with their outer surfaces intact, with the result that different specific names have been applied to the same stems preserved in different ways.

A number of petrified *Lepidodendra* have been described from Europe, but from North America only three or four have been reported. Anatomically, the stems are comparatively simple, a fact in accord with the position of the lycopods in the plant kingdom. The essential features may be summarized as follows:

1. All forms agree in the possession of a single, unbroken vascular cylinder. The primary xylem is a continuous tissue of scalariform tracheids. The protoxylem is exarch, and the development was centripetal.

2. A pith may or may not be present. If present, it apparently represents undifferentiated primary xylem. Some forms have a pith in the main stem but not in the small branches.

3. Secondary xylem is present in some stems. If present, it always forms a continuous sheath of scalariform tracheids and rays around the primary part. In no known instance does the entire xylem cylinder greatly exceed 10 cm. in diameter. No growth rings or other evidence of seasonal activity are manifest.

4. The cortex is thick and consists of three layers. The inner cortex is thin but often preserved. The middle layer has usually disappeared

and the space has become filled with stigmarian rootlets which invaded the tissue after burial, mineral matter, and debris. The outer cortex is often reinforced by a thick layer of periderm, secondarily developed.

(A)

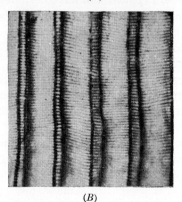

(B)

Fig. 42.—*Lepidodendron Johnsonii.* (A) Cross section of a portion of the vascular cylinder showing the narrow primary xylem layer and the extensively developed secondary xylem cylinder. × 3. (B) Scalariform tracheids of the primary xylem. × 70.

Previous to the rather recent discovery of two new species of *Lepidodendron* in North America, our knowledge of the anatomy of the genus had been based almost entirely upon European material. One of the new American species is *Lepidodendron Johnsonii* (Fig. 42), from the lower Pennsylvanian of Colorado. With the exception of a few details only, the Colorado species is as well understood as any foreign form, and it serves very satisfactorily as a basis for an understanding of the anatomy of the trunks of the genus.

None of the trunks of *Lepidodendron Johnsonii* were found complete, but judging from numerous well-preserved fragments, many of large size, the trunks must have attained a diameter of more than 0.5 m. In the center of the stem there is a large pith 3 or 4 cm. across, which is encircled by a cylinder of primary xylem 1 to 5 mm. thick. The tracheids next the pith are about 0.2 mm. in radial diameter, but diminish very gradually in size toward the outside. Just inside the outer surface they become considerably smaller. The tracheids are scalariform with elongated slitlike openings extending across the walls (Fig. 42*B*). The outer surface of the cylinder is ornamented with low protoxylem ribs from which the leaf traces depart. In cross section these ribs appear as small blunt outwardly projecting points resembling the teeth of a cogwheel. This toothed zone is called the "corona." The protoxylem consists of groups of very small tracheids at the points of the corona teeth.

Surrounding the primary xylem is a layer of secondary wood 3 cm. or more in thickness. The scalariform tracheids are arranged in regular radial series, indicating their origin from a cambium. The secondary xylem resembles the primary xylem except for the arrangement and slightly smaller size of the cells, and the presence of small rays. These rays range from 1 to 20 cells in height although most of them are less than 10. The ray cells are small and contain scalariform secondary wall thickenings similar to those of the ordinary tracheids.

The leaf traces leave the primary cylinder at a steep angle, but they soon bend outward and traverse the secondary xylem horizontally. In a tangential section the traces appear as spindle-shaped vascular bundles about equal in width to two tracheids. These traces continue outward through the phloem, cortex, and periderm, but because of poor preservation of the tissues immediately around the xylem their course cannot be followed satisfactorily through these tissues.

Because of its small size in proportion to the size of the trunk, the xylem cylinder must have served only for conduction, and most of the function of support was taken over by the thick layer of woody periderm around the outside. The plant is, therefore, an excellent example of physiologic specialization. Judging from the size and structure of the tracheids, the xylem must have been a relatively soft tissue that grew rapidly.

Nothing is known of the structure of the tissues between the xylem and the secondary periderm. The intervening space, about 8 cm. in extent, contained thin-walled cells which have disappeared, with the result that the central cylinder of each trunk can no longer be observed in its original position. Often pieces of several cylinders may be found within the large mineral- and debris-filled space surrounded by the periderm shell. Of course in *Lepidodendron* there was never more than one central cylinder within a trunk; the extra ones either floated into the empty cavities through the opening at the top, or the cylinder broke from its own weight into pieces after it was left unsupported by decay of the surrounding tissues. Similarly preserved trunks occur on the island of Arran on the west coast of Scotland.

The periderm of *Lepidodendron Johnsonii* is remarkable for its complexity. The thickness probably reached 20 cm., the most of all known Paleozoic lycopods. The tough tissue is composed of radially seriated cells arranged somewhat as in secondary wood. The cells are of three kinds: fibers, "chambered" cells, and secretory cells. The fibers are thick walled and vertically elongated, with transverse ends as seen in radial section, but pointed in tangential view. The "chambered" cells, which are located mostly in the inner part of the periderm, fit together in radial rows of a dozen to thirty. They are laterally enlarged cells with their flat tangential faces adjacent. The "chambered" cell rows are broadest in the mid-portion but diminish toward either end. These "chambered" cells are so named because some are divided by transverse septa. They have rounded or tapering ends. The secretory cells are arranged in glands, which occur as slightly wavy tangential bands in transverse section. These bands resemble growth layers when seen with the naked eye, but whether they have any seasonal relation is unknown. They probably secreted the waxy material which covered the surface of the trunk, and which is responsible for the frequent excellent preservation of *Lepidodendron* remains.

The leaf cushions and fructifications of *Lepidodendron Johnsonii* are unknown, although impressions of the *Lepidophloios* type are found intimately associated with the silicified trunks.

Lepidophloios Wünschianus, from the Lower Carboniferous of Arran, Scotland, is similar to *Lepidodendron Johnsonii* as pertains to the structure of the main trunk. Its smaller branches are without pith or secondary wood. It has been possible to study the structure of the stele at different levels in the main trunk. Near the base of the tree, the primary xylem is a solid core surrounded by thick secondary wood. At higher levels, the primary wood cylinder becomes larger and a central pith appears. The secondary wood, however, maintains a fairly constant thickness. These facts indicate that the plant was protostelic when young but became medullated as it increased in height.

Also similar to *Lepidodendron Johnsonii* is the Upper Carboniferous species, *Lepidophloios Harcourtii*, which was formerly confused with *L. Wünschianus*. The stem has a large pith, and the corona consists of rather prominent protoxylem points. Most of the material assigned to this species lacks secondary wood. *Lepidodendron brevifolium*, another lower Carboniferous species, differs from *L. Johnsonii* in the absence of a definite corona.

One of the best known species of *Lepidodendron* is the Upper Carboniferous *L. vasculare* (Fig. 43). This form has a solid primary cylinder although in the center the tracheids are mixed with parenchyma. These central tracheids are short cells with reticulate thickenings in the transverse end walls, a character said to be distinctive of the species. Second-

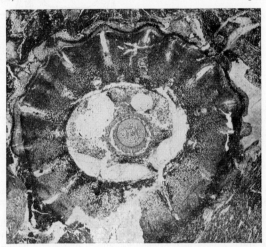

Fig. 43.—*Lepidodendron vasculare (L. selaginoides)*. Complete stem showing the vascular cylinder and the thick cortex. Coal Measures. Lancashire, England. × 3½.

ary wood may or may not be present in small twigs, and sometimes it forms a crescent only partly surrounding the primary core. A considerable amount of secondary periderm was formed. In transverse sections the parenchymatous strips of tissue through which the leaf traces traversed the cortex and periderm are conspicuous features.

The tissues between the stele and the preserved portion of the cortex in *Lepidodendron* are not well known because preservation within this part of the stem is seldom good. The phloem, where preserved, is extremely simple and small in amount, much resembling that of recent lycopods. A pericycle and an endodermis consisting of a layer of radially elongated cells have been observed in *L. vasculare*.

A second American species of *Lepidodendron* to be described in detail is *L. scleroticum*, from Coal No. 6 of the Carbondale formation of the middle Pennsylvanian of Illinois.   In bold contrast to the large stems of *L. Johnsonii*, all the specimens of *L. scleroticum* are small, probably representing the twigs and ultimate branches of a larger plant.   In the smallest twigs, having an over-all diameter of 3 to 10 mm., there is no central pith, the vascular core being protostelic.   Branches with these small solid steles exhibit no secondary development.   In slightly larger stems the protostele develops a parenchymatous interior, and when the stem exceeds 22 mm. secondary wood appears and the pith is well developed.

The distinctive feature of *Lepidodendron scleroticum* is the presence in the middle cortex of nests of sclerotic cells separated by the parenchyma

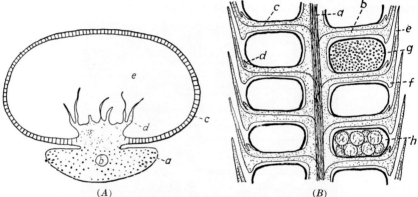

(*A*)   (*B*)

Fig. 44.—(*A*) Idealized transverse section of a sporophyll and sporangium of *Lepidostrobus* cut tangential to the cone: *a*, pedicel; *b*, vascular bundle; *c*, sporangium wall; *d*, nutritive tissue in sporangium cavity; *e*, spore cavity.   Enlarged.   (*B*) Idealized longitudinal section of a portion of *Lepidostrobus* showing relation of sporophylls to axis, and the two kinds of sporangia with their spores: *a*, cone axis; *b*, pedicel; *c*, vascular bundle; *d*, ligule; *e*, terminal bract; *f*, heel; *g*, microsporangium; *h*, megasporangium.   Enlarged.

accompanying the leaf traces.   Such structures had never been previously reported for any species of *Lepidodendron*.

The outer periderm layer arose from a phellogen situated at a depth of 5 to 10 cells beneath the stem surface.   The activity of this phellogen was different from that of modern trees in that it produced more tissue toward the inside (phelloderm) than to the outside (phellem).   The leaf bases have the form of those of *Lepidodendron Volkmannianum*.

Fructifications.—The fructifications borne by the majority of the Paleozoic lepidodendrids belong to the organ genera *Lepidostrobus* and *Lepidocarpon*.   Although ordinarily found detached, it is not infrequent that specimens are found in organic connection with leafy stems, and as a result we are supplied with more details concerning the fructifications of *Lepidodendron* and *Lepidophloios* than with most groups of fossil plants.

Lepidostrobus is an elliptical or cylindrical organ varying up to 6 cm. or more in diameter, and from 3 to 30 cm. or more in length. The cones usually taper slightly at the apex and base. Some were borne on short naked or scaly peduncles and others were sessile.

The general structural features of *Lepidostrobus* may be summarized as follows: The central axis bears closely placed spirally arranged or whorled sporophylls. The structure of the axis is similar to that of a young stem. The sporophyll is peltate, with the upper terminal lobe overlapping the sporophyll above (Fig. 44*B*). Each sporophyll bears a single saclike sporangium attached by its full length to the upper surface. Beyond the sporangium on the sporophyll is a ligule. All known species are heterosporous.

In structural plan the numerous species of *Lepidostrobus* are almost identical throughout the group, the specific differences having to do mostly with relative size and proportion of the parts.

The spirally arranged or verticillate sporophylls stand out from the axis at approximately right angles to it, or they may be deflected slightly downward because of crowding and the weight of the large sporangia. The horizontal stipitate portion bears the sporangium on the upper surface. The stipe terminates in a leaflike lamina or bract that turns upward, overlapping several sporophylls above. The length of this terminal bract differs in different species. In some it is longer than the stipitate portion. There is also a shorter downward prolongation of the lamina known as the "heel," which is completely covered by the lamina of the sporophyll below. Traversing the length of the sporophyll is a single vascular strand that arises from the protoxylem region of the cone axis.

The sporangium is a large saclike structure which is considerably wider than the sporophyll upon which it is borne (Fig. 44*A*). The wall consists of a single layer of prismatic cells. Extending upward into the spore cavity from the base are radiating flaps of tissue, the trabeculae, which served to nourish the developing spores. At the forward end of the stipe just beyond the sporangium is the ligule. This structure is constantly associated with heterosporous lycopods, both living and fossil.

Heterospory seems to have prevailed among all species of *Lepidostrobus*. Although homospory has been reported in one notable instance (*L. Coulteri*), no conclusive evidence was produced showing that another kind of spore was not present either in an unpreserved portion of the cone or in separate cones borne by the same plant.

The existence of homosporous lycopsids in the Carboniferous is proved, however, by an interesting fructification named *Spencerites* from the Coal Measures of Great Britain. In *S. insignis* the cones are rather small, only 8 to 10 mm. in diameter, and the axis bears spirally arranged or whorled sporophylls quite similar to those of *Lepidostrobus*. The chief structural difference lies in the fact that the ovoid or spherical sporangia are not attached to the surface of the pedicel, but rather by a short stalk at the distal end to the fleshy base of the upturned lamina. The spores, which are surrounded by a broad equatorial wing, measure about 140 microns in diameter exclusive of the wing, and are thus intermediate in size between microspores and megaspores. No ligule has been observed.

In *Lepidostrobus* cones with both kinds of spores preserved, the microsporangia are borne in the upper part of the cone and the megasporangia in the lower part The minute microspores are usually less

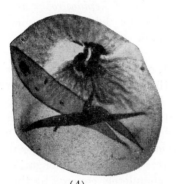
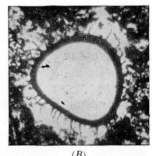

(A)   (B)

Fig. 45.—Isolated lycopodiaceous spores. (A) *Lepidostrobus braidwoodensis*. Carbondale group of Illinois. × 25. (B) *Bothrodendron mundum*. Section of megaspore bearing spiny appendages. Lower Yorkian of Great Britain. × 25.

than 50 microns in diameter and several hundred are formed within a single sporangium. The megaspores by contrast are large, ranging from 300 microns to 3 mm. or more in diameter, and the number per sporangium is distinctly limited. In *L. Veltheimianus* the number of megaspores varies from 8 to 16, and in *L. foliaceus* there appear to be but 4. Four megaspores have also been found in cones attached to stems of *Bothrodendron*. A significant trend in spore development is shown by *L. braidwoodensis* in which the megasporangium contains but one large mature megaspore (Fig. 45*A*). This spore, which slightly exceeds 2 mm. in diameter, is accompanied by the three dwarfed members of the original tetrad. This type of spore development is believed to represent a transitional stage in the evolution of the Paleozoic lycopod fructification from the free-sporing habit of the typical *Lepidostrobus* to the *Lepidocarpon* type in which the single megaspore is retained within a fruitlike or seedlike organ.

The female gametophyte developed within the wall of the megaspore in *Lepidostrobus* in a manner quite similar to that of *Selaginella*.

The megaspores of the Paleozoic Lycopsida are of considerable interest, (1) for purely morphological reasons, and (2) for their possible use in the correlation of coal seams. They display considerable variety in form, and their heavily cutinized exines are frequently ornamented with elaborate appendages (Figs. 7*A*,*B*, and 45). With the aid of an ordinary simple lens these spores can frequently be seen in bituminous coal as small compressed yellowish or amber-colored bodies, and in thin sections of coal they often appear in thin bands parallel to the bedding plane. By macerating the coal with Schulze's reagent or dilute nitric acid they can frequently be removed in large numbers. Macerated coal is probably more useful than thin sections in studying spores because by this method they can be separated completely from the matrix and their surface features observed.

Isolated megaspores of the type produced by the Paleozoic Lycopodiales and Lepidodendrales are usually assigned to the form genus *Triletes*, a name proposed by Reinsch in 1884. Although the accurately drawn figures show conclusively that Reinsch was dealing with lycopod megaspores, he was mistaken concerning their nature. He thought they were algal organisms and that the long appendages were parasites. As Reinsch used the term *Triletes*, it is applicable to more or less triangular, semielliptic or circular bodies, usually compressed, with the surface variously ornamented, and with three converging ridges. In modern terminology, the triradiate ridge distinguishes this type of spore from the *Monoletes* type, which shows but a single mark.

Several attempts have been made to classify lepidodendrid megaspores on the basis of size, shape, and surface ornamentation. Classification of fossil spores of any kind, however, is confusing because of considerable lack of agreement concerning the taxonomic status of the numerous groups or divisions instituted by authors for the reception of the various kinds. Authors have not always explicitly stated whether they intend newly proposed names as taxonomic categories that are subject to the rules of priority or merely as descriptive terms. The most generally followed system of classification is that proposed by Bennie and Kidston in 1886. These authors created three divisions of Reinsch's genus *Triletes*. Division I, *Laevigati*, embraces spores which have a smooth surface or very fine granulations and which are similar to those of *Selaginella Martensii*. The triradiate ridge is prominent and the lines divide the central portion of the spore into three flattened or depressed areas. *Triletes Reinschi* is a widely distributed type that has been reported from Scotland, Poland, the Ruhr Basin, and Illinois. In

Division II, *Apiculati* the outer surface of the spore wall bears mamillate spines. The spores are comparable to those of *Selaginella haematodes* or *Isoetes echinospora*. An excellent example of this division is *Triletes mamillarius*, found in the Carboniferous coal pebbles in the glacial drift at Ann Arbor, Michigan. Division III, *Zonales* (Fig. 7B), includes spores bearing long appendages around the equatorial zone. These appendages form an irregularly reticulated system which produces a winglike expanse of varying extent and which probably facilitated dispersal by the wind. These appendages were thought by Reinsch to be parasites attached to the central body. Sometimes they are nearly as long as the diameter of the spore body itself. The spore body is roundish-triangular, and the lines of the triradiate ridge extend to the equatorial zone. The spore is comparable in some respects to that of *Selaginella caulescens*. Examples are *Triletes rotatus* and *T. superbus*, both from the Carboniferous coal pebbles at Ann Arbor, Michigan, and *T. triangulatus* from Illinois, Germany, Poland, and Scotland.

The fourth spore category described by Bennie and Kidston is the group *Lagenicula* (Fig. 45A), in which the oval or rounded spore has a necklike projection which is lobed into three subtriangular segments. The surface of the spore is smooth or covered with bristlelike hairs or spines. Taxonomically the group is handled differently by authors. Some employ *Lagenicula* as a generic name and others use it as a section or subgenus under *Triletes*. In plants that produced the *Lagenicula* type of megaspore one tetrad only developed within the megasporangium, and from this one spore only matured at the expense of the other three. The three aborted members of the tetrad are present as objects having much smaller diameters than the fully developed spore.

The classification scheme which Bennie and Kidston proposed is admittedly artificial, as indeed must any classification be which treats detached and isolated parts which cannot always be referred to the plant which bore them. Not until a large number of spore forms are found definitely associated with the reproductive organs will it be possible to classify spores into natural groups. Some recent authors have attempted to arrange spores into more natural groupings on the basis of size, extent of triradiate ridge, and general shape of the spore body.

Within recent years a large body of literature on the taxonomy of isolated spore types and their use in stratigraphic correlation has developed. Probably the most thorough work in this field is that of Zerndt on spores of the Carboniferous of Poland. In Germany similar investigations have been carried out by Potonié, Loose, Wicher, and others, and in the United States by several individuals including Schopf and Bartlett. In Great Britain considerable attention has been given the correlative value of microspores. Among these investigators widely differing practices have been followed with respect to spore classification and nomenclature. The literature is too extensive and the whole problem too involved to lend itself to brief summarization, and the individual papers should be consulted by anyone desiring information on this specialized subject.

Lepidocarpon (Figs 46 and 47) is a striking example of the high degree of development attained by the Paleozoic lycopods at the climax of their development. The name is applied to strobili similar in their

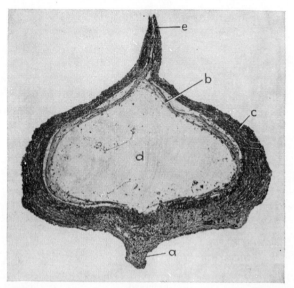

Fig. 46.—*Lepidocarpon sp.* Cross section of the megasporophyll and "seed": *a*, pedicel; *b*, megasporangium; *c*, protective portion of megasporophyll ("integument"); *d*, megaspore with prothallial tissue; and *e*, micropyle. Pennsylvanian. Indiana. × 7½.

general plan of organization to *Lepidostrobus* in that the central axis bears spirally arranged peltate sporophylls, each of which bears a single elongated sporangium attached by its full length on the upper surface. Within the megasporangium only one megaspore tetrad developed, and in most cases it is evident that a single megaspore matured at the expense of the others. The mature megaspore is large and is elongated to correspond somewhat to the size and shape of the sporangium. At maturity the megaspore was retained within the megasporangium, in contrast to the freesporing condition in *Lepidostrobus*, and the entire structure consisting of sporangium and spore was enclosed by lateral outgrowths of the pedicellate portion of the sporophyll. The enclosure was complete except for a narrow, slitlike opening along the top. At maturity the sporophyll and sporangium with its megaspore and contained female gametophyte was shed as a unit. The female gametophyte is made up of thin-walled cells, and aside from being enclosed, the structure closely resembles the gametophyte of certain species of *Lycopodium.* Archegonia were formed in the upper part near the micropylar slit.

Lepidocarpon possesses considerable evolutionary significance in showing that some Paleozoic lycopods had evolved to the point of seed production. The resemblances to a seed are obvious, the main one

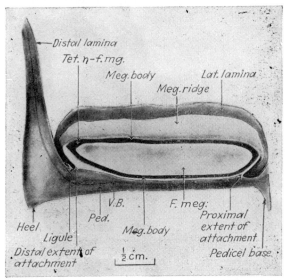

Fig. 47.—*Lepidocarpon ioense.* Longitudinal section. (*After Hoskins and Cross.*) Enlarged.

being the retention of the female gametophyte and megaspore within the megasporangium and the protection of the whole structure by an integumentlike outgrowth of the sporophyll (Fig. 48). Moreover, the microspores gained access to the female gametophyte through a micropylar opening. However, the organ differs in several respects from modern seeds. In the first place, the so-called "integument" is an outgrowth of the sporophyll rather than an ovule, so that in this respect the organ is more nearly analogous to a fruit. In *Lepidocarpon* no true integument, nucellus, or ovule exists. Then also, the micropyle is a narrow slit instead of a circular opening. Lastly, no embryo has been found, a feature shared with other Paleozoic seeds.[1] The best reason for regard-

---

[1] Darrah claims to have found embryos in cordaitean seeds in coal-balls from Iowa, but without figures or a more detailed account the report is hardly credible.

ing the *Lepidocarpon* fructification as a seed is that it functioned as such. The sporophyll outgrowths protected the megasporangium and megaspore, thereby permitting the latter to store a large supply of food, and also provided a receptacle for the microspores which facilitated fertilization.

The microspores of *Lepidocarpon* apparently were borne in strobili of the *Lepidostrobus* type, of which the identity was long a matter of doubt. Quite recently, however, small spores found within the sporangial cavity of *Lepidocarpon magnificum* appear identical with spores present in

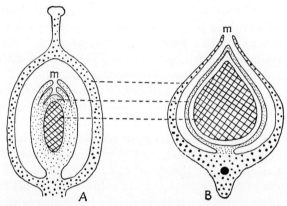

Fig. 48.—Diagrams showing the resemblances and differences between an angiospermous pistil and a lepidocarp. (*A*) Generalized longitudinal representation of pistil (ovary wall coarsely stippled) containing a single orthotropous ovule (finely stippled) with micropyle (*m*) and double integument. Within is the mature megaspore containing gametophytic tissue (cross hatched). (*B*) Lepidocarp consisting of megasporophyll (coarsely stippled) of which the lateral margins form a protective covering around the megasporangium (finely stippled) but leaving a micropylar opening (*m*) along the top. The megasporangium bears a single megaspore which contains gametophytic tissue (cross hatched).

strobili resembling *Lepidostrobus Coulteri*. These spores are about 26 microns in diameter and were produced in large numbers.

Lepidocarpon fructifications have not been found attached to the vegetative parts of the plants on which they were borne, but it is believed that they belong to some of the species of *Lepidodendron* or related genera. In the compression state it is difficult to distinguish *Lepidocarpon* from *Lepidostrobus*, and it is probable that many so-called *Lepidostrobi* are lepidocarps. In *Lepidocarpon mazonensis*, a species recently described from Illinois, the terminal bract of the sporophyll is about three times as long as the horizontal portion that supports the sporangium, and in the detached state the sporophyll would closely resemble such forms as *Lepidostrobophyllum majus* or *L. missouriensis*.

The first species of *Lepidocarpon* to be described was *L. Lomaxi* from the Lower Coal Measures of England.   Before Scott observed the fruits attached to the cone axis, thus proving their lycopodiaceous affinities, petrified specimens had been described as *Cardiocarpon anomalum.* Another well-known form is *L. Wildianum,* from the Lower Carboniferous Calciferous Sandstone at Pettycur, in Scotland.   It differs from *L. Lomaxi* only with respect to minor details.   In 1931, Noé reported *Lepidocarpon* in coal-balls from Illinois, and since then it has been found in considerable abundance in the middle Pennsylvanian of Illinois and Iowa.

The name *Cystosporites* has been proposed for large, isolated, saclike spores believed to belong to *Lepidocarpon.*   These spores are roundish or elongated and may attain a length of 10 mm. or more.   They were produced from tetrads of which one member only reached maturity.   The thin unornamented wall is of peculiar construction.   Instead of appearing granular, as in the case of most spores, it is made up of loosely matted anastomosing fibers, which probably served to facilitate the transfer of food from the outside to the developing gametophyte.

In *Miadesmia* the seedlike habit is further approached in that the megasporophyll forms a complete roof over the megasporangium, leaving only a small micropyle at the distal end.   The sporangial wall is less developed than in *Lepidocarpon,* and in other respects the fructification is decidedly more seedlike.   It is essentially lycopodiaceous, however, as the "integument" is derived from the sporophyll.   The broad terminal lamina probably aided in dispersal.

SIGILLARIA

This genus, the "seal tree," so named because of the appearance of the trunk surface, ranks next to the lepidodendrids in importance among the Paleozoic lycopods.   Like *Lepidodendron,* it is arborescent, although usually less profusely branched.   In some forms the trunk was short and thick.   *Sigillaria reniformis,* for example, had a trunk 6 feet in diameter at the base, but tapered to 1 foot at a height of 18 feet.   In other species the stem was tall and slender.

As in *Lepidodendron,* the leaf scars were persistent, and numerous species have been described, the variations shown on imprints and compressions of the trunks being used as a basis for separation.   *Sigillaria* is usually recognized by the vertical arrangement of the scars, as contrasted with the steep spiral sequence in *Lepidodendron.*   The leaf scars themselves are similar to those of *Lepidodendron,* although the cushion is less prominently raised and in many forms is apparently absent.   The species of *Sigillaria* have been separated into two groups, the *Eu-Sigillariae*, with ribbed stems (Fig. 49*A*), and the *Sub-Sigillariae* (Fig. 49*B*), without ribs. In former times the presence or absence of ribs was supposed to possess considerable taxonomic importance within the genus, but we know now that the two surface patterns occasionally merge and become inseparable on a single trunk. In general, however, the distinction holds.

Partial destruction of the outermost tissues of sigillarian stems often produces changes that are similar to those brought about in *Lepidodendron*. One very characteristic form is *Syringodendron* (Fig. 50) in which

(*A*)   (*B*)

Fig. 49.—(*A*) *Sigillaria scutellata.* (*Eusigillaria.*) Saginaw group, Lower Pennsylvanian. Michigan. (*B*) *Sigillaria ichthyolepis* (*Sub-Sigillaria*). Western Pennsylvania. Natural size.

the leaf scar has disappeared leaving the large and conspicuous pair of parichnos prints. The vascular bundle scar is usually invisible. In *Eu-Sigillariae* the ribs remain, but corresponding conditions in the *Sub-Sigillariae* lack them. The *Syringodendron* appearance is sometimes produced at the bases of large trunks where excessive growth has obliterated the leaf scars.

In the *Eu-Sigillariae*, the ribbed forms, the furrows between the ribs may be straight, with the ribs broader than the leaf scars (*Rhytidolepis* type), or the furrows may be zigzag (*Favularia* type) with a marked transverse furrow between scars on the same rib. Examples of the former are *Sigillaria ovata, S. mamillaris,* and *S. scutellata* (Fig. 49*A*). *S. elegans* is an example of the *Favularia* type.

The *Sub-Sigillariae* are divided into the *Clathraria* and *Leiodermaria* types. *Clathraria* resembles *Favularia* to a great extent in that the low cushions are close together so that the oblique alignment is more pronounced than the vertical one. It has some resemblance to *Lepidophloios*. *Sigillaria ichthyolepis* (Fig. 49*B*) is a familiar example. In *Leiodermaria* there are no raised cushions, and the scars are well separated by the surface of the epidermis, as in *Bothrodendron*. *S. Brardi* is usually cited as an example, although it frequently merges with *Clathraria*.

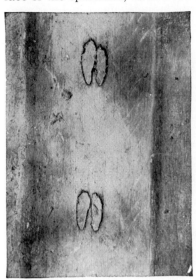

Sigillarian leaves, known as *Sigillariophyllum* when detached, are long grasslike organs similar to *Lepidophyllum*. In some, the vascular bundle is double as in the needles of certain conifers. The leaf margin is enrolled with the stomata in furrows on the lower surface. Lining these furrows are multicellular epidermal hairs, which obviously aided in reducing transpiration.

Anatomy.—Petrified *Sigillariae* are rare, and only one specimen has been reported from North America. The stem is constructed on the same general plan as in *Lepidodendron* although it seems to exhibit features indicative of a greater degree of advancement. As far as is known, a pith is always present, and the primary xylem is a cylinder of scalariform tracheids either continuous (Fig. 51), as in *S. elegans*, or of distinct strands, as in *S. Menardi*. *S. spinulosa* is transitional with the strands partly confluent. The protoxylem is located at the apex of the coarse undulations of the outer edge of the primary wood. Secondary xylem is always present and usually forms a thin layer that conforms to the wavy surface of the primary xylem. The tracheids are scalariform or occasionally pitted, and the rays are small and narrow. A layer of secondary periderm furnished support to the trunk as it did in most species of *Lepidodendron*.

Fructifications.—The fructifications attributed to *Sigillaria* are of two kinds, *Sigillariostrobus* and *Mazocarpon*. *Sigillariostrobus* is similar to *Lepidostrobus* in consisting of a central axis bearing whorled or spirally arranged sporophylls, each supporting a single sporangium on the

FIG. 50—*Sigillaria sp.* Decorticated stem showing the "*Syringodendron*" condition in which the point of attachment of the leaf is represented by two large parichnos scars. Slightly reduced.

adaxial surface. The main distinction between it and *Lepidostrobus* is that the upward pointing lamina at the end of the sporophyll is an acicular bract varying in length in different species. In compressions this pointed structure presents a characteristic appearance quite different from the more obtusely pointed bracts of *Lepidostrobus*.

Sigillarian cones were attached by long peduncles to the trunk or to stout branches just below the leaf-bearing tip portions. They are quite large: *Sigillariostrobus Tieghemi*, for example, grew to a length of 16 cm., and *S. nobilis* to 20 cm. All known specimens are heterosporous.

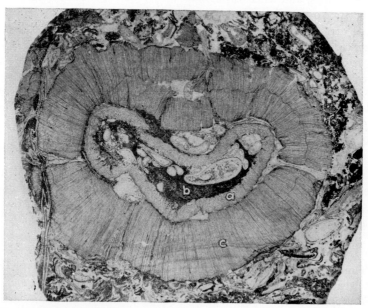

Fig. 51.—*Sigillaria sp.* Transverse section of stem. Coal Measures of Great Britain: *a*, primary wood; *b*, pith containing *Stigmaria* rootlets; and *c* secondary wood. × 2½

Mazocarpon is a fructification supposedly belonging to *Sigillaria* and similar in its general plan of organization to *Sigillariostrobus*. The distinctive feature is the saucer-shaped megaspores, of which there are usually eight, situated in two rows within the megasporangium around a central nutritive pad. At maturity, the entire sporophyll with sporangium attached was shed from the cone axis as a unit. The spores apparently escaped only after disintegration of the sporangium.

Mazocarpon is sometimes regarded as a seedlike organ because, when the megasporangium broke up, fragments of tissue generally adhered to the megaspores, often partially surrounding them. Prothallial tissue has been found within. However, it must be noted that the situation is quite different from *Lepidocarpon*, in which only one megaspore matured and which was completely protected by the sporophyll. In *Mazocarpon* a number of spores matured within the megasporangium, and it seems advisable to regard it as merely showing an advanced condition of heterospory rather than having attained a condition of seed production.

The discovery in 1884 of the fructifications of *Sigillaria* was of far-reaching importance. It not only revealed the true affinities of the genus but it also showed that cryptogamous plants may produce secondary wood. Previous to this, secondary wood was thought to be peculiar to seed plants, and when Brongniart described *Sigillaria Menardi* it was placed in the so-called "gymnospermous dicotyledons." He had held a similar erroneous view concerning the position of *Calamites* in the plant kingdom, not knowing that a plant with secondary wood might also be a spore producer. In 1873 Newberry concluded that the seeds found above the Sharon coal in Ohio belonged to *Sigillaria* because he thought he could account for the fructifications of all the other plants present. The pteridosperms were then unknown and Grand'Eury's announcement of the seeds of *Cordaites* had not appeared. At this time no secondary wood had been observed in *Lepidodendron* and this genus had been correctly placed among the vascular cryptogams. The discovery that *Sigillaria* bore fructifications essentially similar to those of *Lepidodendron* revealed its lycopodiaceous affinities, and also proved conclusively that secondary wood is not peculiar to the seed plants. This paved the way for a better understanding of the relationships of other Paleozoic plants.

FOLIAGE

The detached leaves of *Lepidodendron* and *Lepidophloios* are placed in the organ genus *Lepidophyllum* (Fig. 52), and those of *Sigillaria* in *Sigillariophyllum* or *Sigillariopsis*. Leaves in which the tissues are preserved are often present in abundance in coal-balls and other types of petrifactions, but it is seldom that they can be attributed to any particular stem. Only a few examples are on record of leaves attached to petrified stems although attachment is frequently observed in compressions.

In the detached condition it is usually impossible to distinguish the leaves of the different stem genera, although those of *Sigillaria* are usually longer than those of *Lepidodendron*. The leaves of one species, *Sigillariophyllum lepidodendrifolium*, are reported as having reached a length of 1 m. This is unusual, as lycopodiaceous foliage is typically small, and *Lepidophyllum* does not ordinarily exceed a few centimeters. In shape the leaves of Paleozoic lycopods vary from acicular, needlelike organs similar in appearance to the needles of the longleaf pine, to lanceolate structures more resembling the leaves of *Taxus baccata*.

They are never broad, and attachment is always by the entire width of the base, often at the summit of a persistent cushion. In compressions the simple midrib is usually prominently displayed. Although the same general plan of construction prevails throughout the leaves of all arborescent lycopods, there are notable differences in shape as viewed in cross section and in tissue arrangement.

In practice, the name *Lepidophyllum* is used for both compressions and petrifactions of *Lepidodendron* leaves. *Sigillariophyllum*, however, is used only for compressed sigillarian foliage, and *Sigillariopsis* denotes most petrified forms. *Lepidophyllum* possesses a single xylem strand, but the strand in *Sigillariopsis* is frequently double. Graham, who has recently made a special study of the anatomy of the leaves of Carboniferous arborescent lycopods, proposes to make the single and the double xylem strand the distinguishing feature of the two leaf genera *Lepidophyllum* and *Sigillariopsis* regardless of the plant to which they may belong. This proposal is open to criticism for two reasons. The first one is that on the basis of the single and double strand, some species of *Lepidophyllum* may belong to *Sigillaria*, since it has never been conclusively demonstrated that the sigillarian leaf always possesses a double strand. The double strand does, however, appear to be confined to *Sigillaria*. The second criticism is that we know almost nothing of the structural differences that may be found in different parts of the same leaf. The shape of the leaf changes considerably from the base to the apex, and until serial sections are cut along the entire length of the leaf, we will not know whether the single and double xylem strands are continuous or whether a single strand may divide somewhere between the base and the apex to form two. However, used purely as descriptive categories for leaf sections, the terms may be convenient when defined in this way.

In cross section, the different types of leaves vary considerably in shape. Some forms are round or triangular, and others are flatly rhomboidal. In some species of both *Lepidophyllum* and *Sigillariopsis* there are two large stomatal furrows paralleling the vein on the lower surface. The furrow may be a deep narrow groove, or it may be a rounded or broadened channel with slightly overarching edges. In other forms the stomata are not in furrows, but in bands on the surface.

The epidermis is a single cell layer consisting of small circular and larger rectangular cells. In the stomatal bands the circular cells are absent, and the tissue consists of the regular cells, which may be rectangular or nearly square, and the guard cells. The stomata are oriented in the direction of the length of the leaf. They are not sunken, but are on a level with the other cells of the epidermis. They are usually fairly numerous, and sometimes there are as many or nearly as many stomata as other epidermal cells. The epidermis is supported beneath by a hypodermis of varying thickness. In some species it is a single layer of cells, but in others (in fact, in most species) it may extend inward for a distance of six or eight cells. The hypodermal cells are thick walled, round in cross section, and taper pointed. The tissue is continuous except opposite the stomata, and it may vary in thickness in different parts of the same leaf. Beneath the hypodermis is the mesophyll, a tissue of thin-walled cells that completely surrounds the midrib. It may consist of isodiametric cells irregularly arranged, of cells in chains with large intercellular spaces, or of elongated, palisadelike cells.

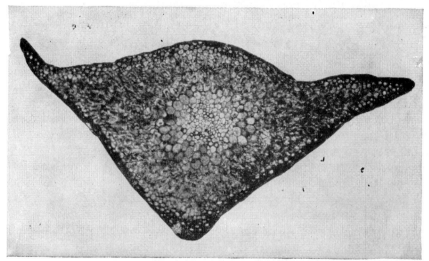

Fig. 52.—*Lepidophyllum sp.*   Transverse section.   Pennsylvanian of Illinois.   × about 50.

The midrib is completely or partly surrounded by a layer of transfusion tissue, which sometimes forms a sheath up to four cells thick. The cells of this layer are rather short and broad with nearly transverse end walls and reticulate secondary thickenings. Occasionally, a few thin-walled cells are interspersed among the thick-walled elements.

Between the xylem and the transfusion tissue is a continuous layer of thin-walled cells that is partly phloem. Since it is usually difficult to identify sieve tubes in fossil material, the extent of the actual phloem is difficult to determine, but it is generally assumed that the phloem occupies the lower side. The xylem is a cylindrical or flattened strand consisting of a few to several dozen scalariform tracheids. The smaller elements, presumably the protoxylem, occupy the two lateral margins, although they sometimes extend downward along the abaxial margin. No secondary tissue has ever been observed.

Rootlike Organs of the Paleozoic Lycopods

The arborescent Paleozoic lycopods possessed an extensive subaerial system of rootlike organs of which the most common form is known as *Stigmaria*. The remains of these are the most abundant plant fossils in Carboniferous rocks, and are often present in large numbers even when other plant remains are absent. Although they occur in all kinds of fossiliferous rocks, including coal-balls, they are most prevalent in the

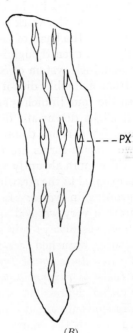

(A)   (B)

Fig. 53.—(A) *Stigmaria ficoides.* Rootstock showing lateral rootlets and rootlet scars. Saginaw coal, Lower Pennsylvanian. St. Charles, Michigan. About ⅓ natural size. (B) Sketch of tangential section through the main stigmarian axis showing the large primary "rays" and the xylem of the appendages (*px*) occupying the upper angles. The lower end of the sketch is toward the apex of the axis. Slightly enlarged.

underclay upon which the coal seams rest. This clay is usually a non-laminated earthy mass which is believed to represent the original top soil in which the coal swamp plants were rooted. The stigmarian remains in the underclay are seldom petrified, and exist mostly as mud-filled cylindrical casts surrounded by a thin crust of carbonaceous material (Fig. 16). These casts permeate the underclay in all directions and sometimes can be traced for distances of many feet.

The majority of stigmariae belong to the large artificial species *Stigmaria ficoides* (Figs. 16, 53, and 54). It seems that this particular type of rootlike organ belonged to many different species and to even different genera of lycopods, but those connected with different plants show few distinguishing characters that enable one to separate them. *S. ficoides* has been observed attached to stumps, showing the surface features of both *Lepidodendron* and *Lepidophloios,* and in some instances to *Sigillaria.*

Some modern authors use the specific name *verrucosa* in place of the more familiar and widely employed *ficoides* on grounds of priority. The former was proposed by Martin in 1804 and the latter by Sternberg in 1820. But since the work of Sternberg is one of those that has been agreed upon as the basis of paleobotanical nomenclature, *S. ficoides* may be validly used even though it is not the older name.

The rootstock system of *Stigmaria ficoides* consists of four main axes arising from a double dichotomy at the base of the trunk. In some instances the four branches spread abruptly away from the trunk base with but slight downward deflection, but in others they slope steeply downward for a short distance before spreading. It is believed that these two growth forms are the result of the habitat of the individual plants. The more abruptly spreading system anchored the plant to soil that was exposed to the air, while the downwardly deflected one supported plants growing in water. In all cases, however, the plants most certainly grew in very wet soil although not necessarily in situations where the surface was always submerged. The individual axes extended for long distances, branched occasionally by a simple dichotomy, and tapered gradually to a diameter of 2 or 3 cm. at the tips.

The casts of *Stigmaria ficoides* are easily recognized by the small circular pitlike depressions ornamenting the smooth or wrinkled surface. The depressions are about 5 mm. in diameter and each is surrounded by a small rim. In the center of the depression, in well-preserved specimens, is a small raised projection. These surface features are arranged according to a definite pattern, which may be described as a spiral or quincunx. They are generally well separated and do not present the crowded appearance such as often characterizes the foliar markings on the surface of the trunks. These scars mark the places of attachment of the lateral appendages or "rootlets" as they are commonly called, but their true homologies are not known with certainty.

The lateral appendages are always much smaller than the main axis and seldom exceed a centimeter in diameter. While still in place in the underclay, they can often be seen spreading out from the axis at nearly right angles to it as slender cylindrical casts or as flattened ribbonlike bands of carbonaceous matter.

Several species of *Stigmaria* have been described on the basis of both external and internal morphology. Of these, however, *S. ficoides* is the most important. A root system slightly different from *Stigmaria* is *Stigmariopsis*, which is believed to belong to the *Subsigillariae*. In *Stigmariopsis* there is the same division into four main axes as in *Stigmaria*, but each axis tapers more abruptly and bears on its lower surface two rows of conical outgrowths, which in turn bear the "rootlets." The surface of the main part is decorated with lengthwise ribs and furrows that give an appearance resembling the pith casts of *Calamites* except for the absence of nodal lines. These ribs are believed to correspond to the ribs on the stem. Little is known of the anatomy of *Stigmariopsis*.

Fig 54.—*Stigmaria ficoides.* Cross section of the xylem cylinder of a large rootstock. In the center is the pith containing several lateral rootlets which had invaded the tissues after death of the plant. The broad bands traversing the xylem are the so-called "rays" through which the lateral rootlets emerge. Coal Measures of Great Britain. × 2½.

The structure of the main axis of *Stigmaria ficoides* is distinctive and is not readily confused with any other plant, although some of the essential features of lepidodendrid and sigillarian stems are present. The outer cortical tissue is a thick resistant layer bounded on the outside by a thin hypodermal zone and having on the inside a band of secondary periderm of radially aligned fiberlike cells. This band originated from a phellogen situated on the inner face. The middle cortex consisted originally of thin-walled cells and in most fossil specimens is represented by a mineral-filled space. The inner cortex is an imperfectly preserved lacunar layer abutting on the phloem.

In the center of the relatively small vascular cylinder is a fair-sized pith from which the cells have mostly disappeared and which may have been hollow during life (Fig. 54). The wood forms a broad zone broken into segments by broad rays.   These segments form a vertical meshwork, and in tangential sections the separating rays show as spindle masses of parenchyma through which the traces supplying the lateral appendages pass.   The wood shows no clear distinction between primary and secondary, the radial alignment of the tracheids extending almost to the pith. However, the innermost xylem elements are smaller and less regular than the others and in radial sections are seen to bear spiral thickenings. The axis therefore contains an endarch siphonostele, an unexpected phenomenon in a lycopodiaceous organ.   The bulk of the xylem consists of radially aligned scalariform tracheids among which there are numerous small secondary rays.

Some differences are shown by other species of *Stigmaria*.   In *S. augustodulensis*, *S. Lohesti*, and *S. dubia*, there is no pith, and the center of the stele is occupied by a solid rod of primary centripetal xylem made up of scalariform tracheids.   A similar situation exists also in the *Stigmaria* of *Bothrodendron mundum*.   In *S. Brardi* and *S. Weissiana* there is a central pith separated from the secondary wood by a well-developed layer of centripetal primary wood.   *S. bacupensis* represents still another type in which there is no pith, and the solid primary wood consists of spiral and scalariform elements showing no evidence of development in any particular direction.   It is most certain that a critical examination of more petrified stigmarian axes would bring to light a still greater variety of structures—there ought to be one for each stem type.   Some of the species mentioned in this paragraph were originally identified as *S. ficoides*, and they were separated only after a careful study of the primary portion.

The lateral "rootlets" of *Stigmaria ficoides* are slender organs a centimeter or less in diameter at the base but tapering gradually toward the apex.   They occasionally fork by dichotomy.   The rootlet is traversed for its full length by a simple vascular strand consisting of a triangular xylem mass with a thin layer of phloem lying along one side. The protoxylem is situated at one of the angles on the opposite side away from the phloem.   The rootlet is therefore monarch and similar to that of some species of *Selaginella*.   A few secondarily formed tracheids are usually present.   Unless misplaced during fossilization, the vascular strand occupies the center of the rootlet.   The strand is surrounded by an inner cortex of thin-walled cells.   The outermost tissues of the rootlet consist of the epidermis and the outer cortex, which are usually well preserved.   The inner cortex, as in the trunks and rootstocks, has usually broken down, leaving an internal cavity.   The vascular strand, instead of lying free within the cortical cavity, is, in most transverse sections, connected with the outer cortex by a narrow bridge of tissue containing a few tracheidlike cells. This structure obviously assisted in the transfer of water from the outer cortex to the vascular strand.

The vascular strand of the appendage is connected to the rootstock along the broad spindle-shaped primary rays that in transverse section break the xylem into segments. The xylem of the appendage bundle lies in the angle of the ray that is directed toward the base of the rootstock, with the protoxylem pointing into the lenticular space left by the decay of the ray parenchyma (Fig. 53*B*).

Within the rootstock the rootlet acquires a cortex of its own, which is confluent with the cortex of the main axis. However, the rootlet appears to break through the outer part of the cortex of the rootstock just as in any ordinary root. The rootlet, therefore, is partly endogenous in origin.

The true morphology of *Stigmaria*, and its relation to the stem, remain, even after more than a century of research, one of the great unsolved problems of paleobotany. When the prevalence of *Stigmaria* is considered this may seem strange, but there are difficulties attending its interpretation that are not easily overcome. For instance, the underclay in which the majority of *Stigmariae* occur parts irregularly, making it difficult to follow the axes for any distance except under occasional circumstances when they are exposed on quarry floors or mine roofs. Then, it is only under exceptional circumstances that even the slightest clues are provided concerning the stems to which they were attached. In the relatively few specimens where attachment has been observed, the surface features of the stumps were not always sufficiently preserved to permit accurate identification. Another difficulty lies in the fact that the structure of *Stigmaria* is different from that of the roots of most other plants, either ancient or modern. From the purely functional standpoint *Stigmaria* served as the root of the plant of which it was a part. That it was an organ of anchorage is plainly obvious, and there is no reason to question its capacity for absorption. A problem closely correlated with its structure and function is that of development, and pertaining to this there has been great diversity of opinion. Some writers of the early part of the last century (Artis, Lindley, and Hutton, and Goldenberg in particular) regarded *Stigmaria* as an independent growth of unknown affinity and unassociated with any kind of aerial stem. Others thought it to represent a succulent dicotyledon of the *Cactus* or *Euphorbia* type, although Brongniart suggested in 1849 that it might be the root of *Sigillaria*.

One of the most elaborate explanations of the morphology and development of *Stigmaria* was formulated about 1890 by Grand'Eury, who believed that both *Stigmaria* and *Stigmariopsis* are rhizomes and not true roots. He concluded that the spores of *Lepidodendron* and *Sigillaria* germinated in water and developed into long branched axes capable of an independent existence. Under certain conditions, probably in shallow water, the rhizomes sent out bulblike outgrowths that grew into erect stems. In the next stage the developing trunks became basally enlarged and sent out four primary roots that grew obliquely downward and sent out appendages. These then served to anchor the trunk, and after the decay of the parent rhizome the plant carried on an independent existence.

Grand'Eury's theory of the nature and development of the rootlike organ of the arborescent lycopods was based principally upon *Stigmariopsis* stumps standing *in situ*, but he applied the same explanation to *Stigmaria*. His theory has difficulties that are not readily explained. It seems almost unbelievable that the parent rhizome and the roots that developed later should have the same structure. One would expect them to show sufficient differences to distinguish them when found in close association. British writers of the same period held rather firmly to the belief that *Stigmaria* is a true root, and their views were based upon a number of excellent stumps discovered when railway construction commenced in Great Britain about 1840. At about the same time similar stumps were described from Cape Breton Island in Nova Scotia, some of which are figured by Sir William Dawson in his "Geological History of Plants" and "Acadian Geology."

Modern research has thrown little additional light on the *Stigmaria* problem and the remains are generally ignored by present-day paleobotanists. When all the accumulated evidence is brought together, the salient facts are that *Stigmaria* and *Stigmariopsis* functioned as the roots of the Paleozoic arborescent lycopods, and that those attached to different kinds of plants were much alike in structure. The likeness is so great that the vast majority of them have been lumped together into one species. On purely morphological grounds *Stigmaria* cannot be regarded as a true root, and probably not as a rhizome. Rhizomes, especially those that produce extensive vegetative growth, usually bear numerous aerial stems, but in *Stigmaria* multiple attachment has not been observed. Moreover, the connection between *Stigmaria* and the trunk offers good evidence that the former is an outgrowth from the base of the trunk rather than being the parent organ from which the stem was initiated. *Stigmaria* is in all probability a highly specialized and modified branch system which, from the standpoint of function and position, has become transformed into a root. The possibility that the appendages of the main stigmarian axis are homologous with leaves has been suggested. When their function and position is considered, this interpretation may seem needlessly farfetched, but the orderly arrangement that they present is far more remniscent of a foliar whorl than of the irregular distribution of lateral roots. Moreover, similar modifications of the basal foliage occur in the aquatic fern *Salvinia*, in which the lowermost leaves have become transformed into long filiform segments somewhat resembling roots although of questionable function.

OTHER ANCIENT LYCOPODS

Mention has already been made of the fact that the names *Lycopodites* and *Selaginellites* are often given fossils resembling the modern genera *Lycopodium* and *Selaginella*. The distinction between *Lycopodites* and *Selaginellites* is based upon the kinds of spores produced. *Lycopodites* is the term for plants which may resemble either *Lycopodium* or *Selaginella* but are known to be homosporous. *Selaginellites*, on the other hand, is for heterosporous forms. *Lycopodites* has been used for vegetative *Lycopodium*-like twigs even in the absence of spores, some of which have later been shown to be coniferous twigs. The smaller branches of some species of *Bothrodendron* resemble *Lycopodites*. *Lycopodites* and *Selaginellites* range from the Lower Carboniferous (or possibly even from the Upper Devonian) to Recent.

Pinakodendron is a Carboniferous lycopod that somewhat resembled *Bothrodendron* in the absence of raised leaf cushions. It differs from it in the presence of a network of raised lines between the leaf scars. The outstanding feature of the genus, however, is that no strobili were produced, but certain portions of the stem became fertile as in *Baragwanathia* or in some of the modern species of *Lycopodium*. The sporophylls are long and narrow, and on the stem just above the point of leaf attachment the imprint of a megaspore tetrad is visible. There is no pronounced distinction between the sporophylls and the vegetative leaves.

Omphalophloios is a middle Pennsylvanian lycopod of somewhat uncertain affinities, although it is probably a relative of *Lepidodendron*. It has rhomboidal leaf cushions that change considerably in shape on axes of different size, but they bear a general resemblance to those of *Lepidodendron* except for the absence of the median keel. Its characteristic feature is the leaf scar. Slightly above the middle of the scar is a small triangular print, and above this is an oval or subcordate area outlined by a low rim. Within this area is an oval print. In the absence of attached leaves or preserved structure, it is impossible to explain the markings on the scar. Although *Omphalophloios* probably represents a stem, it may be a rhizomelike structure similar to *Stigmaria*. *Omphalophloios* originally was described from material from the Des Moines series of Missouri, but similar forms have been observed elsewhere.

Pleuromeia is a genus of Triassic plants somewhat resembling *Sigillaria*. The plant possessed a straight, upright, unbranched stem sometimes exceeding 1 mm. in height. Terminating the trunk was a large strobilus of spirally arranged, oval or reniform sporophylls. The plant was heterosporous. Linear or lanceolate leaves were borne on the upper part of the trunk beneath the strobilus. The base of the trunk was divided by dichotomies into four, or sometimes more, short fleshy lobes which were upturned at the ends, and which were covered with small slender rootlets.

The internal tissues of *Pleuromeia* are not well preserved, and comparisons with other plants are based mostly on the external form. Probably the most striking feature of the plant is the resemblance between the lobed basal portion and the rhizomorph of the living *Isoetes*. In *Isoetes* the growing point is located in a median groove and successive rows of rootlets are produced in a rather orderly succession. It is probable that *Pleuromeia* produced its rootlets in a similar manner.

The double dichotomy at the base of the lepidodendrid trunk where the four stigmarian axes depart also resembles the basal lobing of *Pleuromeia* and *Isoetes*. The structure of the rootlets is similar in all three, although *Stigmaria* differs in that it apparently had an apical growing point. It is improbable that in *Stigmaria* new rootlets were initiated within the grooves between the large rootstocks in the same manner as in *Isoetes*.

PHYLOGENETIC CONSIDERATIONS

It is beside the purpose of this volume to give a complete analysis of the relationships of the ancient lycopods, and therefore mention will be made of a few salient features only.

The climax of lycopod development is revealed in the arborescent genera of the Carboniferous. These plants became large trees with secondary wood and extensive root systems. All forms were spore bearers but heterospory became firmly established, and in some genera evolution had proceeded to the point of production of seedlike organs.

The lepidodendrids and sigillarians represent highly specialized trends within the Lycopsida, which lacked the plasticity necessary to enable them to adapt themselves to the more adverse climates of the late Permian and Triassic. For this reason we find them all but disappearing from the fossil record at the close of the Paleozoic. There existed, however, along with the arborescent genera small herbaceous lycopods that grew near the ground in protected places, and these were able to cope more successfully with the changing environment. Much less is known of them than of their more spectacular relatives, but some of the more or less indifferently preserved fossils that are sometimes indiscriminately called *Lycopodites* and *Selaginellites* may be the actual forerunners of the Recent genera.

The division of the lycopods into the Ligulatae and Eligulatae extends to the Paleozoic forms. As far as we know all the lepidodendrids and sigillarians were ligulate and heterosporous, and the two characteristics seem to accompany each other in the ancient as well as in the Recent genera. Reports of homosporous *Lepidostrobi* have never been verified, and it is altogether probable that all members of this cone genus were both ligulate and heterosporous. *Spencerites*, on the other hand, is a homosporous, eligulate organ, but the plant that bore it is unknown. It is undoubtedly lycopodiaceous but not a member of the Lepidodendraceae.

According to the systems of classification generally followed in botanical textbooks, the plants grouped in this volume in the Psilopsida, Lycopsida, Sphenopsida, and Filicinae constitute the Pteridophyta, and the lycopods and scouring rushes are termed the "fern allies." However, we know from anatomical and historical evidence that the ferns and lycopods are not close relatives, but that they represent separate lines of development as far into the past as it is possible to trace them. The true fern allies are not the lycopods and scouring rushes, but the seed plants.

The Lycopsida agree with the Psilopsida in showing many primitive features in the vascular system, but in other respects there are some pronounced differences. In the lycopsids the sporangium is always associated with a foliar organ and is not a part of the main shoot as in *Rhynia* or *Psilophyton*. In the living psilotalean genus *Tmesipteris* the synangia are situated on short shoots that have been variously interpreted, some investigators maintaining that they are sporophylls, while others explain them as cauline. Situated beside the synangia on the short shoots are expanded leaflike structures, the morphology of which has been a matter of dispute. However, those who have most recently studied *Tmesipteris* seem inclined to agree that the short shoot that bears the synangium in a terminal position is not homologous with the lycopod sporophyll. Another distinction is that many lycopods possess organs definitely differentiated as roots, and the whole plant body shows a degree of differentiation scarcely more than suggested among the Psilopsida. No heterosporous psilopsids are known.

References

ANDREWS, H.N., and E. PANNELL: Contribution to our knowledge of American Carboniferous floras, II. *Lepidocarpon, Ann. Missouri Bot. Garden,* **29,** 1942.

ARNOLD, C.A.: *Lepidodendron Johnsonii* sp. nov., from the lower Pennsylvanian of central Colorado. *Univ., Michigan Mus. Paleontology, Contr.* **6** (2), 1940.

BANKS, H.P.: A new Devonian lycopod genus from southeastern New York, *Am. Jour. Botany*, **31**, 1944.

BARTLETT, H.H.: Fossils of the Carboniferous coal pebbles of the glacial drift at Ann Arbor, *Michigan Acad. Sci. Papers*, **9**, 1928 (1929).

BENNIE, J., and R. KIDSTON: On the occurrence of spores in the Carboniferous formation of Scotland, *Royal Phys. Soc. Edinburgh Proc.*, **9**, 1886.

BERTRAND, C.E.: Rémarques sur le *Lepidodendron Harcourti* de Witham, *Travaux et mém. faculté de Lille*, **2**, 1891.

GRAHAM, R.: An anatomical study of the leaves of the Carboniferous arborescent lycopods, *Annals of Botany*, **49**, 1935.

HIRMER, M.: "Handbuch der Paläobotanik," Munich, 1927.

HOSKINS, J.H., and A.T. CROSS: A consideration of the structure of *Lepidocarpon* Scott based upon a new strobilus from Iowa, *Am. Midland Naturalist*, **25**, 1941.

KIDSTON, R.: Carboniferous lycopods and sphenophylls, *Nat. History Soc. Glasgow Trans.*, **6**, 1901.

KISCH, M.: The physiological anatomy of the periderm of fossil Lycopodiales, *Annals of Botany*, **27**, 1913.

KRÄUSEL, R., and H. WEYLAND: Die Flora des deutschen Unterdevons, *Preuss. geol. Landesanstalt Abh.*, neue Folge, **131**, 1930.

———, and H. WEYLAND: Pflanzenreste aus dem Devon, IV, *Protolepidodendron. Senckenbergiana*, **14**, 1932.

LANG, W.H., and I.C. COOKSON: On a flora, including vascular plants, associated with *Monograptus*, in rocks of Silurian age, from Victoria, Australia, *Royal Soc. London Philos. Trans.*, **B 224**, 1935.

PANNEL, E.: Contribution to our knowledge of American Carboniferous floras, IV. A new species of *Lepidodendron*, *Ann. Missouri Bot. Garden*, **29**, 1942.

REED, F.D.: *Lepidocarpon* sporangia from the Upper Carboniferous of Illinois, *Bot. Gazette*, **98**, 1936.

SCHOPF, J.M.: Two new lycopod seeds from the Illinois Pennsylvanian, *Illinois Acad. Sci. Trans.*, **30** (2), 1938.

———: Spores from the Herrin (No. 6) coal bed in Illinois, *Illinois Geol. Survey, Rept. Inv. 50*, 1938.

SCOTT, D.H.: On the structure and affinities of fossil plants from the Paleozoic rocks, IV. The seed-like fructification of *Lepidocarpon*, *Royal Soc. London Philos. Trans.*, **B 194**, 1901.

———: "Studies in Fossil Botany," 3d ed., pt. I, London, 1920.

SEWARD, A.C.: "Fossil Plants," Vol. 2, Cambridge, 1910.

WALTON, J.: Scottish Lower Carboniferous plants: the fossil hollow trees of Arran and their branches. *Royal Soc. Edinburgh Trans.*, **58**, 1935.

WHITE, D.: *Omphalophloios*, a new lepidodendroid type, *Geol. Soc. America Bull. 9*, 1898.

———: A remarkable fossil tree trunk from the Middle Devonic of New York, *New York State Mus. Bull. 107*, 1907.

WILLIAMSON, W.C.: On the organization of fossil plants from the coal measures, III. Lepidodendraceae (continued), *Royal Soc. London Philos. Trans.*, **162**, 1872.

CHAPTER VI

ANCIENT SCOURING RUSHES AND THEIR RELATIVES

The geological history of the scouring rushes and their relatives is a close parallel to that of the lycopods. Their earliest representatives appeared during the Devonian, the group became a conspicuous element of the flora of the Carboniferous, and their subsequent decline has reduced them to a mere vestige of their former status. The decline, in fact, has been even more severe than that endured by the lycopods, as is shown by the presence in the Recent flora of but 1 genus with some 25 species.

With the possible exception of one or two Middle Devonian forms of which the affinities are still somewhat conjectural, the stems of all members of the scouring rush group (the Sphenopsida) are marked by distinct nodes at which are borne whorls of leaves, and the stem surface between the nodes is ornamented with lengthwise extending ribs and furrows. Coupled with these characteristics of the vegetative phase of the plant, most genera possess especially modified fertile stalks, or sporangiophores, which, although they are variously interpreted, are probably homologous with the sporophylls of the lycopods. In the surviving genus *Equisetum* the leaves are vestigial organs forming a low toothed sheath around the stem and generally incapable of photosynthetic activity. The individual plants are small, limited to a few feet in height, and show no appreciable size increase due to cambial activity. The Paleozoic genus *Calamites*, on the other hand, had green leaves that were often several centimeters long. There was extensive cambial activity and the plants often became large trees.

The categorical ranking of this group of ancient and modern plants varies among the several systems of classification that have been in use during the last half century. In some it is rated as an order (Articulatales or Equisetales) but others treat it as a class (Articulatineae or Equisetineae), as a phylum (Arthrophyta), or as a division (Sphenopsida). Divisional ranking seems most satisfactory because the very pronounced characteristics that all its members possess in common serve not only to unify it but also to segregate it on an equal level from all other divisions. The term "phylum" which is approximately equivalent to "division" has not come into general use in botany.

The division Sphenopsida consists of five orders or classes, depending upon the categorical rank assigned them. They are the Hyeniales, the

129

Pseudoborniales, the Sphenophyllales, the Calamitales, and the Equisetales. The exact time of the first appearence of the Sphenopsida is of course unknown, but there is some evidence that primitive members of the division flourished beside the Psilophytales and ancient lycopods of the early Devonian. This belief is supported by the discovery in the Lower Devonian of two very small plants, *Climaciophyton* and *Spondylophyton*, which consist of stems bearing leaflike organs in whorls at regular intervals. Nothing is known of the internal structure or of the fructifications of either of these plants, and since we are acquainted only with the external aspect of the vegetative parts there is no proof that they belong to the Sphenopsida or that they are not ancient thallophytes or members of some extinct group as yet unknown. Their assignment to the Sphenopsida is therefore tentative.

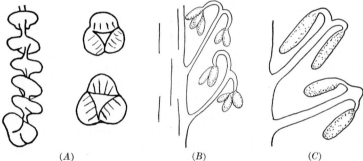

(A)   *(B)*   *(C)*

Fig. 55.—(A) *Climaciophyton trifoli...um.* (*After Kräusel and Weyland.*) × 4; (B) *Hyenia elegans.* Fructification. (*After Kräusel and Weyland.*) × 3. (C) *Calamophyton primaevum.* Fructification. (*After Kräusel and Weyland.*) × 7.

Climaciophyton (Fig. 55A) is from the Lower Devonian Wahnbach beds of Germany. It is a small plant with a three-angled stem not more than a millimeter in diameter, which at rather short intervals is encircled by whorls of three short, rounded, broadly attached, leaflike appendages which appear to be quite thick. *Spondylophyton* is from the Lower Devonian of Wyoming. It is a plant apparently of cespitose habit with small unbranched axes 1 or 2 mm. in diameter bearing whorls of two to four leaflike appendages. These organs are 5 to 8 mm. long and are dichotomously split to the base into truncate linear segments. The whorls are from 4 to 6 mm. apart except near the stem tips where they may be closer.

HYENIALES

The Hyeniales consist of the two genera *Hyenia* and *Calamophyton*. Both are from the Middle Devonian. They are small plants with stems bearing short forked appendages in whorls of three or more.   Along the lower portions of the plant these appendages are sterile, but at the tips they bear small terminal sporangia on the recurved ends.   The Hyeniales are placed within the Sphenopsida solely on external characters because nothing is known of the internal structure of any of the members.   The position of the sporangia suggests affinity with the Psilophytales.   Too little is known of any of them to enable us to assert positively that a phylogenetic sequence is represented, but it is possible that the Hyeniales constitute a side line of the psilophytalean complex from which the later sphenopsids evolved.

Hyenia is the largest genus with three species.   *H. elegans* is from the upper Middle Devonian of Germany, and *H. sphenophylloides* was found in a bed of similar age in Norway.   *H. elegans*, the best known of the species, has an upright stem less than 1 cm. in diameter, which branches in a digitate manner.   The straight slender branches bear whorls of narrow, repeatedly forked leaves, which are from 1 to 2.5 cm. long.   *H. Banksii*, from the Hamilton group of Orange County, New York, differs somewhat in habit.   It has an upright stem that bears spreading lateral branches of only slightly smaller diameter.   The leaves are 7 to 11 mm. long and are usually two but sometimes three-forked.   As in the other species, the stems are not jointed.   The leaf whorls are spaced at intervals of about 8 mm.   The number of leaves per whorl is uncertain.

Fructifications have been described only for *Hyenia elegans* (Fig. 55B). The upper portions of the branches are transformed into spikes bearing short-forked sporangiophores arranged the same as the leaves.   No leaves are present among the sporangiophores, the entire stem tip being fertile.   The bifurcated tips are recurved, and each bears several oval sporangia.   The spores have not been observed.

Calamophyton has two species, *C. primaevum* from the same locality and horizon in Germany as *Hyenia elegans*, and *C. Renieri* from the Middle Devonian of Belgium.   The genus resembles *Hyenia* in having an upright stem that divides into several smaller branches, but the stems and branches are jointed with short internodes.   The nodes of the smaller branches bear whorls of small, wedge-shaped, apically notched leaves. The fertile portions of the stem are similar to those of *Hyenia* in consisting of fertile spikes of forked sporangiophores, but only two sporangia are present on each sporangiophore (Fig. 55C).

The digitate branching of the main stem of the Hyeniales is different from that of any of the other members of the Sphenopsida.   In *Equisetum* and in the Carboniferous genus *Calamites*, the branches are attached to the nodes, often in pronounced whorls around the main central axis, which extends upward as a straight shaft.   However, the whorled leaf arrangement of the Hyeniales is definitely a sphenopsid character.

PSEUDOBORNIALES

This order is founded upon the single genus and species *Pseudobornia ursina* (Fig. 56), which was discovered in 1868 in the Upper Devonian of Bear Island in the Arctic Ocean.   The remains are not petrified, but consist of well-preserved compressions, which reveal stems of varying dimensions.   The largest of these, which were as much as 10 cm. in

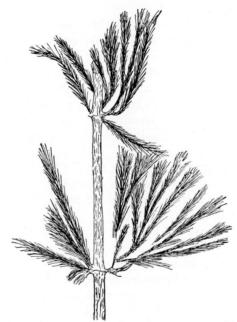

Fig. 56.—*Pseudobornia ursina.   (After Nathorst.)*   About ½ natural size.

diameter, are believed to have grown more or less horizontally and to have borne upright leafy stems of smaller size.   All the stems are distinctly jointed, and the smaller upright ones bear at the joints whorls of large, deeply incised leaves.   Each joint appears to have borne four leaves.   Each leaf is divided by repeated dichotomies into several divisions, each of which is in turn deeply cut in a pinnate fashion into numerous fine segments.   The fructifications are lax spikes up to 32 cm. long, which bear whorled sporophylls (sporangiophores) resembling reduced foliage.   The sporangia, which are borne on the lower side, contain spores believed to be megaspores.   Because we know nothing of the internal structure of *Pseudobornia,* and have scant knowledge of its fructifications, the relationship of this plant to the other members of the Sphenopsida is somewhat problematic, but the form of the leaves is suggestive of affinities in the direction of the Sphenophyllales and *Asterocalamites* of the Calamitales.

SPHENOPHYLLALES

Sphenophyllum, the principal genus of the Sphenophyllales, was a small plant that probably produced a dense mat of vegetation on the muddy floor of the coal-swamp forests.   Some of the species may have produced trailing stems of considerable length, which grew over other low plants or fallen logs.   The reclining habit of the plant is indicated by the slenderness of the stem and the frequent insertion of adventitious roots at the nodes.   The occurrence on the same plant of shallowly and deeply notched leaves has been interpreted by some investigators as an indication that the plant was aquatic, with the deeply cut leaves growing beneath the water.   However, on some stems it may be seen that the two kinds of leaves are interspersed, showing that there is no apparent relation between leaf form and habit.   The stem of *Sphenophyllum* was probably an aerial organ for the most part, although since the plant grew in a swampy environment it might have occupied habitats where partial submergence was frequent.

Compressions of the leafy stems of *Sphenophyllum* are widely scattered throughout the shales of the Carboniferous coal-bearing formations, and petrified stems are often present in coal-balls.   The genus first appeared in the Upper Devonian, the oldest species being *S. subtenerrimum* from Bear Island.   It is most abundant in the Upper Carboniferous (Pennsylvanian) and Lower Permian, but survived into the Triassic, where it disappeared.

The name *Sphenophyllum* is derived from the shape of the leaf.   The leaves (Figs. 57 and 58), are always small, usually less than 2 cm. in length but occasionally more.   They are arranged in verticels of 6 or 9, although sometimes there may be as many as 18.   The broad apex of the leaf is variously notched or toothed, depending to some extent upon the species, although there is considerable variation in this respect within some of the species.   *Sphenophyllum emarginatum* (Fig. 57*B*) has shallow rounded teeth.   In *S. cuneifolium* (Figs. 57, *D* and *E*) *S. saxifragaefolium*, *S. bifurcatum*, and *S. majus* (Figs. 57*A* and 58*A*) the teeth are pointed, and often there is a deep central cleft which divides the terminal portion of the leaf into lobes which in turn may be secondarily notched.   The leaves of *S. myriophyllum* (Fig. 57*C*) are divided by three deep notches, which extend nearly to the base.   In some species, such as *S. speciosum* or *S. verticillatum*, the leaf apex is somewhat rounded and very shallowly

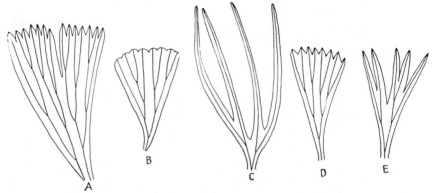

FIG. 57.—Leaf forms of *Sphenophyllum*. (*After Zeiller.*) ✕ about 3. (*A*) *S. majus.* (*B*) *S. emarginatum.* (*C*) *S. myriophyllum.* (*D*) and (*E*). *S. cuneifolium.*

(*A*)   (*B*)

FIG. 58.—(*A*) *Sphenophyllum majus.* Westphalian B. Belgium. (*After Renier and Stockmans.*) (*B*) *Sphenophyllum oblongifolium.* Lawrence shale, Virgil series, Pennsylvanian. Eastern Kansas. Natural size.

toothed. Each *Sphenophyllum* leaf is provided with several slightly diverging veins, one of which terminates within each apical tooth. All the veins are derived from a single vein which enters the base of the leaf and which divides by repeated dichotomies.

The stems of *Sphenophyllum* seldom exceed 5 mm. in diameter. They are slender, with nonalternating ribs extending the length of the internodes. The strobili are elongated lax inflorescences borne terminally on the lateral branches.

Structurally preserved stems of *Sphenophyllum* are frequently found in coal-balls and other types of petrifactions (Fig. 59). Early during the development of the stem a phellogen layer was initiated rather deep

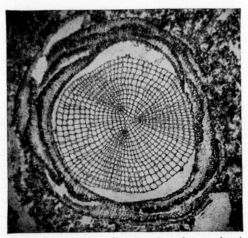

Fig. 59.—*Sphenophyllum plurifoliatum.* Cross section of stem showing the three-angled primary xylem mass, the secondary xylem and periderm layers.   × about 9.

within the cortex and this produced successive layers of corky tissue. Ultimately the entire cortex was utilized, and periderm formation continued within the secondary phloem as in modern woody stems. The stele consists of a central column of xylem surrounded by phloem. The latter, however, was of delicate construction and hence seldom preserved. The primary xylem, which occupies the center of the stele, is a solid three-angled rod of conducting cells without pith and with the protoxylem located at the angles. The external position of the protoxylem indicates that the metaxylem developed centripetally, a type of development rare in stems, except in the Lycopsida, but characteristic of roots. A cross section of a stem usually shows three protoxylem points, although in *S. quadrifidum*, a French species, each point is said to be a double one, making the stele hexarch instead of triarch. The protoxylem tracheids are small, but inwardly the tracheids of the metaxylem become larger and in some species they bear multiseriate bordered pits.   There is no conjunctive parenchyma, and the primary xylem therefore bears considerable resemblance to the exarch protostele of a root.

The three-angled primary xylem rod is surrounded by secondary wood.   The first-formed secondary wood was deposited within the shallow embayments between adjacent protoxylem points but later it extended completely around the primary axis.   In young stems it conformed to the shape of the primary wood but later, as it increased in extent, it became round in cross section.

The secondary xylem consists of tracheids and rays.   The tracheids are large, are arranged in regular radial sequence as a result of their origin from the cambium, and bear multiseriate bordered pits on their radial walls.   In most of the species in which the stem anatomy has been investigated the rays show the usual structure, but in *Sphenophyllum plurifoliatum* they consist merely of short horizontal cells which connect vertical strands of parenchyma occupying the spaces between the corners of the tracheids.

The roots of *Sphenophyllum* are adventitious organs attached at the nodes of the stem.   They are small, and structurally they resemble the stem except that there are two protoxylem points instead of three. Secondary growth took place as in the stem.   The phloem is usually better preserved, and a broad periderm layer is usually present.

The Sphenophyllales reproduced by spores as did all other known members of the Sphenopsida.   The spores were borne in sporangia produced in strobili of various sizes although they are usually not more than 1 cm. in diameter.   The length varies more than the diameter.   In some species the strobili are lax structures not strongly differentiated from the vegetative shoots, but in others they are compact organs borne in either a lateral or a terminal position.

Detached sphenophyllean strobili are assigned to the organ genus *Bowmanites*.   The name *Sphenophyllostachys* has also been used but the other was proposed first.   The cone axis is jointed, reminiscent of the stem on which it is borne, but the internodes are typically shorter. Instead of bearing leaves, the nodes support sporangiophores usually accompanied by sterile bracts.   These sterile bracts are borne in verticils and are often joined laterally with each other for a part of their length with the terminal portions free.   In some the union extends for more than half the length of the bract but in others, of which *Bowmanites Kidstoni* is an example, they are united only at their bases with the free terminal part split in a manner resembling the bifurcation of the vegetative leaves.   Attached to the axis between these verticils of coalescent bracts are the sporangiophores, which differ markedly in length in different species. Sometimes they are fused for a short distance to the upper surface of the bract so that in the compressions they appear to arise from it. In *B. majus* the stalks are borne in such close proximity to the bracts that they were long undetected, and in the older texts the small cluster of four sporangia is shown sessile on the upper surface (Scott's "Studies," Fig. 51). In *B. Dawsoni* (Fig. 60) the coalescent bracts form a cup around the cone axis with the upturned tips free. There are several sporangiophores for every bract and they are of different lengths so that the sporangia are arranged within two or more concentric whorls within the cups of coalescent bracts. Each sporan-

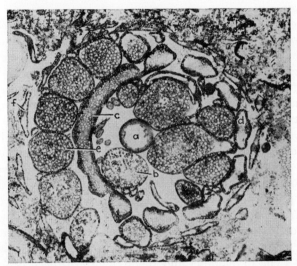

Fig. 60.—*Bowmanites Dawsoni*. Slightly oblique transverse section of strobilus: *a*, axis; *b*, sporangium of higher whorl; *c*, whorl of basally coalescent bracts which become free on opposite side at *d*; *e*, sporangium of whorl below; *f*, free tips of bracts constituting a still lower whorl. × 10.

giophore bears a single oval sporangium terminally on its upwardly curved tip. *B. Römeri* is similar to *B. Dawsoni* except that each sporangiophore bears a pair of terminal sporangia. In both species the number of sporangia borne within a single verticil is large. In *B. fertile* the bracts are not coalescent, but they depart from the nodes as six pairs of slender stalks. Immediately above and between each pair a sporangiophore arises as a single stalk that soon branches into 14 to 18 slender pedicels. Each pedicel is a peltate structure with two backwardly directed sporangia. Homospory is generally assumed to have prevailed in the majority of species but heterospory has been demonstrated in at least one species and has been suspected in others. In the supposed homosporous forms the spores vary from about 75 to 125 microns in diameter exclusive of the *perispore*, which is an irregular or wrinkled layer covering the exine. In *B. delectus* some of the sporangia bear small spores averaging about 80 microns in diameter, but other sporangia in the same strobili contain spores nearly ten times as broad. Of the latter, only a few spores, probably about 16, are present within a sporangium. Not all of the megaspores developed to maturity. Some of them are aborted and have developed to about one-third the dimensions of the fully formed ones. *B. delectus* is a sphenophyllaceous fructification in which heterospory is clearly expressed, but the plant which produced it is unknown. Its affinities are evident from the whorled arrangement of the sporangia.

The trend in the Sphenophyllales appears to have been toward strobilar complexity, and unlike the lycopods there is no indication of seed production. Cryptogamic methods cf reproduction seem to characterize the order as they do within the entire sphenopsid group.

Boegendorfia semiarticulata is a sphenopsid from the Upper Devonian of Germany, which has some resemblance to *Sphenophyllum*. Its lateral branches depart from a central stem at nodes spaced 0.5 to 2 cm. apart. The small wedge-shaped leaves are deeply cut into two to five segments, and attached to some of the nodes are small spikelike structures that appear to be strobili. The affinities of *Boegendorfia* are expressed only by superficial characters and its exact position within the Sphenopsida is questionable.

Cheirostrobus is the name given a very interesting calcified strobilus from the well-known Lower Carboniferous plant-bearing beds at Pettycur, Scotland. The plant that bore it is unknown. The cone is large, being as much as 4 cm. in diameter and 10 cm. long. At each node there is a whorl of about 12 so-called "sporophylls," which, as in *Sphenophyllum*, are superimposed in successive whorls. Each sporophyll is divided just beyond its base into two lobes, the upper being fertile and the lower sterile. The whole sporophyll is also split laterally in a digitate fashion into three portions so that altogether there are six parts, three above and three below. Both the sterile lobe and its fertile complement extend out at right angles from the axis. The tip of the sterile part is provided with an upturned bract that overlaps the corresponding portion of the sporophyll in the whorl next above. The fertile lobe above each sterile part is a peltate organ bearing four closely packed elongated sporangia, which extend back to the axis. The spores average 65 microns in diameter and are all of one kind.

The division of the sporophyll of *Cheirostrobus* into a superior sporangiophore complex and an inferior sterile bract is quite like the situation in *Sphenophyllum*, which is believed to be its closest relative. There is also considerable resemblance to *Palaeostachya*, one of the fructifications of *Calamites*, in having four sporangia borne in a peltate manner. Although the plant to which *Cheirostrobus* belongs is unknown, this fructification nevertheless shows the extent of strobilar complexity attained by some of the Paleozoic Sphenopsida. It is a synthetic type combining characteristics of the Sphenophyllales and the Calamitales.

CALAMITALES

The genus *Calamites* occupies a position among the Sphenopsida that is comparable to *Lepidodendron* among the lycopods, and from the standpoint of bodily dimensions and structural complexity it marks the culmination of the entire class of plants. Its remains are extremely abundant and they are almost always present in one form or another wherever Carboniferous plants are found.

CALAMITES

In *Calamites* the segmented construction of the plant body is strongly expressed in the stems, the branches, and the fructifications. Although it has been possible to reconstruct several types, it is only rarely that the various parts of the plants are found connected. The plant body was of frail construction, which resulted in the branches and strobili readily becoming detached from the main trunk, and consequently most of our information on the complete plant has been assembled part by part from painstaking comparisons of separate parts and from occasional instances where organs have been found attached. Separate generic names have been given all parts of the plant including the pith casts, leaf whorls, and strobili. The name *Calamites* was originally used for the pith casts in 1784 by Suckow, who was one of the first to suspect the true relationships of the fossils. Previous writers believed them to be reeds or giant grasses such as bamboos or sugar cane. Although the name was originally applied to the pith casts, it has through common usage come to be understood as applying to the plant in its entirety. Other names, such as *Arthropitys* for one type of stem and *Annularia* for one of the common leaf forms, now refer to different organs of the one natural genus.

Three subgenera of *Calamites* have been distinguished in the mode of branching. In *Stylocalamites* the branches are few and irregularly scattered. This group contains the well-known *Calamites Suckowii* (Fig. 61). Subgenus *Calamitina* bore branches in whorls but only at certain nodes. All the internodes are short, but the branch-bearing nodes surmount internodes much shorter than the others. To this group belong *Calamites Sachsei*, *C. Schützeiformis*, (Fig. 63A) and *C. undulatus*. In the third group, *Eucalamites*, branches are borne at every node. There are two forms of *Eucalamites*, one typified by *C. cruciatus* (Fig. 63*B*), which has many branches per whorl, and the other by *C. carinatus*,

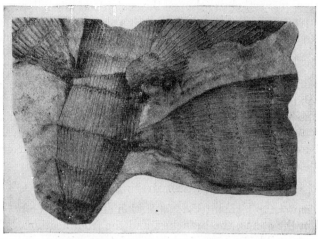

Fig. 61.—*Calamites Suckowii.* (*Stylocalamites.*) Pith cast of main stem and three branches. (*After Grand 'Eury.*) Reduced.

Fig. 62.—*Calamites Cistii.* Nodal region of pith cast showing alternation of ribs and the scars of the infranodal canals × 2.

which has a pair of opposite branches at each node. *Calamites* is therefore a large and diversified genus that may ultimately be divided into several when more information concerning the various forms is forthcoming.

It is difficult to form satisfactory estimates of the size of *Calamites* because of the fragmentary nature of the remains. A trunk 10 m. long was found in a British coal mine, and medullary casts 15 cm. in diameter are occasionally found and specimens twice as large are known. It is likely that the plants were tall in proportion to the diameter of the trunk because the casts usually taper little or none at all over distances of several feet. A height of 15 or 20 feet is probably a reasonable estimate

(A)   (B)

Fig. 63.—(A) *Calamites Schützeiformis* (Calamitina). × 1½. (B) *Calamites cruciatus* (*Eucalamites*). (*After Renier.*) Reduced.

for an average tree of most species although trees 30 to 40 feet tall certainly existed.

Considerable information concerning the manner of growth of *Calamites* has been gathered from the upright casts exposed in the sea cliffs at Joggins, Nova Scotia, which were described many years ago by Sir William Dawson. At this place Dawson saw numerous casts standing upright in a layer of sandstone about 8 feet thick, and some of them (which presumably were casts, judging from Dawson's description) were as much as 5 inches in diameter and showed no decrease in passing from the bottom to the top. At one place he reports 12 casts within a distance of 8 feet along the face of the cliff.

The trunks displayed at Joggins, and similar ones at other places, indicate that many of the species of *Calamites*, particularly those belonging to the subgenus *Stylocalamites*, were gregarious plants, which grew in dense stands on inundated flats.   As the individual plant grew it developed an upright stem that was anchored to the muddy soil by adventitious roots arising from the lowermost nodes.   Branches arising near the base turned upward, and they themselves were supported at the basal bend by additional roots.   This method of branching probably repeated itself indeterminately in some species, thus producing a tangled or matted mass of roots and horizontal stems that impeded the movements of sediments and caused quick burial of these parts.   The rapid accumulation of sand and mud made possible the entombment of long lengths of trunks.

The habit of growth of *Stylocalamites* was probably not greatly different from that of some modern species of *Equisetum*, particularly *E. hyemale*, and the general aspect of a calamitean forest may be visualized by imagining a clump of scouring rushes in which the individual plants are enlarged to the proportions of small trees.   In other calamitean species, especially those of the subgenus *Eucalamites*, the habit was probably more arborescent.   The central trunk probably arose from a horizontal rhizome and the smaller branches spread outward to form a symmetrical crown.   As for habit, these may be compared with the sterile shoots of the common horsetail rush, *E. arvense*.   However, similarities in habit between the ancient *Calamites* and the modern species of *Equisetum* does not of necessity imply close affinity.

Anatomy.—A prominent anatomical feature shared by *Calamites* and most species of *Equisetum* is the large pith which broke down during life with the resulting formation of an extensive central cavity which extends uninterrupted from one internode to the next.   When the stem became covered with sediment some of it infiltrated into the pith cavity and hardened with little distortion, forming the cast.   After exposure by decay or carbonization of the surrounding tissues the casts reveal on their surfaces the topography in reverse of the inner surface of the vascular cylinder.   The resemblance between the markings on these casts and those on an *Equisetum* stem was noted years ago, an observation that led at the time to correct surmises concerning affinities.   The similarity, however, amounts only to a superficial resemblance because the markings on the casts bear no relation to the surface of the stem.   The stems of *Calamites* were ribbed in some instances and smooth in others, but the surface ribs do not correspond to those on the casts.

The ribs on the calamitean pith cast are the counterparts of the furrows between the inwardly projecting primary wood strands on the inner face of the woody cylinder, and the intervening furrows are the imprints of the primary wood strands themselves. The ridges on the casts are usually from 1 to 3 mm. broad and extend the full length of the internodes. The crest of the ridge is usually rounded, although in some species it is nearly flat. The nodes are clearly marked on the surface of the cast and the nodal line is represented by a zigzag furrow (Fig. 62). For the most part, but not always, the ridges and furrows of adjacent internodes alternate with each other. On the upper end of each ridge of most species there is a small oval elevation that indicates the position of the *infranodal canal*. These so-called "canals" are not empty passageways but strips of thin-walled tissue which extended from the pith along dilated upper portions of large rays which merge with the pith between the primary wood bundles. The position of these infranodal canal scars aids in determining the upper and lower ends of the casts because of their position below the nodal line.

It is sometimes possible to determine the mode of branching of a calamite from the position of the branch scars on the pith casts. The branches came off just above the nodes, and at the base of the branch the pith narrowed rather abruptly to a narrow neck of tissue connecting with the pith of the main stem. This connection is often seen in casts of the familiar *Calamites Suckowii* (Fig. 61). Some of the older figures of the casts show them turned wrong way around because it was originally thought that the tapered end was the apex. The main stem was tapered at the apex and at the root-bearing base but less abruptly.

Petrified stems of *Calamites* are frequent in coal-balls and other Carboniferous petrifactions. Anatomically the stems closely resemble those of *Equisetum*, and were it not for the presence of secondary wood in *Calamites*, the two would be virtually indistinguishable in the fossil condition. In transverse section the large pith cavity is usually filled with crystalline material, but the extreme outer part of the pith, being more resistant than the inner part, is often preserved as a narrow zone of thin-walled cells lining the medullary cavity (Fig. 64). During development, when the tissues differentiated from the terminal meristem, a ring of a dozen or more collateral bundles formed around the pith, and in many young stems these bundles show as very slightly separated strands. In the majority of stems examined a continuous cylinder of secondary wood surrounds the primary system. On the inner face of the secondary sheath the primary strands project inwardly as slightly rounded blunt wedges (Fig. 64B). At the point of each wood wedge there is a conspicuous open space which careful examination of well-preserved stems has shown to be identical with the carinal canal of the *Equisetum* bundle. This canal or cavity is the *protoxylem lacuna*, which formed from the breaking down of the first formed xylem elements during development of the primary vascular skeleton of the stem.   Within these cavities in *Calamites* disorganized fragments of annular and spiral

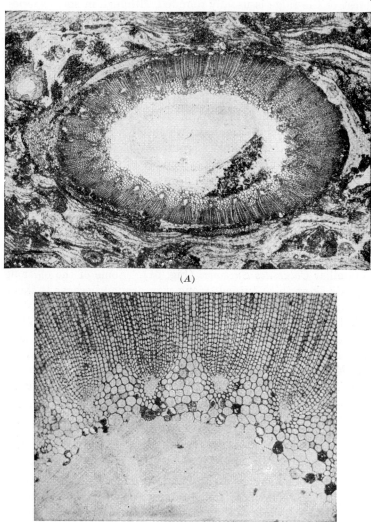

(A)

(B)

Fig. 64.—(A) *Calamites sp.*   Transverse section of the xylem cylinder of a small stem. × 10.   (B) *Calamites (Arthropitys) sp.*   Inner portion of stem showing the carinal canals in the primary xylem and their relation to the secondary xylem.   × about 20.

tracheids have been observed.   External to the canal, the intact primary wood consists of scalariform and pitted tracheids that merge with the pitted elements of the secondary wood.

The secondary wood of *Calamites* consists of tracheids and rays. The tracheids bear their bordered pits on the radial walls and are essentially like those of the gymnosperms. The rays, on the other hand, show considerable diversity in structure, and have served as the basis for generic separation of calamitean stems. *Arthropitys* (Fig. 64*B*), the simplest of calamitean stems, has narrow rays seldom more than one cell wide. In *Arthrodendron* and *Calamodendron* the rays are complex structures. In both, the broad primary rays beginning at the pith between the wood bundles extend far into the secondary wood dividing it into segments, which in turn contain secondary rays of the ordinary type. In *Arthrodendron* the broad primary rays contain zones of vertically elongated fibrous elements interspersed with ordinary ray parenchyma; and in *Calamodendron* the fibrous portions are segregated to the margins with a band of parenchyma in the middle. *Arthropitys* is the most common type of stem and the others are rare. It has been established beyond doubt that that they all belong to *Calamites* but the special names are rarely used except when it is necessary to refer to some particular stem type.

The insertion of the leaf trace is similar in *Calamites* and *Equisetum* although the point of departure is not so constant in the former. The traces usually depart at the nodes from short slanting connections between the primary bundles across the nodal lines. No leaf gaps are formed and in all cases the trace is a simple bundle.

Foliage.—The leaves of *Calamites* are found mostly on the smallest branches although they often remained attached after the stems had attained considerable size. They are often present at the same nodes that support larger branches. The leaf types most commonly encountered are *Annularia* and *Asterophyllites*. Another, *Lobatannularia*, is rare, having been reported only from the Permian of China.

Of the three leaf types, *Annularia* (Figs. 65 and 66*A*) probably exhibits the greatest diversity in form. *Annularia* leaves are arranged in whorls around the stem, but in most compressions the plane of the whorl lies parallel to the stem. Whether this position existed during life is somewhat uncertain although we are reasonably sure that the plane was highly oblique if not quite coincident with that of the stem. The number of leaves in a whorl varies, but usually there are from 8 to 13. The leaves are linear, lanceolate, or spatulate with a single vein, and range in length from 5 mm. or less in the very small *A. galioides* to several centimeters in the familiar and widely distributed *A. stellata*. In many species the midrib is prolonged at the tip into a short but sharp mucro. In some leaf whorls the individual leaves are of equal length, but often those lying nearest the stem are slightly shorter than those extending

Fig. 65.—*Annularia stellata.* Leaf whorls. Middle Pennsylvanian. Henry County, Missouri. Natural size.

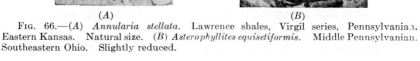

(*A*)   (*B*)

Fig. 66.—(*A*) *Annularia stellata.* Lawrence shales, Virgil series, Pennsylvania. Eastern Kansas. Natural size. (*B*) *Asterophyllites equisetiformis.* Middle Pennsylvanian. Southeastern Ohio. Slightly reduced.

away from it.   In some compressions the leaves of a whorl appear to be joined basally into a sheath, but whether this is a constant feature throughout the genus is doubtful.

Little has been found out about the internal structure of *Annularia*. In transfer preparations made from compressions Walton saw in the cells of some of the leaves dark elongated inclusions, which are probably of the same nature as the starch masses found in petrified specimens of *Asterophyllites*.

Lobatannularia (Fig. 67) is similar to *Annularia* except that the leaf whorl is markedly eccentric and with a gap on that side of the whorl toward the base of the branch.   The leaves near the gap are shorter than the others, and there is a pronounced tendency for all the leaves to bend forward.

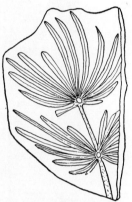

Leaves of *Asterophyllites* are slender and needlelike and form a whorl around the shoot. The number per whorl varies, but there are usually from 15 to 30.   *Asterophyllites equisetiformis* is the most common species (Fig. 66*B*).

Asterophyllites has been found petrified. Each leaf has a small vascular strand surrounded by a starch-bearing bundle sheath. A palisade layer connects this sheath with the epidermis, which contains stomata.   It is evident that the leaves of *Calamites* possessed chlorophyll and carried on photosynthesis, a function which in the Recent members of the Sphenopsida has become relegated to the stem.

Fig. 67.—*Lobatannularia ensifolia*. Upper Shihhotse series, Lower Permian. Central Shansi, China. (*After Halle.*)   Reduced.

The roots of *Calamites* were produced adventitiously at the nodes of the stems between the leaves.   When petrified, they are described as *Astromyelon* and when compressed as *Pinnularia*.   Structurally, the calamitean root bears considerable resemblance to the stem.   The pith did not become hollow during life, hence no casts were formed, and the primary wood strands lack the carinal canals that are such conspicuous features in the stem.   Large air spaces are often present in the cortex. Unlike the stem, the roots are not noded and the branching is irregular.

Fructifications.—The fructifications of the Calamitales are of interest not only on anatomical and morphological grounds but also for the bearing they have on the problem of the origin and evolution of the Sphenopsida. At least five types have been described.   Two of them, *Calamostachys* and *Palaeostachya* (Figs. 68*A* and *B* and 69) have been found as petrifactions but compressions only are known of the others.   All agree in the possession of a central articulated axis bearing whorls of sporangiophores, and some bear intercalated whorls of sterile bracts. The four types, *Calamostachys*, *Palaeostachya*, *Macrostachya*, and *Cingularia* are attributed to *Calamites*. The fifth, *Pothocites*, is the fructification of *Asterocalamites* and is discussed in connection with that genus.

Calamostachys is a large and widely distributed organ genus that ranges from the Lower Coal Measures into the lower Permian. *Cal-*

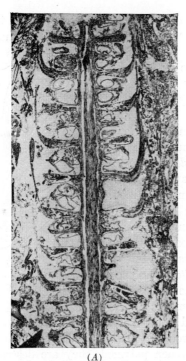

(A)   (B)

FIG. 68.—(A) *Calamostachys Binneyana*. Longitudinal section of strobilus showing alternating whorls of sporangiophores and bracts. × 5.
(B) *Macrostachya sp.* Compression of strobilus. Saginaw group. Grand Ledge, Michigan. ¾ natural size.

amostachys Binneyana and *C. Casheana*, both Coal Measures types, are probably the best known species, and will serve as a basis for discussion of the genus.

The cone axis of *Calamostachys Binneyana* contains a central pith surrounded by three or more separate vascular bundles, the number being one-half that of the sporangiophores in a whorl. Unlike the stem bundles, the cone bundles do not alternate at the nodes but pass continuously along the axis. The axis supports successive whorls of sterile bracts and sporangiophores. The sterile bracts are joined for some distance out from the base but the tips are free. The sporangiophore whorls are situated midway between the bract whorls (Fig. 68*A*), and the number of sporangiophores is about one-half that of the bracts. In *C. Binneyana* there are usually about 6 sporangiophores and 12 bracts, but the number varies in different parts of the cone. *C. tuberculata*, a similar species, has more bracts and sporangiophores, sometimes as many as 32 and 16, respectively.

The sporangiophores are cruciate organs borne at right angles to the axis. The four-armed shield at the end of the stipe bears on its inner side four elongated saclike sporangia, which extend back to the cone axis. A tangential section cut through the cone will therefore show the small stipe in cross section surrounded by four rather large sporangia. The sporangial wall is usually one cell thick.

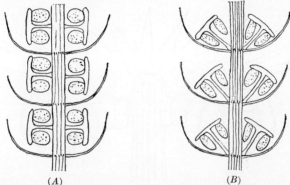

(*A*)   (*B*)

Fig. 69.—Diagrams showing the relation of the sporangiophores and sterile bracts in *Calamostachys* and *Palaeostachya*. (*A*) *Calamostachys*. (*C*) *Palaeostachya*.

The spores of *C. Binneyana* are numerous, are all of one kind, and measure about 90 microns in diameter. They were produced in tetrads and frequently the members are unequally developed, one or more members of the set being larger than the others. However, there is no evidence that megaspores and microspores were formed. The larger ones simply developed to a larger size at the expense of the others.

Calamostachys Casheana is similar in structure to *C. Binneyana* except that it is heterosporous. Some of the sporangia contain numerous small spores only slightly smaller than those of *C. Binneyana*, but in other sporangia there are fewer spores which are about four times as large.

Palaeostachya is organized similarly to *Calamostachys* except that the peltate sporangiophores are situated in the axils of the sterile bracts (Fig. 69*B*). One very interesting anatomical feature is that the vascular bundle supplying the sporangiophore passes vertically upward to about the middle of the internode (the point where the sporangiophore is attached in *Calamostachys*) and then it bends abruptly downward and enters the sporangiophore at the base. The explanation given for the peculiar course of the bundle is that the *Palaeostachya* cone is a derivative of the *Calamostachys* type and that the sporangiophore has moved from a position midway along the internode to an axillary one.

Macrostachya (Fig. 68*B*) is the name given compressions of large calamitean cones bearing successive whorls of bracts and sporangiophores probably attached as in *Calamostachys*. The internal anatomy is not well understood, and since its chief distinguishing feature seems to be size, the name may not designate a type that is morphologically distinct. The best known American species is *Macrostachya Thomsoni*, from the middle Pennsylvanian of Illinois. The cone is 21 cm. long by about 2.5

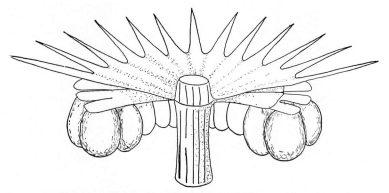

Fig. 70.—*Cingularia typica*. (*After Hirmer.*) Enlarged.

cm. in diameter, and the whorls, which consist of 30 to 36 bracts, are 5 to 7 mm. apart. This species apparently is homosporous although the spores are said to be from 350 to 400 microns in diameter and therefore comparable in size to megaspores. *M. infundibuliformis*, the most common European species, is similar, but differs in having fewer bracts per whorl and in possessing two kinds of spores.

Cingularia, a fructification from the middle Upper Carboniferous, is a lax cone in which the horizontally extending sporangiophore is divided into an upper and a lower lobe. In *Cingularia typica*, the most familiar species, the upper lobe is a sterile bract. The lower portion is split apically into two short lobes and each bears two inferior sporangia (Fig. 70). The inferior position of the sporangiophores and sporangia is a marked departure from that in other Sphenopsida or other vascular cryptogams. In *C. Cantrilli* both the upper and the lower lobes bear sporangia, and no sterile bracts or lobes are present within the cone.

ASTEROCALAMITES

This genus, from the Upper Devonian and Lower Carboniferous of Europe, is one of the oldest members of the Calamitales. The pith casts, which are similar in general appearance to those of *Calamites*, are characterized by nonalternation of the ribs at the nodes. The leaves differ strikingly from those of *Calamites*. They are 5 cm. or more long and instead of being simple are forked several times. In this respect the Middle Devonian genus *Hyenia* is resembled. Detached strobili probably belonging to *Asterocalamites* are described as *Pothocites*. The cone axis bears whorls of peltate sporangiophores but only a few sterile bracts are present, and in some specimens they have not been seen at all. In this respect it resembles *Equisetum* more than *Calamostachys* or *Palaeostachya*.

RESEMBLANCES AND DIFFERENCES BETWEEN CALAMITES AND EQUISETUM

The essential points of resemblance are (1) the jointed stems with leaf whorls at the nodes, (2) the prominent carinal canals (protoxylem lacunae) in the primary wood, and (3) the production of sporangia on sporangiophores. The main differences are (1) *Calamites* has large leaves, which apparently contained chlorophyll, (2) *Calamites* had considerable secondary wood, and (3) the presence in calamitean strobili of whorls of sterile bracts between the sporangiophore whorls.

EQUISETALES

Fossil stems that resemble *Equisetum* are usually assigned to *Equisetites*. A few stems referable to *Equisetites* have been found in the Carboniferous, of which *Equisetites Hemingwayi* from the Middle Coal Measures of Yorkshire is the most completely preserved. This species has slender jointed stems, and attached at some of the joints are broadly oval strobili made up of six-sided peltate sporangiophores. The sterile bracts characteristic of calamitean strobili are lacking. The strobilus, therefore, is essentially equisetalean but borne in a lateral position. *Equisetites mirabilis* is older, being from the Lower Coal Measures, but as its cones are unknown, the affinities cannot be established with certainty.

Equisetites is frequently found in rocks of Mesozoic and Cenozoic age but the remains show little to distinguish them from Recent species. The *Equisetum* type of plant is therefore an ancient one, and this genus may be regarded as the oldest of living vascular plants.

There were several other equisetalean genera in existence during the late Paleozoic and early Mesozoic. One of these is *Phyllotheca*, found principally in the Permian but extending into the Lower Cretaceous. Its leaves are united basally into a sheath but terminate as slender narrow teeth. The peltate sporangiophores that are borne between the leaf whorls bear numerous sporangia. Our knowledge of the plant as a whole is rather meager.

Neocalamites is a Triassic genus that somewhat resembles *Calamites* in habit. Its whorled leaves are arranged in a manner similar to those in *Annularia* but they are free at the base and not joined as they usually are in *Annularia*. It is probably a direct descendant of *Calamites*.

Schizoneura is another equisetalean genus of which rather little is known. The jointed longitudinally furrowed stems bear leaf sheaths at each node that are split into two large multinerved segments on opposite sides of the stem. No fructifications have been found. The genus ranges from the late Permian to the Jurassic.

EVOLUTION OF THE SPHENOPSIDA

Of all those vascular plant groups that have been continuously in existence since the late Paleozoic, the Sphenopsida have come nearest the point of extinction. A possible exception might be the Psilopsida if the Psilotales are descendants of the Psilophytales. Represented by a single genus and about 25 species, the living Equisetales are a mere vestige when the group is viewed in the light of its history. The Lycopsida, by contrast, are represented by several hundred species.

The Sphenopsida was a diversified as well as a large group during the Paleozoic. The class was made up of several divergent lines, the largest being the Sphenophyllales and the Calamitales. The Equisetales are closely related to the Calamitales and the probabilities are that both rose from some common source during the pre-Carboniferous or that the Equisetales developed as an offshoot of the older calamitean stock. The Sphenophyllales are more remote and stand out as a distinct order as far into the past as they can be traced. They probably emerged from some primeval sphenopsid stock before the Equisetales and Calamitales became separate. *Sphenophyllum* is older than *Calamites*. It has been found in the Upper Devonian. The Pseudoborniales probably represent an extremely ancient branch of the sphenopsid complex of relatively short duration which became extinct at an early date.

A possible explanation of the diversification within the Sphenopsida may be that during early Devonian times, or maybe even before, the primordial sphenopsid stock emerged as a branch of the psilophytalean complex. The oldest evidence to support this is found in the Middle Devonian Hyeniales. These plants had retained the dichotomous habit, which was so strongly entrenched within the Psilophytales, but the fertile branches became shortened to mere side shoots with bifurcated recurved tips.   In the original condition all these modified side shoots bore sporangia, but gradually the lower ones became sterile.   Sterilization progressed upward, ultimately resulting in a leafy shoot bearing at its tip a spikelike cluster of fertile stalks.   Therefore both the leaf and the sporangiophore of the Sphenopsida may be considered as having been derived from the bifurcated fertile appendages of the Hyeniales and hence are homologous structures.   A semblance of the original bifurcation is retained in the form of the sporangiophores of the different Sphenopsida.   They are peltate and cruciate in *Equisetum* and *Calamites* and more or less bifurcate in *Sphenophyllum*.   The leaf has undergone more change than the sporangiophore.   In *Sphenophyllum* the cuneiform leaf, which is often split apically, reflects the original bifurcate organ as does also the leaf of *Asterocalamites* and *Pseudobornia*.   In the leaves of *Calamites* and *Equisetum*, on the other hand, simplification has been carried out to the extreme, there remaining practically no trace of the original forked condition.

The morphology of the sterile bracts in the cones of certain of the sphenopsids presents a special problem that has received several explanations.   One is that the bracts and sporangiophores are lobes of dorsally divided sporophylls.   In *Cingularia Cantrilli* both lobes have remained fertile, whereas in *C. typica* the upper lobes are sterile.   Then in *Bowmanites* and *Palaeostachya* the upper lobe is the fertile one, and in *Calamostachys* the fertile lobe, or sporangiophore, has moved upward about midway between adjacent whorls of sterile lobes.   This view, however, seems unlikely when the course of the vascular bundles supplying the sporangiophores is considered.   Other investigators have held that either the bracts are sterilized sporangiophores or that the sporangiophores are fertile leaves, and thus homologous with true sporophylls.   A more recent proposal is that the sterile bracts have intruded the cone from below, and in the cones of the *Calamostachys* and *Palaeostachya* types have become regularly intercalated between the sporangiophore whorls.   The various arguments for and against these several theories cannot be discussed, but considerable significance should be attached to the fact that in the oldest of the Calamitales, *Asterocalamites*, the sterile bracts are irregularly dispersed and sometimes absent, thus approaching the *Calamostachys* condition only in part.   If the last mentioned explanation of the presence of sterile bracts in the calamitean cone is accepted, then *Equisetum* has either lost its sterile members or (and this seems more probable) it never had them.   The sporangium of *Equisetum* would then be a very primitive organ derived from the fertile branch tip of the Psilophytales. Consequently the fructifications of the Calamitales would be the derived forms.

In conclusion, it may be stated that the sphenopsid stock contains several divergent lines which are interrelated only in so far as they may be assumed to have had a common origin. The Sphenopsida are entirely remote from the lycopods, ferns, or seed plants. Although the sphenopsids and the lycopods are essentially alike in their cryptogamic methods of reproduction, the similarities in the life cycles of the two groups are certainly to be accounted for as a result of parallel development.

THE NOEGGERATHIALES

This group of plants, which ranges from the Permo-Carboniferous to the Triassic, is of uncertain position in the plant kingdom, and its inclusion following the Sphenopsida is mainly a matter of convenience. The Noeggerathiales show more points of resemblance with the Sphenopsida than with any other class although they are not embodied within it as it is ordinarily defined. Some authors establish the Noeggerathiales as a group of vascular plants distinct from the lycopods, sphenopsids, and ferns and would give them equal rank. This, however, seems unjustified in view of our meager knowledge of its members.

The two genera making up the Noeggerathiales are *Noeggerathia* and *Tingia*. In *Noeggerathia* the leafy shoot bears two rows of obovate leaves spread in one plane but each leaf is slightly oblique to the axis. The branch was originally thought to be the compound leaf of a cycad or fern. Each leaf has several dichotomizing veins that terminate at the distal margins. In *Noeggerathia foliosa*, the species with the best preserved cones, the latter are terminal on the vegetative shoots and are made up of sporophylls that are shorter and relatively broader than the foliage leaves. They are narrowed at the point of attachment and the arrangement is the same as that of the leaves. About 17 sporangia are situated on the upper surface in longitudinal and transverse rows. Two kinds of spores were produced. The microspores, which measure approximately 100 microns in diameter, are thin walled with slightly roughened coats. The megaspores, which are about eight times as large, are round, thin walled, and have smooth coats. The spores, therefore, are quite similar to sphenopsid spores in general.

In *Tingia* the leaves are longer than those of *Noeggerathia*. In some leaves the margins are nearly parallel and the apex is irregularly cut. The leaves are arranged in four rows with those in two of the rows slightly shorter than the others. The cones are borne in pairs on bifurcated branches, but whether they are terminal or lateral is undecided. The cone consists of tetramerous whorls of long, free sporophylls with upturned tips. The quadrilocular sporangia are supported upon the upper surface of the horizontal part of the sporophyll a short distance away from the axis, and there is some evidence that they are borne on a short stalk. This latter feature has considerable value in determining the affinities of the genus because its presence would strengthen the assumption of proximity to the Sphenopsida. Preservation, however, is inadequate to show with certainty the attachment of the sporangium.

It must be admitted that there are few genuine sphenopsid characters in *Noeggerathia* and *Tingia*. There is no suggestion of articulation in either the stems or the reproductive organs, and of the two genera, *Noeggerathia* is less of a sphenopsid than *Tingia*. It is mainly the suggestion of a short stalk beneath the sporangium of *Tingia* that impels comparison with the Sphenopsida, although as mentioned the spores are not unlike those of other sphenopsids. Those who maintain that the Noeggerathiales should be placed within a class by themselves are probably right, but our present knowledge of them is insufficient to place them on a par with other groups.

References

BROWNE, I.M.P.: Notes on cones of the Calamostachys type in the Renault and Roche collections, *Annals of Botany*, **39**, 1925.

————: A new theory of the morphology of the calamarian cone, *Annals of Botany*, **41**, 1927.

————: The Noeggerathiae and Tingiae. The effect of their recognition upon the classification of the Pteridophyta: an essay and a review, *New Phytologist*, **32**, 1933.

DARRAH, W.C.: A new Macrostachya from the Carboniferous of Illinois, *Harvard Univ. Bot. Mus. Leaflets*, **4**, 1936.

HALLE, T.G.: Palaeozoic plants from central Shansi, *Palaeontologia Sinica*, ser. A, **2**, 1927.

HARTUNG, W.: Die Sporenverhältnisse der Calamariaceen, *Abh. Inst. Paläobot. u. Petr. Brennst.*, **3** (6), 1933.

HEMINGWAY, W.: Researches on Coal-Measure plants. Sphenophylls from the Yorkshire and Derbyshire coal fields, *Annals of Botany*, **45**, 1931.

HIRMER, M.: "Handbuch der Paläobotanik," Munich, 1927.

HICKLING, G.: The anatomy of *Palaeostachya vera*, *Annals of Botany*, **21**, 1907.

HOSKINS, J.H., and A.T. CROSS: Monograph of the Paleozoic cone genus Bowmanites (Sphenophyllales), *Am. Midland Naturalist*, **30**, 1943.

KIDSTON, R., and W.J. JONGMANS: "Monograph of the Calamites of Western Europe," The Hague, 1915.

KRÄUSEL, R., and H. WEYLAND: Beiträge zur Kenntnis der Devonflora, II. *Senckenberg. naturf. Gesell. Abh.*, **40**, 1926.

LACEY, W.S.: The sporangiophore of *Calamostachys*, *New Phytologist*, **42**, 1943.

LECLERCQ, S.: Sur une épi fructifère de Sphénophyllales, I. *Ann. soc. géol. Belg.*, **58**, 1935; II. *ibid.*, **59**, 1936.

————: À propos du *Sphenophyllum fertile* Scott, *Ann. soc. géol. Belg.*, **60**, 1936.

SCOTT, D.H.: "Studies in Fossil Botany," 3d ed., Vol. 1, London, 1920.

SCHULTES, R.E., and E. DORF: A sphenopsid from the Lower Devonian of Wyoming, *Harvard Univ. Bot. Mus. Leaflet,* **7,** 1938.

SEWARD, A.C.: "Fossil Plants," Vol. 1, Cambridge, 1898.

THOMAS, H.H.: On the leaves of *Calamites* (Calamocladus section), *Royal Soc. London Philos. Trans.,* **B 202,** 1911.

WALTON, J.: On the factors which influence the external form of fossil plants, with descriptions of the foliage of some species of the Paleozoic equisetalean genus *Annularia, Royal Soc. London Philos Trans.,* **B 226,** 1936.

WILLIAMSON, W.C., and D.H. SCOTT: Further observations on the organization of fossil plants from the Coal Measures, I. *Calamites, Calamostachys* and *Sphenophyllum, Royal Soc. London Philos. Trans.,* **B 185,** 1894.

CHAPTER VII

PALEOZOIC FERNLIKE FOLIAGE

The prevalence of fernlike foliage in the Carboniferous rocks many years ago won for that period the name "Age of Ferns," and the notion that the Carboniferous flora was predominantly filicinean persisted until the beginning of the present century. Following the epoch-making discovery that some of this foliage belonged to seed plants rather than to cryptogams, the question arose and considerable discussion ensued concerning the relative abundance of ferns and seed plants in the Carboniferous. Suggestions were even made that ferns might not have existed at all during this time, but we know now that the Carboniferous rocks contain the remains of many kinds of ferns, most of which have long been extinct. However, we have no infallible methods of separating fern and seed-fern foliage unless fructifications are present or diagnostic anatomical structures are preserved.

The first successful attempt to classify the various forms of Paleozoic fernlike foliage was made by Brongniart who in 1828 published his "Prodrome d'une histoire des végétaux fossiles," in which he instituted a number of provisional genera based upon shape, venation pattern, and mode of attachment of the pinnules. At about the same time Sternberg, in a series of volumes, described additional genera. In 1836, Göppert, in a work entitled "Die Fossilen Farnkräuter," endeavored to classify fossil fernlike foliage according to its resemblance to recent genera, and he proposed many such names as *Gleichenites, Cyatheites, Asplenites, Aneimites*, etc. Many of Göppert's genera were the same as those previously described by Brongniart and Sternberg and consequently have been abandoned on grounds of priority, although a few, as *Aneimites*, for example, are still used. Many years later Stur and Zeiller based a few additional generic names upon the mode of branching of the fronds, but this method of classification is of limited use because of the usual fragmentary nature of the material. During the latter half of the last century many spore-bearing fructifications were found in association with certain of the leaf types, and some authors, Stur, Zeiller, Weiss, and Kidston in particular, adopted the practice of substituting the name of the fructification for that proposed by earlier authors for the vegetative fronds. Where the fructifications are known, a classification based

157

upon them is preferable to that based upon vegetative characters alone because it more nearly approaches a natural one.   Nevertheless, those names originally proposed for sterile parts are still freely used where necessary.

The artificial genera proposed by Brongniart and subsequent authors are of limited biological value because they often include unrelated species, but they are sometimes useful in stratigraphic correlation. Certain forms, or an assemblage of forms, may be characteristic of a

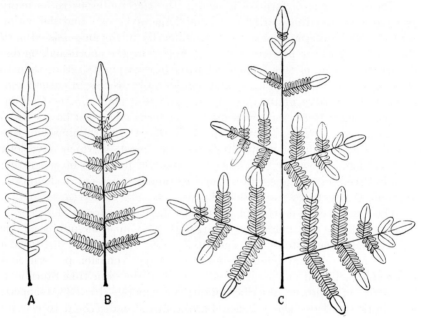

A   B   C

Fig. 71.—Diagrams showing the manner of division of the frond.   Monopodial type.
(*A*) Simply pinnate frond.   (*B*) Bipinnate frond.   (*C*) Tripinnate frond.

given formation or of a lesser stratigraphic unit, and it was mainly on the basis of fernlike foliage that David White discovered distinct floral zones in the Pottsville (lower Pennsylvanian) series of the Eastern United States.   In like manner, Robert Kidston was able to separate the four major divisions of the Coal Measures of Great Britain, and in continental Europe similar divisions of the Westphalian have been made.   Quite recently W.A. Bell has successfully subdivided the Carboniferous rocks of Nova Scotia by similar means.

The architectural plan of the leaf, or *frond*, of the Paleozoic ferns and seed-ferns is similar to that of many of the recent ferns.   The frond was attached to the main stem of the plant by a stalk or *petiole* of varying length and sometimes of considerable size.   In order to support the weight of the large and much branched frond, the petiole was usually provided with considerable supporting tissue in the form of subepidermal strands, which often show as ribs or furrows on the compression surfaces. Externally the petioles were either smooth or covered with emergences such as glandular hairs or trichomes. The branching of the frond was various. In some fronds the petiole forked near the top into equal branches, and sometimes a second dichotomy produced four branches in two pairs (Fig. 72). In others the petiole continued as a central shaft to the apex and bore its subdivisions in a pinnate manner (Fig. 71). This part of the stalk (or its divisions if dichotomously forked) is the

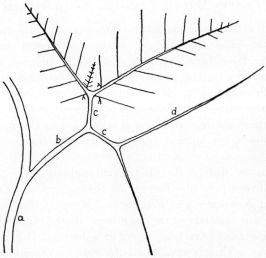

Fig. 72.—Diagram of the four-parted frond of *Mariopteris*. Dichotomous type: *a*, primary rachis; *b*, secondary rachis; *c*, tertiary rachis; *d*, pinna rachis. (*After Corsin.*)

rachis. The *pinnae*, which are the subdivisions of the rachis, may be opposite, subopposite, or alternate upon the rachis and they are all arranged in the same plane as the rachis. The plane of the individual pinna, however, is sometimes at right angles to that of the rachis as in the case of those Paleozoic ferns belonging to the Zygopteridaceae. In others the plane of the pinna is the same as that of the frond. The pinna bears its subdivisions upon the *pinna rachis*, which of course is a branch of the main rachis. If the main rachis bears pinnules only and is without pinnae, the frond is *simply pinnate* (Fig. 71A), but if the pinnules are borne in pinnae the frond is *bipinnate* (Fig. 71B). Further subdivisions make the frond *tripinnate* (Fig. 71C) or *quadripinnate*, as the case may be. The first major subdivision of the frond, regardless of whether there are further subdivisions, is a *primary pinna*, and the subsequent divisions produce *secondary* and *tertiary pinnae*. The pinnules themselves may be variously notched and divided. The tendency is for the frond to produce its more complex divisions near the base and to become simpler toward the tip where the pinnae gradually become so reduced in size that they become replaced by ordinary pinnules. The frond, or any of its subdivisions, is usually terminated by a simple pinnule that is often longer or larger than the others.

IMPORTANT FORM GENERA OF PALEOZOIC FERNLIKE FOLIAGE

I. Pinnules Triangular or with Nearly Parallel or Slightly Convex Margins, and Broadly Attached

The midrib is distinct and continues most or all of the way to the apex.

Alethopteris.—The pinnules are usually fairly large (up to 5 cm. in length), longer than broad with a rounded or pointed apex, usually decurrent especially on the lower side, with a deep midrib that in most species extends to the apex. The lateral veins are distinct, are simple or fork one or more times, and depart at a steep angle but curve abruptly outward and pass straight or along a broad arc to the margin. Those veins in the decurring portion arise directly from the rachis (Figs. 73G and 105A).

Alethopteris is widely distributed throughout the Coal Measures. It is almost always present wherever fernlike plants occur but is seldom as plentiful as *Neuropteris* or *Pecopteris*. Most species are believed to belong to or are related to the seed-fern *Medullosa*. The seeds and pollen-bearing organs have been found attached in a few instances.

Lonchopteris.—This is similar to *Alethopteris* except that the veins form a network. It is restricted to a few horizons in the Upper Carboniferous, and in North America has been found only in the Sydney coal field of Nova Scotia.

Callipteridium.—This form genus is similar to *Alethopteris* and is sometimes confused with it. The pinnules differ from those of *Alethopteris* in being more blunt and not at all or only slightly decurrent. It also bears small accessory pinnae on the rachis between the normal pinnae. Some of the small forms resemble *Pecopteris* except that some of the veins enter the pinnule directly from the rachis. It is less common than either *Alethopteris* or *Pecopteris*.

The foliage included in *Callipteridium* probably belongs to more than one genetic group. Some forms tend to merge with *Pecopteris* and *Alethopteris* on one hand and with *Odontopteris* on the other. *C. Sullivantii*, a species that is common in the lower Allegheny series of Illinois and Missouri, is included by some authors in *Alethopteris*.

FIG. 73.—Diagrams showing the pinnule characteristics of the most important Paleozoic fernlike leaf types. (A) and (B) *Odonopteris*. (C) and (D) *Sphenopteris*. (E) *Neuropteris*. (F) *Mixoneura*. (G) *Alethopteris*. (H) *Alloiopteris*. (I) and (J) *Pecopteris*. (K) *Mariopteris*.

Megalopteris.—The frond is cut into large, long, strap-shaped pinnules that are sometimes bifurcated nearly to the base. The midrib is strongly developed, and the rather fine lateral veins after forking one or more times ascend to the margin at a steep angle or bend sharply and pass at right angles to it (Fig. 74*G*).

Because of the fragmentary condition in which *Megalopteris* is usually found, it is sometimes confused with *Taeniopteris*. *Taeniopteris*, however, has a simple undivided leaf and a distinct petiole, and occurs in the late Pennsylvanian, the Permian, and the early Mesozoic. *Megalopteris* has been found only in North America where it occurs most abundantly in the Conequenessing stage of the upper Pottsville. The plant is probably a pteridosperm. No seeds or other reproductive structures have been found attached but it is often associated with large broad-winged seeds of the *Samaropsis Newberryi* type.

Pecopteris.—The pinnules are usually small with parallel or slightly curved margins and are attached by the whole width at the base or are sometimes slightly contracted. A single vein enters the base and extends almost to the tip. The lateral veins are rather few, are simple or forked, and straight or arched. The pinnules stand free from each other or they may be partly joined laterally (Figs. 14, 73*I* and *J*, and 74*D*).

Pecopteris embraces a large and artificial assemblage of fronds belonging to several natural genera. Many have been found bearing fernlike fructifications such as *Asterotheca*, *Scolecopteris*, or *Ptychocarpus*, and others bear seeds of pteridosperms and other organs believed to be pollen-producing structures. Some authors dispense with the name *Pecopteris* for those forms in which the fructifications are known and apply the names of the fructifications. The pecopterids extend from the early Pennsylvanian into the Permian, but they are most abundant in the middle and late Pennsylvanian and early Permian.

Mariopteris.—The coriaceous lobed or unlobed pinnules are either broadly attached or more or less constricted. They may be free from each other, or decurrent and joined, and the margin may be undulate or toothed. A rather pronounced midvein extends nearly to the apex of the pinnule. The lateral veins are forked, and depart from the midvein at an acute angle. The basal pinnule on the posterior side of the pinna is distinctly larger than the others and is rather prominently two-lobed (Figs. 72, 73*K*, and 74*H*).

The pinnules of *Mariopteris* do not differ greatly from those of *Pecopteris*, and indeed the genus was founded by Zeiller upon certain pecopterid forms in which the subdivisions of the fronds appeared distinctive. It also contains some types originally contained in *Sphenopteris*.

Fig. 74.—Foliage of Paleozoic ferns and pteridosperms. (*A*) *Eremopteris sp.* Saginaw group, Michigan. (*B*) *Sphenopteris artemisaefolioides.* Saginaw group, Michigan. (*C*) *Sphenopteris obtusiloba.* Saginaw group, Michigan, (*D*) *Pecopteris candolleana.* Coffeyville formation, Oklahoma. (*E*) *Linopteris muensteri.* Enlarged pinnule showing the neuropteroid form and anastomosing venation. Carbondale group, Morris, Illinois. (*F*) *Neuropteris plicata.* Carbondale group, Braidwood, Illinois. (*G*) *Megalopteris sp.* Apical portion of frond showing apices of four large segments. Pottsville series, Logan County, West Virginia. (*H*) *Mariopteris sphenopteroides.* Des Moines series, Henry County, Missouri.

163

The morphology of *Mariopteris* has been variously interpreted but its essential structural features are as follows: A geniculate axis, which may be as much as 2 cm. in breadth, gives off alternately arranged (some authors claim spirally arranged) branches, which divide into four by two successive dichotomies. Each dichotomy is at a wide angle, and the two innermost branches of the second dichotomies are longer than the other two. These four ultimate branches in turn bear the primary pinnae, which may themselves in turn be bipinnate, tripinnate, or even quadripinnate. The larger portions, however, are naked, no pinnules being borne upon the subdivisions below those of the second dichotomy. Most species of *Mariopteris* are believed to be pteridosperms although no fructifications have been found attached.

Alloiopteris.—The pinnules are small, alternately arranged, decurrent, and more or less united laterally. The vein entering the pinnule branches two or three times and the veinlets pass at a steep angle to the apex where they terminate usually in three more or less prominent teeth (Fig. 73*H*).

In the fertile condition this leaf type is known as *Corynepteris*. It is rare but widely distributed, and is believed to represent the frond of the fern *Zygopteris*.

II. Pinnules Rounded or Tapering or with Parallel Sides, or Oval or Tongue Shaped

Midrib, if present, is often indistinct and usually breaks up into separate arched veins at one-half or two-thirds of the distance to the apex; margin usually entire.

Neuropteris.—The pinnules are constricted basally to a narrow point of attachment. The base is rounded to cordate and often slightly inequilateral. One vascular strand enters the pinnule from the rachis, which gives off arched, dichotomously forked veins (Figs. 73*E*, 74*F*, 105*B*, 106, and 179*A*, and *B*).

Neuropteris is probably the most abundant and widely distributed of the Carboniferous fernlike leaf types, and it ranges from the Lower Carboniferous to the late Permian. It is most abundant in the Pennsylvanian, especially the lower and middle parts. It is one of the leaf forms of the Medullosaceae and seeds have been found attached to several specimens. The so-called "neuroalethopterids" resemble the true neuropterids except that the midrib is distinct and extends (as in *Alethopteris*) almost to the tip of the pinnule. Examples are *Neuropteris Elrodi*, *N. Schlehani*, and *N. rectinervis*.

Cyclopteris.—The pinnules are large and rounded or auriculate. The veins radiate outwardly from the point of attachment. *Cyclopteris* is not a distinct morphological type, the name being applied usually to the rounded basal pinnules often present on the fronds of *Neuropteris* or other genera.

Linopteris (Dictyopteris).—The pinnules are similar to those of *Neuropteris* but the veins form a network as in *Lonchopteris*. The genus is widely distributed but is rather rare (Fig. 74*E*).

Odontopteris.—The often decurrent pinnules are attached by a broad or slightly narrowed base, and the arched veins enter directly

Fig. 75.—Adolphe Théodore Brongniart. 1801–1876.

from the rachis. In some forms the veins in the central portion are more or less aggregated into a midrib (Figs. 73*A*, and *B*).

Odontopteris ranges from the lower Pennsylvanian into the Permian, and some species are important "index fossils." The members of the group are believed to be pteridosperms.

Mixoneura.—This is a group of foliage types that presents characters intermediate between those of *Neuropteris* and *Odontopteris*. It resembles the former in having narrowly attached pinnules, but one, or more, of the basal veins passes directly from the rachis into the pinnule. Toward the tip of the pinnae the pinnules tend to become more odontopteroid. It is probably derived from the *Neuropteris* type (Fig. 73*F*).

Lescuropteris (*Emplectopteris*).—This form resembles *Odontopteris* except that the veins form a network. Seeds have been found attached. It is a rare type from the late Pennsylvanian and Permian.

III. Pinnules Narrow at the Base, Often Shortly Stalked, and Usually Toothed or Lobed

Sphenopteris.—The pinnules are contracted at the base and often attached by a short stalk. They are usually small, oval or oblong in outline, and lobed or toothed or sometimes cut into narrow acute or obtuse lobes. The midvein is straight or flexuous. The lateral veins depart at an acute angle and dichotomize a few times, and then pass either singly or in groups to the tips of the lobes of the pinnule (Figs. 73*C*, and *D*, 74*B*, and *C*).

Sphenopteris is one member of a large and diversified group of leaf types that first appeared in the Devonian and became widely spread during the Carboniferous. It represents a primitive type of foliar organ. In the lower portion of the Coal Measures, *Sphenopteris* is represented by many species, and although it persists as a prominent member of the flora of the Middle and Upper Coal Measures, it becomes subordinate in numbers to the neuropterids and pecopterids. Some of the sphenopterids bore fernlike fructifications and others produced seeds. *S. Hoeninghausi* is the frond of *Calymmatotheca Hoeninghausi*, a pteridosperm, but others belong to the fern genera *Oligocarpia* and *Urnatopteris*.

Diplothmema.—This genus was proposed by Stur for fronds bearing *Sphenopteris* foliage but in which the leaf stalk below the lowest pinnae is equally forked. In some forms each branch forks a second time much after the manner of *Mariopteris*. It is therefore impossible to distinguish sharply between *Diplothmema* and *Mariopteris*, but in general *Mariopteris* bears pecopteroid pinnules whereas in *Diplothmema* they are usually sphenopteroid. Where the forking of the frond cannot be observed, neither of these criteria is always easy to apply although the coriaceous character of most forms of *Mariopteris* is sometimes diagnostic. Lesquereux segregated certain species of *Diplothmema* into the genus *Pseudopecopteris*, but the name is not widely used.[1]

Rhodea.—The pinnules of this genus are divided by dichotomous divisions into narrow segments in which the single vein is bordered by a very narrow band of the lamina. *Rhodea* bore fructifications of various types including *Telangium*, *Urnatopteris*, and *Zelleria*. It is most common in the Lower Carboniferous but extends into the Upper Carboniferous.

---

[1] For a recent discussion of the genus *Diplothmema* and the use of the name, the reader is referred to a posthumously published paper by David White cited in the bibliography.

Rhacopteris.—The pinnules of this form are semiflabelliform or rhomboidal and are entire or shallowly or deeply cut into slender lobes. They stand out from the rachis at a wide angle. There is no midvein, and the ultimate veins spread equally throughout the lamina and terminate in the marginal lobes. The genus is characteristic of the Lower Carboniferous.

Aneimites.—The pinnules are fairly large (sometimes as much as 4 cm. long) and pyriform or ovate. They often have convex basal margins, which converge into a stout footstalk. The apex is entire or rounded or it may be cleft into two or more rounded segments separated by a sinus of varying depth. The dichotomously forked veins radiate evenly throughout the pinnule. Forms in which the pinnules are rather distinctly three-lobed are sometimes assigned to *Triphyllopteris* (Fig. 178*B*), but this genus is not readily distinguishable from *Aneimites*.

Aneimites is characteristic of the Lower Carboniferous. Several species of *Triphyllopteris* have been named from the Pocono and Price formations of Pennsylvania and West Virgina.

Eremopteris.—In this genus the pinnules are decurrent, lanceolate to spathulate or rhomboidal, toothed or entire, forwardly pointing, and coalescent toward the tip of the pinna but tending to become more divided and open toward the base. The lower inferior pinnule of each pinna is axillary and deeply cut. There is no midrib and the venation is dichotomous (Fig. 74*A*).

Eremopteris is differently interpreted by authors, some including in it species normally retained within *Sphenopteris*. It differs from a true *Sphenopteris* by the lack of a midrib in the pinnules. The frond is large and much branched, and is usually preserved as fragments. Its fructifications are unknown but it is probably a pteridosperm.

In the foregoing pages those form genera of Paleozoic fernlike foliage are described which are either in most common usage or are distinctive for some reason or other. The others, of which there are several, are rare or are restricted to a few horizons or are segregates of better known types. For the most part the types described are characteristic of the Mississippian and Pennsylvanian of North America or of the Lower and Upper Carboniferous of Europe. When spore-bearing organs are found attached to a frond belonging to any of these leaf genera, the usual practice is to replace the form genus name with that of the fructification if it has been previously named. Seed-bearing fronds, on the other hand, have usually retained the name of the frond (as for example, *Alethopteris Norinii* and *Neuropteris heterophylla*), or the name of the stem if it happens to be known.

Of the hundreds of species of pteridophylls that have been described from the Carboniferous, relatively few have been found bearing fructifications. The discovery of the fructifications assists greatly in determining affinities.

References

BERTRAND, P.: Bassin houiller de la Sarre et de la Lorraine, I, Flore Fossile, 1er fasc., Neuroptéridées, *Études des Gîtes Minéraux de la France*, Lille, 1930; 2me fasc., Aléthopteridées, *ibid.*, 1932.

BRONGNIART, A.: "Histoire des végétaux fossiles ou recherches botaniques et géologiques sur les végétaux dans les divers couches du globe," Paris, 1828–1844.

CORSIN, P.: Bassin houiller de la Sarre et la Lorraine, I, Flore Fossile, 3me fasc., Marioptéridées, *Études des Gîtes Minéraux de la France*, Lille, 1932.

CROOKALL, R.: "Coal Measure Plants," London, 1929.

GOEPPERT, R.H.: Systema filicum fossilium, *Nova Acta Acad. Caesareae Leopoldina natur. Curios*, suppl. 17, Breslau, 1836.

JANSSEN, R.E.: Leaves and stems of fossil forests, *Popular Sci. Ser., Illinois State Museum*, **1**, 1939.

KIDSTON, R.: Fossil plants from the Carboniferous rocks of Great Britain, *Great Britain Geol. Survey Mem.*, **2** (1), 1923; *ibid.*, **2** (2), 1923–1925.

LESQUEREUX, L.: Description of the coal flora of the Carboniferous in Pennsylvania and throughout the United States, *Pennsylvania 2d Geol. Survey Rept. P*, text, 1880, atlas, 1879.

NOE, A.C.: Pennsylvanian flora of northern Illinois, *Illinois State Geol. Survey Bull.* 52, 1925.

STERNBERG, G.: "Versuch einer geognostische-botanische Darstellung der Flora der Vorwelt," Leipzig, 1821–1838.

STUR, D.: "Beiträge zur Kenntnis der Flora der Vorwelt," Vol. I, Die Culm-Flora des märisch-Schlesischen Dachschiefers, Wien, 1875; Vol. II, Die Culm-Flora der Ostrauer und waldenburger Schichten, Wien, 1877.

WHITE, D.: Fossil flora of the Lower Coal Measures of Missouri, *U. S. Geol. Survey Mon.* 37, 1899.

———: (A posthumous work assembled and edited by C. B. Read): Lower Pennsylvanian species of Mariopteris, Eremopteris, Diplothmema, and Aneimites from the Appalachian region, *U. S. Geol. Surv. Prof. Paper* 197 C, 1943.

ZEILLER, R.: Flore fossile du bassin houiller de Valenciennes, *Étude des Gîtes Minéraux de la France*, Paris, Atlas, 1886, Texte, 1888.

———: "Eléméntes de Paléobotanique," Paris, 1900.

CHAPTER VIII

THE ANCIENT FERNS

The ferns have a long history that has been traced to the middle of the Devonian period. The oldest ferns are only partly distinct from their psilophytic forebearers, but in the late Devonian we find them standing out against the landscape as prominent elements of the flora. Ferns increased markedly in species and in numbers of individuals during the early Carboniferous, and they had become a diversified plant group before the close of the Paleozoic coal age. Some of the Paleozoic ferns appear to be related to modern families. The Filicinae were more successful than the contemporaneous lycopods and scouring rushes in their struggle against the adverse climate of the late Permian, and they survived into the Mesozoic in large numbers. Not until the Jurassic do they show much tendency toward a large-scale decline. Ever since the close of the Paleozoic the ferns have been the most successful vascular cryptogams and today they constitute the largest class. The Recent flora contains approximately 175 genera and 8,000 species as compared with a total of 700 lycopods and 25 scouring rushes.

Early paleobotanical writers believed that ferns outnumbered other plants during the late Paleozoic, a notion which has not completely disappeared even at the present time. Many individuals still regard the finely divided fossil fronds found on coal-mine dumps as ferns although it was demonstrated more than four decades ago that many of these ancient fernlike plants bore seeds. Although the emphasis once placed on ferns in the literature was excessive, they were, nevertheless, important plants of the late Paleozoic. Because of their long history, and the fact that their anatomical and reproductive features indicate relationships in many directions, the ferns have entered strongly into phylogenetic controversy. In the most modern systems of classification in which all phases of the structure and life histories of plants are taken into account, the ferns are classified along with the gymnosperms and flowering plants in the Pteropsida, which is believed to be a more natural group than the Pteridophyta to which they are traditionally assigned.

Because of the general similarities in the life cycles of ferns and the Bryophyta, the two are closely associated in morphological thought. This association is the result of resemblances in the structure of the sex organs, the production of spores which germinate outside the sporangium, and the well-defined alternation of generations.   In addition to these, the ferns and bryophytes often occupy similar habitats.   On the other hand, there are numerous differences of such an obvious and fundamental nature that any close phylogenetic relation between them is precluded.   The resemblances are superficial and the result of parallel development.

ANATOMICAL CHARACTERISTICS OF FERNS

The ferns exhibit more variety in body structure than any other group of vascular plants of comparable rank.   Some of them have simple vascular systems, whereas others have them built along complicated lines.   The primary xylem is exarch or mesarch (usually the latter) and consists for the most part of scalariform tracheids.   In *Pteridium* some of the largest tracheids lose the closing membranes of the scalariform pit openings and thus take on the characteristics of vessels.   The stele may be a protostele (*Gleichenia*), a siphonostele (*Adiantum* and *Dicksonia*), or a dictyostele (*Cyathea* and *Pteridium*).   Of these the protostele is the most primitive, and the dictyostele of the *Pteridium* type, in which the bundles are not arranged in a circle, is the most advanced.   The leaf traces are usually large, and they present a variety of appearances in cross section.   The leaf gaps are usually conspicuous as large breaks in the continuity of the vascular cylinder.   In the protostelic and siphonostelic types the phloem is present as an unbroken zone completely surrounding the xylem core, and in the dictyostelic forms it may surround all the xylem (*Osmunda*) or it may ensheath each individual strand (*Pteridium*).   Internal phloem occurs in several genera (*Dicksonia*).   Secondary xylem is extremely rare in ferns, occurring in living forms only in the genus *Botrychium*.   Instead of the production of secondary vascular tissues as a means of supplying the larger plant body, the fern vascular system breaks up into separate strands or adds to its bulk by the development of folds and accessory strands.   Thus in the tree ferns belonging to the Cyatheaceae the stele in many species changes gradually from a siphonostele in the basal part of the trunk to a dictyostele of large dimensions accompanied by medullary strands in the higher parts. In others, especially in the Polypodiaceae, Matoniaceae, and Marattiaceae, inward projections of the stele may develop into accessory strands, or in some cases into complete internal cylinders.

The fructifications of modern ferns are synangia or individual sporangia borne singly or in sori.   An annulus may or may not be present. The fructifications are usually borne upon the under surface of the unmodified pinnules, although in many ferns they are marginal, and occasionally (as in *Onoclea*) the fertile foliage may be slightly modified by a reduction of the lamina. They are never on sporangiophores or on the upper surface. The spores are small and numerous. The exines are wrinkled or variously marked, and the triradiate tetrad scar is usually visible. Heterospory is present in living ferns only in the Marsiliaceae and the Salviniaceae. This brief generalization of recent ferns may serve to some extent as a basis for understanding the ancient forms.

THE OLDEST FERNS

The abundance of ferns and fernlike plants in the Carboniferous naturally arouses curiosity concerning their earliest history. It is

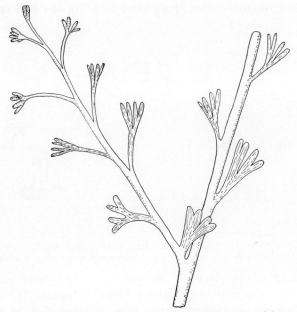

Fig. 76.—*Protopteridium minutum.* Branch system showing the ultimate subdivisions flattened into leaflike organs. *(After Halle.)* × about 2.

possible that they developed at some time during the late Silurian or early Devonian as an offshoot from the psilophytic complex. When first suggested, this was merely a supposition not based upon any known links between the ferns and Psilophytales, but now four genera are known from the Middle Devonian that appear to combine some of the characteristics of the two groups. One of these is *Protopteridium* (Fig. 76), from Belgium, Scotland, Eastern Canada, Bohemia, and China. The plant consists of sympodially branched axes upon which the ultimate branches are dichotomously divided, and some of the latter are flattened in such a manner as to suggest a primitive frond. Oval sporangia are borne at the ends of some of the smallest subdivisions and on one side of the sporangium there is a band of specialized cells along which the sac apparently split when the spores were mature. The upper Middle Devonian Hamilton formation of western New York has yielded three petrified stems, which also appear to occupy an intermediate position. The simplest of these, and the one apparently closest to the Psilophytales, is *Arachnoxylon* (Fig. 77*A*), a plant represented by small stems with a xylem strand only a few millimeters in diameter. The xylem is deeply lobed with six projecting points. Near the extremity of each lobe is a small island of parenchyma tissue completely immersed in the xylem, which resembles the "peripheral loop" of the Paleozoic Zygop-

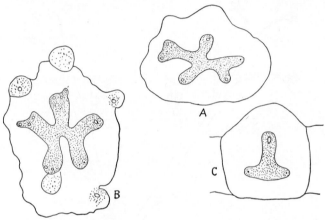

Fig. 77.—Middle Devonian fernlike axes. (*A*) *Arachnoxylon Kopfi.* × about 12. (*B*) *Iridopteris eriensis.* × about 7. (*C*) *Reimannia aldenense.* × about 10.

teridaceae and Cladoxylaceae. No leaf traces have been observed, an indication that the plant was leafless, and this feature, in combination with the general simplicity of the axis, is suggestive of psilophytic affinities. *Reimannia* (Fig. 77*C*), has a slightly more complicated stem in which the xylem strand is T-shaped in cross section. Small leaf traces depart from the three xylem arms but nothing is known of their form or structure. The most advanced of the three psilophytic ferns from the Hamilton formation is *Iridopteris* (Fig. 77*B*), in which a five-lobed xylem cylinder gives off spirally arranged leaf traces. The xylem is peculiar in that each lobe contains near the tip two parenchymatous areas placed side by side. This genus is decidedly less psilophytic than either *Arachnoxylon* or *Reimannia*, and were it not for the fact that the three together constitute a series of forms of increasing complexity, *Iridopteris* might be looked upon as quite apart from the psilophytalean line. As the matter stands, *Arachnoxylon*, *Reimannia*, and *Iridopteris* furnish anatomical evidence of psilophytalean ancestry of the ferns, whereas *Protopteridium* indicates a similar relationship from the standpoint of habit.

The Upper Devonian has not yielded many ferns, but the few that have been found reveal further departures from the Psilophytales. *Asteropteris noveboracensis* was described many years ago by Dawson from the Portage beds (Upper Devonian) of central New York. The stem has a deeply stellate symmetrical xylem strand with 12 lobes from which leaf traces depart. The trace sequence shows that the fronds were produced in whorls with the successive verticils superimposed. Near the tip of each xylem arm is the diagnostic parenchymatous area, which indicates affinity with the Zygopteridaceae. The traces themselves are also of the zygopterid type in that soon after departure each trace becomes constricted in the middle with two lateral loops, with the long axis of the trace section lying at right angles to the radius of the main stem. At this stage the trace resembles that of *Clepsydropsis* described on a later page. At a higher level, however, each trace has two loops at each end. The main point of interest in connection with *Asteropteris* is that it was the first Devonian plant to be discovered in which the tissues were sufficiently preserved to reveal definite fernlike affinities, and consequently it was long known as the oldest fern. We know now, however, that fernlike plants were in existence during the Middle Devonian, and it is possible that they exist undiscovered in still older rocks.

All the early fernlike plants of which we possess information on the internal structure have a lobed or fluted xylem strand which bears some resemblance to that found in the Carboniferous Zygopteridaceae. The fluted xylem therefore appears to be an ancient feature that developed from the simple psilophytalean strand before the decipherable history of the ferns began. Not all the Paleozoic ferns, however, show the fluted xylem because in the Botryopteridaceae, a family that existed nearly contemporaneously with the Zygopteridaceae, a round or nearly round strand is present. It is believed that the Botryopteridaceae attained its state of development through simplification and that it is not as primitive as the other family.

An ancient plant believed to be a fern and occurring in both the Devonian and the Carboniferous is *Cladoxylon*. The oldest species is *Cladoxylon scoparium* from the Middle Devonian of Germany. Other occurrences of the genus have been reported from the Upper Devonian Portage shales of New York and the New Albany shale of Kentucky. Our knowledge of the German form is most complete. The largest stems, which are almost 2 cm. in diameter, branch somewhat irregularly in a crude dichotomous fashion which recalls *Hyenia* or *Calamophyton*. Clothing the smaller branches are small, deeply cut, wedgeshaped, aphlebialike leaves. On the upper extremities of some of the branches the appendages are fertile and bear at their tips small oval sporangia. The stems are polystelic, a transverse section showing a number of elongated xylem strands, some of which are small and simple, and others which are larger and bent into various shapes.

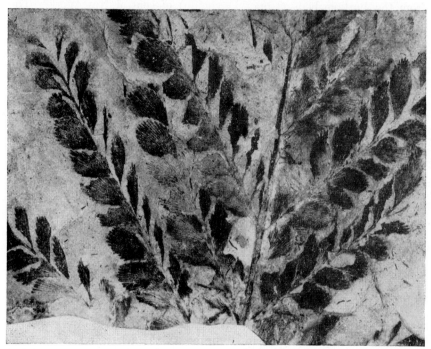

Fig. 78.—*Archaeopteris latifolia.* Upper Devonian. Port Allegany, Pennsylvania. Natural size.

The most widely distributed fern in the Upper Devonian is *Archaeopteris*. Its exact position in the plant kingdom was long a matter of uncertainty but the latest evidence indicates that at least some of the species are ferns and it is probable that all of them are. The genus is known mostly from frond compressions and it is possible, though not probable, that the *Archaeopteris* type of frond could have been borne on more than one kind of plant.

Although *Archaeopteris* is decidedly fernlike in general appearance, some paleobotanists have maintained that it is a seed plant even though seeds have neither been found attached to it nor closely associated with it. A couple of accounts exist of seedlike objects in intimate association with *Archaeopteris,* but it is just as possible that they could be spore-bearing organs as seeds. Too little is known of the anatomy to shed much light on the question of affinity, and it was not until some recently discovered compressions of exceptionally well-preserved sporangia were studied that the exact situation became clear.

Archaeopteris is sometimes used as an "index fossil" of the Upper Devonian, which means that its presence is accepted as substantial proof of the age of the rocks in which it is found. It does occur, however, very sparingly in the Middle Devonian.

Fig. 79.—*Archaeopteris macilenta.* Upper Devonian. Factoryville, Pennsylvania. (*From specimen in United States National Museum.*) Lacoe Collection. Natural size.

In North America *Archaeopteris* occurs throughout the Upper Devonian shales and sandstones of Catskill and Chemung age in northeastern Pennsylvania and adjacent portions of New York and in rocks of similar age in eastern Maine and Eastern Canada. In Pennsylvania and New York the most common species are *Archaeopteris Halliana, A. latifolia* (Fig. 78), *A. macilenta* (Fig. 79), and *A. obtusa.* The plants from Maine and Canada are similar, and may not be specifically distinct, but they are known (in part) as *A. Jacksoni, A. Rogersi,* and *A. gaspiensis.* In Canada the most prolific *Archaeopteris* locality is the paleontologically famous "Fish Cliffs" on the north shore of Scaumenac Bay along the southern border of the Gaspé Peninsula, where Dawson collected vegetative and fertile fronds at about the middle of the last century. In addition to plants, the locality has yielded several forms of Devonian fish, one of which is *Eusthenopteron*, an organism in which the fins had developed to the extent that a limited amount of motility on dry land was possible.

Well-known European species of *Archaeopteris* are *Archaeopteris hibernica*, the type species from County Kilkenny, in Ireland, and *A. Roemeriana* from Belgium, Germany, and Bear Island in the Arctic Ocean. *A. hibernica* occurs sparingly in Pennsylvania and is closely related to *A. latifolia* but differs in having larger pinnules. *A. Roemeriana* is very similar to *A. Halliana*.

Little is known of the habit of *Archaeopteris* other than that it produced large fronds which sometimes reached a meter or more in length. It was probably a shrub rather than a tree, but whether the main stem was upright or prostrate is unknown, although there is some evidence that the fronds were attached above the ground and did not arise from subterranean parts. It probably produced a low dense bushy growth along streams and embayments where there was abundant soil moisture.

The fronds of *Archaeopteris* are compound like those of many ferns. The rather strong straight rachis bears two rows of suboppositely placed pinnae, all spread in the same plane. The wedge-shaped pinnules, also in two rows, vary from 1 to 5 cm. or more in length, are attached by a narrow base, and broaden to a rounded apex. They may or may not overlap, depending upon the species. There is no midvein. The single vein entering the base divides by several dichotomies and the branches pass straight or with slight curvature to the distal margin. The margin may be smooth (*A. Halliana*), toothed (*A. latifolia*), or deeply cut (*A. macilenta*).

We know little of the vascular system of *Archaeopteris* because only compressions have been found. However, the thick carbonaceous residues of some of the leafstalks reveal the presence of vascular strands similar to those of ferns.

The sporangia of *Archaeopteris* are borne on specialized fertile pinnae. These may make up a large part of the frond or only a small portion of it. Frequently they are limited to the lower portion of the rachis, but they are always similar in size and arrangement to the vegetative ones. The sporangia themselves are oval or linear exannulate spore cases either sessile or shortly stalked borne singly or in clusters on greatly reduced pinnules. Sometimes a fertile pinna may bear a few vegetative pinnules at the base, thus suggesting a homology between the two types.

The spores and sporangia of several species of *Archaeopteris* have been studied but the spore condition is well known only in *A. latifolia*

(Fig. 80). In this species there are two kinds of sporangia, slender ones measuring approximately 0.2 mm. in width by 2 mm. in length, and broader ones about 0.5 mm. by 2 mm. in comparable dimensions. The smaller ones contain numerous (100 or more) very small spores about 30 microns in diameter, whereas the others hold 8 to 16 spores, which are at least 10 times as broad. The exines of both types are smooth or very slightly roughened and are heavily cutinized. These obviously are microspores and megaspores. From the descriptions (by Johnson) of the sporangia of *A. hibernica* from Ireland, it is probable that a similar spore condition is present in this species. Although no mention is made of spores of two sizes the sporangia are described as "locular," but the supposed locules are in all probability the outlines of large spores similar

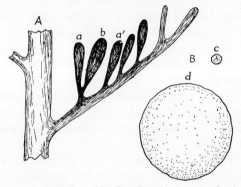

Fig. 80.—*Archaeopteris latifolia.* (*A*) Fertile pinna: *a* and *a'*, microsporangia; *b*, megasporangium. (*B*) Sketches showing relative sizes of *c*, microspores and *d*, megaspores. Enlarged.

to the megaspores of *A. latifolia*, and the sporangium is in all likelihood a simple undivided spore case.

The spores produced by *Archaeopteris latifolia* indicate beyond all reasonable doubt that this species is a fern and not a seed plant. It does, however, indicate a trend in the direction of seed plant evolution, and it may constitute the necessary heterosporous stage in the derivation of the seed from the terminal sporangium of the Psilophytales. Morphologically it may stand between the homosporous ferns and the pteridosperms, or it may be a more direct connection between the pteridosperms and the Psilophytales. In either case, however, it appears to be a "missing link" in the evolution of the seed plants. It is possible that some species of *Archaeopteris* had reached the point of seed production, but this is entirely conjectural. Small paired cupules resembling seed husks have been found associated with the foliage of *A. latifolia* in Pennsylvania, and before the heterosporous condition was observed in this plant, these cupules were thought to be its probable seeds. At present the evidence that *Archaeopteris* reproduced by cryptogamic methods is much stronger than the evidence for any other method.

A plant probably related to *Archaeopteris* but nearer the psilophytic level is *Svalbardia*, with one species *S. polymorpha*, from beds belonging either to the upper Middle Devonian or lower Upper Devonian of Spitzbergen. The genus is founded upon branches and branch systems of rather small size in which the internal structure is not preserved. Branching is lateral and opposite to subopposite. The ultimate vegetative branches are slender filiform ramifications bearing fan-shaped pinnulelike appendages split apically by two or four dichotomies into slender segments. These organs are about 2.5 cm. long. The venation is dichotomous and each segment receives one vein.

The fertile branches are panicles 3 to 4 cm. long. Branchlets depart from the axis of the panicle at nearly right angles to it, and they bear near the middle on the upper side as many as 12 sporangia on simple or divided stalks. The branchlet tips are sterile and curved forward. The sporangia are pyriform or cylindrical, are 0.5 to 0.7 mm. wide by 1.5 to 2 mm. long, and have thin walls. The spores, which appear to be all of one kind, are slightly oblong and measure 60 to 70 microns in greatest dimension. The exine is nearly smooth and there is a distinct triradiate tetrad scar.

The size of *Svalbardia* is unknown, but judging from the fragments of the branch systems, the plant was probably a small shrub not exceeding 2 m. in height.

THE LATER PALEOZOIC FERNS

During the late Paleozoic (Lower Carboniferous to Permian) a few ferns appeared that can be assigned with varying degrees of certainty to modern families. For example, the structurally preserved stems of the Permian genera *Thamnopteris* and *Zalesskya* show essential characteristics of the Osmundaceae. In older rocks the fructification *Oligocarpia* is provisionally aligned with the Gleicheniaceae, and *Senftenbergia*, another fructification, shows certain essential features of the Schizaeaceae. Throughout the Carboniferous (including the "Permo-Carboniferous") there are numerous fructifications and a few stems that have been compared with the Marattiaceae. Some other modern families have been reported from the Paleozoic but final investigations have in every instance shown the evidence to be inconclusive.

The late Paleozoic ferns consist mainly of representatives of a large group, the Coenopteridales. The Coenopteridales first appeared in the Devonian and probably became extinct before the onset of the Triassic.

If they did outlive the Paleozoic they did so in such reduced numbers or in such modified form that they have escaped recognition in the later periods.

The Coenopteridales

The name Coenopteridales is applied to a large assemblage of structurally preserved stems and frond parts, which often occur in abundance in coal-balls and other types of petrifactions of late Paleozoic age. The taxonomic treatment of the group varies, and consequently it is impossible to present any standardized classification of its members. Some of the generic names refer to stems, but others apply to leafstalks of which we have no knowledge or are uncertain of the stems to which they were attached. The Coenopteridales have also been called Palaeopteridales, Primofilices, and Renaultificales, the latter in honor of the great French paleobotanist Renault. No two of these names as originally proposed were used in exactly the same way, but their applications are similar. They all refer to Paleozoic ferns that cannot be assigned to Recent families. In this book the group is tentatively divided into three families, the Botryopteridaceae, the Zygopteridaceae, and the Cladoxylaceae. Some authors describe additional families based for the most part upon single species, but it is felt that the inclusion of more family names than those given would serve no useful purpose. There are numerous Paleozoic plants belonging to other groups that are difficult to classify, and the list of *plantae incertae sedis* is large and will always be so. Bertrand, who proposed the name Renaultifilicales for those ferns here described under the Coenopteridales, recognized two subgroups. The first is the *Inversicatenales*, which conforms in general to the Botryopteridaceae, and the second is the *Phyllophorales*, which is made up principally of the Zygopteridaceae and the Cladoxylaceae. He assigns each group ordinal rank.

Botrypoteridaceae.—This family of Paleozoic ferns contains the simplest although not necessarily the oldest of the Coenopteridales. The stem contains a simple protostele and the petiole bundle is a single strand that is often but not always deeply indented or curved in cross section. In all genera except *Tubicaulis* the pinnae are spread in the same plane as the rachis, a feature that distinguishes the family from the Zygopteridaceae.

The Botryopteridaceae are typified by *Botryopteris*, the best known genus of the family. It contains five species which range from the Lower Carboniferous to the Permian. The slender stem, which is but a few millimeters in diameter, is much branched and bears spirally arranged fronds. The main stem has a small mesarch protostele. The phloem completely surrounds the xylem, and no secondary tissues were formed.

As seen in cross section the xylem strand in the leafstalk is distinctive. It is solid, and has one to three protoxylem points on the adaxial side.

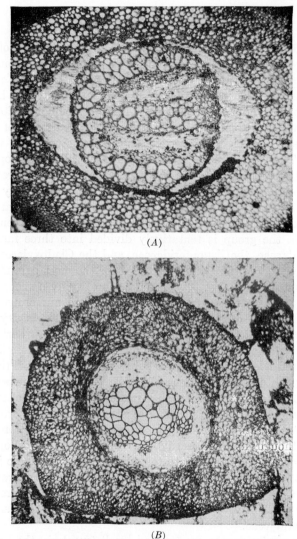

(A)

(B)

Fig. 81.—(A) *Botryopteris forensis*. Transverse section of petiole. McLeansboro formation. Illinois. × about 20. (B) *Botryopteris hirsuta*. Transverse section of petiole. Coal Measures. England. × about 30.

In some species the strand is deeply indented, giving the appearance of a trident (*B. forensis*, Fig. 81A), but in others the protoxylem is situated on short points (*B. hirsuta*, Fig. 81B), or on the smooth surface (*B.*

antiqua). The fronds forked at least once, and the branches bore pinnae of the first order. In *B. forensis* the ultimate frond segments were of the *Sphenopteris* type, but in the other species the pinnules are unknown.

Botryopteris forensis is one of the few petrified Paleozoic fern species to which sporangia have been actually found attached (Fig. 82). The sporangia are oval or pyriform, measuring 1.5 to 2 mm. in length and 0.7 to 1 mm. in diameter. They are clustered and short-stalked, and occur in tufts on the branches of the fertile rachis. The annulus is a broad, oblique band of thick-walled cells limited to one side of the sporangium. The spores, which are small and all of the same kind, were discharged through a split opposite the annulus.

Botryopteris bore its roots adventitiously along the stem among the leaves. They are diarch and closely resemble those of living ferns.

Anachoropteris, a Middle Carboniferous genus belonging to the Botryopteridaceae, is known mainly from structurally preserved leafstalks. The xylem strand in cross section is strongly incurved with revolute margins. Apparently the curvature is away from the stem, and the protoxylem groups (of which there are two or more) are located on the adaxial or convex side. The fructification is a synangium of four sporangia borne at the end of a vein on the incurved margin of the

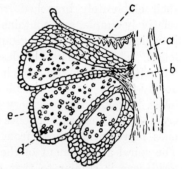

Fig. 82.—*Botryopteris forensis.* Cluster of three sporangia attached to fragment of fertile rachis: *a*, fertile rachis; *b*, short stalk bearing sporangia; *c* annulus belonging to sporangium of adjacent cluster; *d*, numerous spores in sporangial cavity; *e*, sporangial wall. (*After Renault.*) × about 30.

small pinnule. Each sporangium contains a number of small spores. The entire synangium is bounded by a thick layer of cells that caused the sporangia to open at maturity.

Tubicaulis is sometimes made the type of a distinct family. It consists of three species, one from the Lower Coal Measures, and two from the Permian. The stem has a simple protostele with centripetally developed protoxylem. Surrounding the vascular cylinder is a thick cortex that produced a stem of considerable size. Within this broad cortex are the spirally arranged leaf traces in various stages of departure. The xylem of the trace is outwardly curved with the protoxylem on the adaxial side. *Tubicaulis* differs from the typical botryopterids in that the plane of the pinnae is horizontal to the main rachis instead of parallel to it, and because of this feature it is placed within the Zygopteridaceae by some authors.   It suggests a common origin of the Botryopteridaceae and the Zygopteridaceae, and illustrates one of the essential difficulties in the formulation of a consistent classification of the ancient ferns.

Zygopteridaceae.—This group of ferns is both older and more complex than the Botryopteridaceae, and it shows possible connections with the Psilophytales through the Middle Devonian genera *Arachnoxylon*, *Reimannia*, and *Iridopteris*.   As was pointed out, *Arachnoxylon* possesses a deeply lobed xylem strand without evidence of departing traces, but near the outer extremity of each there is a clearly defined peripheral loop.   The other two genera have leaf traces while at the same time they retain much of the essential psilophytalean simplicity.

Externally the members of the Zygopteridaceae are characterized by their elaborately branched fronds.   The stem, where it is known, is relatively simple, and contains a shallowly or deeply lobed xylem strand in which the protoxylem groups occupy positions near the ends of the lobes.   The stele is usually described as a protostele although the central part often contains tracheids intermixed with parenchyma.   In the more deeply furrowed forms the pithlike tissue in the center extends into the lobes.   Around the periphery the tracheids form a compact tissue.   In cross section the arms or lobes appear to be of unequal length, the longer ones representing those from which the traces are about to depart. The spiral arrangement of the fronds exhibited by most of the Zygopteridaceae is similar to that of the Osmundaceae, and the cross section through a well-preserved stem will reveal the complete sequence of trace transformation.

The main stalk of the zygopterid frond has been termed the *phyllophore*.   It differs from an ordinary leafstalk or petiole in that it is intermediate in position and structure between the stem and a typical leafstalk.   It possesses axial symmetry instead of bilateral symmetry and bears its subdivisions in two or four rows.   These subdivisions (the pinnae of the first rank) are attached so that they lie in a plane at right angles to that of the phyllophore axis.   The pinnae of the first rank are often divided into secondary pinnae which are again turned at right angles to the axis on which they are borne.   Modifications of this basic plan are often exhibited by particular genera of the family.

The xylem strand of the main leafstalk (the phyllophore) presents a variety of appearances in cross section and the figures presented are often striking (Fig. 83).   In most genera it is thinnest along the middle and thickest at the edges.   In such forms as *Clepsydropsis*, *Metaclepsydropsis*, *Diplolabis*, and *Asteropteris* the trace is somewhat dumbbell- or hourglass-shaped in cross section, with the peripheral loops occupying prominent positions at the ends (Figs. 83C and D).   Where a pinna trace departs from the phyllophore the loop opens to the outside and faces a similar concavity in the departing strand. The gap thus formed closes gradually above the axil produced by the departing trace except in some cases where the sequence of departure is too close to permit closure before another trace originates. In others, such as *Ankyropteris* and *Etapteris* (*Zygopteris*) the figure recalls the letter H (Figs. 83*A* and

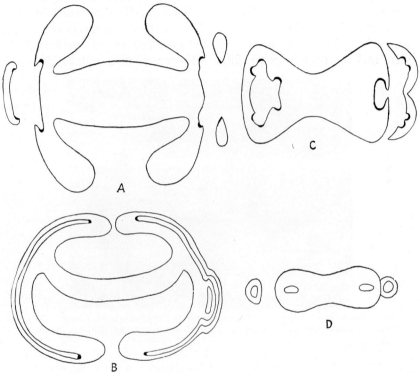

Fig. 83.—Zygopterid phyllophore types. (*A*) *Etapteris Scotti.* (*After Bertrand.*) (*B*) *Ankyropteris westphalensis.* (*After Hirmer.*) (*C*) *Metaclepsydropsis duplex.* (*After Hirmer.*) (*D*) *Clepsydropsis antiqua.* (*After Bertrand.*)

B), with the crossbar lying tangential to the circumference of the stem. *Stauropteris* is different from any of the foregoing. In this genus the xylem strand is a deeply invaginated cruciform structure, which at places breaks into separate strands (Fig. 84).

The oldest genus to be assigned to the Zygopteridaceae is the Upper Devonian *Asteropteris* previously described. Another ancient type is *Clepsydropsis*, which ranges from the late Devonian to the Permian. In *Clepsydropsis* the phyllophore trace is narrowly elongated in cross section, the rounded ends contain a prominent peripheral loop, and it is slightly constricted in the middle.   It is regarded as the ancestral type from which others such as *Ankyropteris* and *Zygopteris* were derived.

Clepsydropsis leafstalks have been found attached to more than one kind of stem.   *Asterochlaena*, one of the stems, is a Permian genus with a very deeply lobed xylem cylinder.   The tips of the long lobes are divided into two or three small lobes from which the traces depart in a spiral sequence around the stem.   The center of the stellate xylem contains a mixed pith in which the tracheids are mostly short and wide, and a narrow radial band of small-celled tissue extends from this mixed pith into each of the xylem arms.   The phloem follows the outline of the xylem and the deep bays are filled with parenchyma.   *Asterochlaena* is

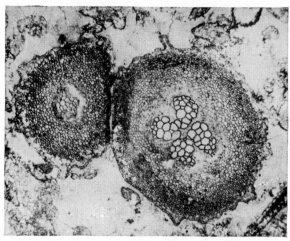

Fig. 84.—*Stauropteris oldhamia.*   Transverse section of phyllophore showing four-parted xylem strand and lateral branch.   Coal Measures.   Great Britain.   × 16.

peculiar in the possession of the most elaborate stele found in any of the Zygopteridaceae but in combination with traces of the relatively simple *Clepsydropsis* type.

Austroclepsis, from Australia, and probably of Carboniferous age, has clepsydropsid phyllophores although its stem is more like that of *Ankyropteris*, which is to be described shortly.   In this fern the repeatedly forked stems are bound together by a heavy growth of adventitious roots into a mass that resembles the false stem of the Cretaceous genus *Tempskya*.   The Devonian genus *Asteropteris* also has clepsydropsid phyllophores which, however, arise in verticils instead of spirals.

Our knowledge of *Zygopteris*, the type genus of the Zygopteridaceae, centers mainly around *Z. primaria*, from the lower Permian of Saxony. Until recently the plant was very incompletely known because the specimen used in the original description showed only the leafstalks, and these had been imperfectly described. It so happened that the type specimen, along with another block which may even represent the basal part of the same plant, was cut into several pieces and distributed among several European museums. These pieces were recently reexamined by Sahni who has given the first account of the stem on which the leafstalks were borne. When the several slabs were assembled into approximately their original positions, a tree fern of some size was revealed with a central stem about 1.5 cm. in diameter surrounded by a mantle of leafstalks and adventitious roots having a diameter of about 20 cm. The xylem cylinder consists of a small central rod of scalariform tracheids surrounded by a layer of secondary wood. This is one of the rare examples of a fern axis, either living or fossil, with secondary wood. However, a stem of the same type but lacking leaf-stalks discovered previously in the Lower Coal Measures had been named *Botrychioxylon* in allusion to its resemblance to the recent *Botrychium*, but had the stem of *Zygopteris primaria* been known, this genus would never have been founded. In addition, it was found that the leafstalks of this plant are identical with those previously known as *Etapteris*, of which several species exist. It appears, however, that no stems are known for any of the other species of *Etapteris*, and while the presumptive evidence may be strong that all of them were originally borne on stems of the *Botrychioxylon* type, actual proof is lacking, and for that reason the names *Botrychioxylon* and *Etapteris* are retained as organ genera for detached stems and leafstalks of the types to which the names were originally applied. But on grounds of priority *Zygopteris* is the correct name for the combination whenever any of the parts are found together.

The petioles of *Zygopteris primaria* are cylindrical and up to 2 cm. in diameter, and because no compressions have been found the form of the frond must be inferred entirely from the internal anatomy. The frond was evidently bipinnate and each pinna was deeply divided down the middle. The two halves of the pinna apparently gave off pedately branched subdivisions on which the lamina was borne, but the exact form is not clearly shown in the pertified material.

As stated, *Zygopteris primaria* bore leafstalks of the *Etapteris* type of which the structure is probably best expressed in another species, *Etapteris Scotti* (Fig. 83*A*), a common fossil in the coal-balls of the English Coal Measures. The vascular strand is surrounded by a three-layered cortex of which the innermost delicate layer has mostly disappeared. The xylem strand has an H-form as afore-mentioned, with a straight, rectangular median band and stout, somewhat flexed lateral arms. Two external protoxylem groups lie in shallow depressions on opposite sides of the figure at the ends of the median band. The pinna traces arise as a pair of strands departing laterally from the position of the protoxylem groups.   They soon fuse to form a flat slightly curved bar, which then divides to produce two crescentric strands.   No peripheral loop is formed in this species as the sinus left by the departing traces does not close as in other members of the zygopterid group.

A few fructifications are attributable to *Zygopteris*.   One is *Corynepteris*, borne on foliage known in the sterile condition as *Alloiopteris*.   The fertile pinnules may be very similar to the sterile ones or they may be slightly modified.   The sporangia are large, ovate, and sessile, and are grouped into spherical sori of five or six around a central point.   The annulus is a broad band that extends up along the edge of the sporangia, where they touch, and around and over the apex.   Dehiscence probably took place around the edges where the sporangia are in contact, or along the inner surface.   The sorus bears some resemblance to *Ptychocarpus*, and the suggestion has been made that it may indicate some connection between the Zygopteridaceae and the Marattiaceae.

In *Notoschizaea*, a fructification similar to *Corynepteris*, a broad annulus extends over the entire outer (abaxial) face of the sporangium, and the vertical dehiscence line extends down the center.   *N. robusta*, the only species, occurs in the middle Pennsylvanian of Illinois.

Slightly curved, elongated sporangia have been found on a petiole known as *Etapteris Lacattei*.   These sporangia are 2.5 mm. long, and occur in tufts of three to eight on short branched pedicels.   Extending along each side of the sporangium is a band of cells with thickened walls, which constitutes a vertical annulus of a unique type.   The spores are about 80 microns in diameter and resemble those of ordinary ferns.

Ankyropteris, with eight species, is similar in age to *Zygopteris*.   *A. Hendricksi*, from the early Pennsylvanian of the Ouachita Mountains of Oklahoma, is the only one from North America.   *Ankyropteris* resembles *Zygopteris* except that no secondary xylem is present in the stem.   The xylem core is an angular structure with about five lobes of unequal length, which result from the spiral departure of the foliar traces.   In the center, and extending into the lobes, is a mixed pith similar to that of some other forms.   The phyllophore trace in cross section is characteristic.   In *A. Hendricksi* it is H-shaped, and in *A. westphalensis* (Fig. 83B) the free ends are inwardly curved so that the structure resembles a double anchor.   In all species the pinna traces depart opposite the ends of the crossbar of the strand.   The peripheral loop is a narrow strip of parenchyma situated along the lateral extensions of the xylem, and one peculiarity of the genus is that when the pinna traces depart the loop remains closed.   The phloem completely surrounds the xylem and conforms to its outline.

There are several other well-defined generic types assigned to the Zygopteridaceae which are known only from the leafstalks. The Lower Carboniferous *Metaclepsydropsis* was probably a procumbent plant in which the fronds were attached to a horizontal rhizome. The rectangular xylem strand is slightly constricted in the middle (hourglass-shaped) and near each end is a large round or oval loop (Fig. 83C). The loop opens to form a large gap when a pinna trace departs. The pinna traces are given off alternately from the opposite poles, and immediately after departure the crescent-shaped trace curves inward and soon divides laterally to form two strands. The pinnule traces arise from these.

Stauropteris is a rather primitive type that is sometimes segregated into a separate family. It has three species, two from the Lower Carboniferous and one from the Upper Carboniferous. The branching of the frond is distinctive. The main rachis bears alternate branch pairs, and each individual branch is in turn divided in a similar manner. All together the main rachis is branched into rachises of three orders and the same four-ranked arrangement of the lateral branches is maintained throughout. The ultimate ramifications bear no laminae but each is terminated by a sporangium. These features suggest affinity with the Psilophytales. The xylem of the main rachis and of the branches is a cruciate mass that is often broken into two or more separate strands (Fig. 84). The terminal sporangia are of the exannulate eusporangiate type, and the spores escape by an apical pore.

A genus that is interpreted as transitional between the Osmundaceae and the Zygopteridaceae is *Grammatopteris*, from the lower Permian of France and Saxony. The plant had a small upright trunk equipped with a slender protostele surrounded by a cortex and a thick armor of leaf bases and adventitious roots. The leaves depart from the stem in a close spiral, and in a single transverse section a series representing the change in form of the leaf trace from the moment it departs from the xylem cylinder until it enters the petiole base can be observed. At first the xylem of the trace is a small elliptical strand but farther out it lengthens and becomes a thin band. This is quite different from the strongly curved trace strand of the Osmundaceae, and more nearly approaches the condition in the Zygopteridaceae. Probably the most outstanding osmundaceous character is the incision of the surface of the xylem, which give a shallowly lobed effect. The traces, however, depart from between the lobes, and not from them. The sinuses between the lobes are therefore regarded as rudimentary leaf gaps. The genus, however, is believed to be closer to the Zygopteridaceae than to the Osmundaceae, but it may belong to an offshoot of the Zygopteridaceae from which the osmundaceous line developed.

Cladoxylaceae.—Cladoxylon, the principal genus of this group, is one of the oldest plants to be attributed to the ferns. Its affinities have been variously interpreted. The presence of secondary wood is suggestive of a place among the pteridosperms where it was assigned by Scott and other authors, but the appearance of the primary xylem and the clepsydropsid traces are features more in common with the Zygopteridaceae. Some investigators have placed it in a group by itself (the Cladoxylales) below the ferns in the evolutionary scale and somewhat transitional between them and the Psilophytales.

The anatomical structure of *Cladoxylon* is peculiar, a fact responsible for the doubt and uncertainty concerning its systematic position. The

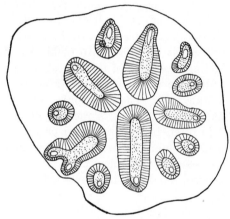

Fig. 85.—*Cladoxylon taeniatum.* (*After Bertrand.*) × about 3.

stem is polystelic (Fig. 85). In cross section there are a number of separate radially elongated steles, some straight, but others which are curved, sharply bent, or irregular. These steles form an anastomosing system vertically. Each stele consists of a central oval or radially elongated solid primary xylem strand surrounded by its own zone of secondary wood. Often the secondary xylem is thicker on the side toward the interior of the stem than on the other side. The peripheral loop is situated at the outer extremity of each primary xylem arm from where the leaf trace departs. The secondary wood consists of pitted tracheids and small rays. The genus is somewhat restricted, ranging from the Middle Devonian to the lower part of the Lower Carboniferous. It therefore represents an ancient line that became extinct before the major development of the Carboniferous flora took place, and probably left no immediate relatives.

OTHER PALEOZOIC FERNS

Gleicheniaceae.—The existence of this family in the Carboniferous is suggested by *Oligocarpia*, a fructification borne on certain species of *Pecopteris* and *Sphenopteris*. The fructification consists of a sorus of four to six sporangia placed in a circle on the lower surface of the pinnule at the ends of the veinlets. An annulus of two or more rows of cells extends along the outward margin of each sporangium. The spores produced within the sporangium are small, smooth, and numerous.

The resemblance between *Oligocarpia* and the recent genus *Gleichenia* has long been known, but the affinities have been questioned because of the structure of the annulus. In *Gleichenia* the annulus is strictly uniseriate. Professor Bower calls attention to the resemblance between *Oligocarpia* and those members of the Marattiaceae in which the sporangia are arranged in a circle.

Oligocarpia is found principally in rocks of middle and late Coal Measures age. Undoubted members of the Gleicheniaceae (*Gleichenites* and *Mertensites*) occur in the Triassic, and from there the family extends in unbroken sequence to the present.

Osmundaceae.—The fossil record of this family begins with two late Permian genera *Thamnopteris* and *Zalesskya*. The two are much alike, and a brief description of the former will suffice.

The stem of *Thamnopteris* is a protostele surrounded by a relatively thick cortex and a much wider zone of persistent leaf bases. The tracheids in the center of the solid xylem strand are short and wide, apparently being modified for water storage. The outer ones served for conduction. The xylem is surrounded by a continuous layer of phloem, a feature that links this genus with the Osmundaceae.

The spirally arranged leaf traces of *Thamnopteris* originate on the outside of the xylem cylinder, and they first show in cross section as oval strands with centrally located protoxylem points. Slightly higher, a parenchymatous area forms in the protoxylem region, which later breaks through on the adaxial side of the strand. In the leaf base the xylem narrows to a horseshoe-shaped band with the concavity toward the stem. Within a single transverse section of a stem surrounded by leaf bases the entire sequence may be seen. The plant was probably a small tree.

Schizaeaceae.—The existence of this modern fern family in the Carboniferous seems fairly well established by the occurrence in both the late Lower Carboniferous and the Upper Carboniferous of the fructification *Senftenbergia* (Fig. 86), with oval or barrel-shaped sporangia borne in two rows on the lower surface of the pinnules of certain species of *Pecopteris*. Each sporangium is attached by a short narrow stalk, and around the apical part is a circular, bandlike annulus two to four cells in width. The spores are numerous, have smooth walls, and measure about 45 microns in diameter. These sporangia differ from those of certain of the living species of *Aneimia* and *Schizaea* only with respect to minor details. In *Schizaea* the annulus is only one cell wide but otherwise the resemblance to *Senftenbergia* is close. If the affinities of

FIG. 86.—*Senften-bergia ophiodermatica.* (*After Radforth.*) Enlarged.

Senftenbergia are correctly interpreted, the Schizaeaceae becomes one of the oldest of surviving fern families.

Marattiaceae.—There exist throughout the rocks of the Upper Carboniferous and Lower Permian numerous spore-bearing organs, which show considerable resemblance to the fructifications of the marattiaceous ferns. In the Marattiaceae the sporangia are grouped into linear or circular clusters, and in many genera lack an annulus and are joined into synangia. In a synangium the spores occupy locules in the multichambered organ. The sporangial (or synangial) wall is several cells thick, and spore dispersal is through a pore or slit. The Marattiaceae belong to the eusporangiate group of ferns in which the sporangium develops from a superficial cell group. This feature, however, is of little use in the study of fossil ferns because the manner of development of the sporangia cannot be studied.

Those Paleozoic fern fructifications resembling the Marattiaceae are borne mostly upon *Pecopteris* foliage. The fertile pinnules are but little if at all modified, and the sporangial clusters or synangia are produced on the under surface, usually in rows on either side of the midrib. As in modern genera, they are situated over a veinlet or between the veinlet tip and the margin.

Ptychocarpus (Figs. 87 and 88*B*) is a fructification of the Middle and Late Coal Measures age, which is produced on the foliage of *Pecopteris unitus*. It is common in both Europe and North America, being abundant in the plant-bearing nodules at Mazon Creek and in the strip mines near Braidwood in Illinois (Fig. 87). The fructification is an exannulate synangium circular in cross section, but when viewed laterally has the form of a truncate cone that narrows slightly toward the apex. The entire structure is 1mm. or less in diameter, and consists of five to eight sporangia, that are united to each other for their full length at the sides. In the center is a vertical column of tissue to which the sporangia are attached by their inner surfaces. The synangia are arranged in a single row on each side of the midrib of the pinnule. The method of dehiscence is unknown but it is believed that the spores escaped through a terminal pore. The sporangium is surrounded by a layer of delicate tissue and the whole assemblage is embedded in an envelope of lax parenchyma, which makes up the wall of the synangium.

Ptychocarpus bears a strong though probably superficial resemblance to the modern marattiaceous fern *Christensenia* in which the synangium is circular. An important difference, however, is that the *Christensenia* synangium is hollow in the center, and spore dispersal takes place through a dehiscence line on the upper face of each locule.

FIG. 87.—*Ptychocarpus (Pecopteris) unitus.* Portion of fertile pinna showing two-ranked synangia. Mazon Creek, Illinois. × about 2.

Asterotheca (Fig. 88C), also borne on *Pecopteris* foliage, is a synangium somewhat resembling *Ptychocarpus* except that the sporangia per synangium are fewer, and they are united only at the base, being free for the greater part of their length. It differs from *Scolecopteris*, the next fructification to be mentioned, in being entirely sessile. It is therefore evident that the three fructifications *Ptychocarpus*, *Asterotheca*, and *Scolecopteris* are essentially similar, and differ only with respect to the extent of attachment between the individual sporangia and the length of the central receptacle.

The sporangia of *Asterotheca* usually number four or five. The synangium is situated on a veinlet or between the veinlet tip and the pinnule margin. There is usually one row on each side of the midrib.

Asterotheca has several species of which some of the most familiar are *A. arborescens*, *A. cyathea*, *A. Miltoni*, *A. oreopteridea*, and *A. hemiteliodes*. It ranges throughout the Coal Measures (Pennsylvanian) and into the Permian, being most abundant in the Middle and Upper Coal Measures. There is some evidence that the foliage which produced *Asterotheca* was borne on trunks of the *Psaronius* type.

Scolecopteris (Fig. 88A) is similar to *Asterotheca* except that the synangia are situated upon a short pedicel. The cluster consists of four or sometimes five sporangia, which are pointed at the apex and joined only at the extreme base. They therefore stand free from each other for the greater part of their length.   The genus contains but three species, *S. elegans* from the lower Permian of Saxony, *S. minor* from the middle Pennsylvanian of Illinois, and *S. Oliveri*, from the Permo-Carboniferous of Autun, in France.   The species commonly called *S. polymorpha* is sessile and has been transferred to the genus *Acitheca*.

Cyathotrachus* is a Coal Measures fructification that resembles *Ptychocarpus* except for its taller shape and in having a cup-shaped depression in the center similar to the living *Christensenia*.   Another difference is that the synangia are short stalked.   It is known only from

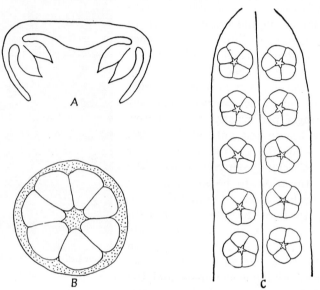

FIG. 88.—Paleozoic exannulate fructifications. (*A*) *Scolecopteris;* (*B*) *Ptychocarpus;*
(*C*) *Asterotheca.*

isolated specimens but in association with *Pecopteris* foliage.   *C. alatus* is from the British Coal Measures and *C. bulbaceus* is from the middle Pennsylvanian of Illinois.

In contrast to the circular synangium of the type just described, the name *Danaeites* has been applied to linear synangia, which somewhat resemble the modern genus *Danaea*.   *D. sarepontanus* is a fossil species in which the synangia consist of 8 to 16 sporangia in two contiguous rows on the lower surface of pecopteroid pinnules.   Dehiscence is by an apical pore, and the entire structure is slightly embedded within the tissue of the leaf.   *Danaeites* is a rare type, and was reported many years ago from the upper Middle Carboniferous of the Saar Basin.   It has not been sufficiently figured to be accepted as conclusive evidence of the implied relationship with the recent genus *Danaea*.

Previous to the discovery of the pteridosperms all fossil fernlike foliage and fructifications were supposed to represent true ferns, and those sporangia provided with an annulus were assigned to the Leptosporangiatae and exannulate ones to Eusporangiatae (Marattiaceae). In the light of present knowledge it still seems probable that fossil annulate fructifications represent nonmarattiaceous ferns, although the possibility must be admitted that some pteridosperms might have had microsporangia which are indistinguishable from the sporangia of leptosporangiate ferns. No annulate pteridosperm microsporangia, however, have yet been discovered. The discovery of the pteridosperms revealed a problem concerning the assignment of fructifications similar to that pertaining to foliage. We are handicapped by not knowing much about the microsporangiate organs borne on particular pteridosperms.

The problem of placing fossil exannulate sporangia in their proper plant groups is still more acute. It is frequently impossible to be absolutely certain whether a given fructification belongs to some ancient marattiaceous fern or to a pteridosperm, and Kidston, in commenting upon such familiar forms as *Asterotheca* and *Ptychocarpus*, says that the evidence is about as strong for placing them in one group as in the other. The majority of paleobotanists have been prone to consider the clustered type, such as *Asterotheca* and *Ptychocarpus*, marattiaceous and singly borne ones as probably being pteridospermous. This practice, however, is open to criticism for two reasons. In the first place it assumes that no singly borne exannulate sporangia were present on ferns during the Paleozoic, and second, there is evidence that amounts to practical certainty that some pteridosperms had synangial microsporangiate organs. *Archaeopteris*, according to the latest available information, is an example of a fern with separate exannulate sporangia, and *Aulacotheca*, *Crossotheca*, and *Whittleseya* are synangiate organs evidently belonging to pteridosperms. However, none of those synangia believed to belong to pteridosperms bear much resemblance to the fructifications of the living Marattiaceae.

Our information being as imperfect as it is, it is impossible to cite the exact differences between the fructifications of the Paleozoic ferns and pteridosperms. The manner in which the fructifications are borne may have some meaning in certain groups of plants. In the Lyginopteridaceae and the Medullosaceae, two families of pteridosperms, the fructifications, where known, are borne either on reduced frond segments or as replacements of ordinary pinnules on the pinna rachis. In no known instances in these families are either the microsporangia or the seeds produced on the surfaces of unmodified pinnules as are the sporangia of most ferns. From this it may logically follow that exannulate sporangia on reduced pinnules or pinnules modified in other ways are pteri-dosperms (*Archaeopteris* being an exception), although it would not be safe to reason that all spore-bearing organs borne on unmodified foliage belong to ferns. Some pteridosperms (*Sphenopteris tenuis*, *Emplectopteris triangularis*, and *Pecopteris Pluckeneti*) carried their seeds on unmodified foliage, and it is possible that their microsporangia were similarly situated.

Psaronius.—This genus of Paleozoic tree ferns is often compared to the Marattiaceae but in all probability belongs to a separate family. Its foliage is believed to be of the *Pecopteris* type but neither the foliage nor the fructifications have been found attached so the systematic

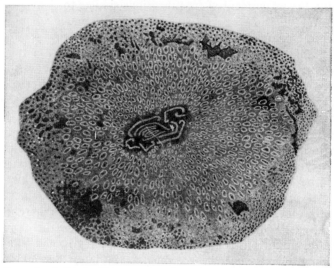

Fig. 89.—*Psaronius Gutbieri.*   Transverse section of stem showing the central vascular portion and the enclosing root mass.   (*After Corda.*)

position of the tree has never been determined with finality. Some investigators have suggested that it might have borne seeds, but this is merely a supposition not in accord with the evidences of affinity supplied by anatomical structure. The largest trees reached an estimated height of 10 m. or more, and the silicified trunks frequently exceeded 0.5 m. in diameter. The fronds, which were borne in a crown at the summit of the trunk, were arranged in two or more rows. Trunk compressions showing the frond scars are known as *Megaphyton* if the scars are in two ranks and *Caulopteris* if there are more than two rows. The scars are large oval imprints sometimes 10 cm. or more in height, and bear on the surface a horseshoe-shaped impression of the frond trace. Emerging from the trunk below the crown of leaves are numerous adventitious roots, which form a mantle of increasing thickness toward the base. The trunk portion below the leaf zone therefore resembles a slender inverted cone. There was no secondary growth in the stem, and the root sheath was mainly responsible for the support of the crown.

The morphology of the root sheath has been interpreted differently by different authors. It was originally thought to be cortical with roots growing vertically through it as in the living *Angiopteris* and some palms. The other, and most generally accepted explanation, is that the root mantle is extracortical, as indicated by its structure and the fact that no leaf traces pass through it. The roots are embedded within a compact tissue of hairlike structures, which arise at least in part from the roots themselves. These hairs may be seen arising from the roots, mostly on the side toward the outside of the trunk, and pressing firmly against the inner surface of the roots external to them.

Fig. 90.—*Psaronius lacunosus.* Transverse section of stem showing arrangement of the vascular bundles. Conemaugh series. Ohio. *Drawn from photograph by Blickle.* Slightly less than ½ natural size.

The individual roots are small, usually less than a centimeter in diameter. The xylem strand is six-pointed in cross section in most species and with only a few exceptions is composed entirely of primary tissue.

The vascular portion of the *Psaronius* trunk is usually only a few centimeters in diameter, and hence small in proportion to the size of the complete trunk. It is surrounded by a well-defined zone of sclerenchyma that is continuous except for breaks where leaf traces emerge. These breaks correspond with the arrangement of the leaves as revealed on the stem surface—thus in the *Distichi* the breaks occur in two series on opposite sides (*Megaphyton*), whereas in the *Tetrastichi* and the *Polystichi* (*Caulopteris*) there are four or more than four series, respectively. The interior of the stelar portion consists of ground parenchyma in which are embedded numerous concentric vascular bundles which form a complicated system but in which the arrangement corresponds definitely with the attachment of the fronds. The bundles are of various sizes, but in general they are elongated in a tangential direction. The xylem consists of scalariform tracheids with or without an intermixture of parenchyma (depending upon the species) and the protoxylem is more or less centrally located. Phloem completely surrounds the xylem. Some of the bundles, mainly those opposite the "leaf gaps" in the peripheral sclerenchyma, are a part of the leaf trace system. Those toward the outside, which are called the reparatory steles, are the ones which depart at the lowest levels as leaf traces, and those toward the interior will supply leaves at successively higher levels. Alternating with the reparatory steles are the so-called "peripheral steles," which give rise to the adventitious roots. They also fuse with the reparatory steles below the places where the latter pass out as traces.

Psaronius ranges from the Lower Coal Measures to the Triassic, but the greatest number of species occurs in the Rotliegende (Lower Permian). The oldest petrified species is *P. Renaulti*, from the Lower Coal Measures of Lancashire. Anatomically this species is relatively simple, with a single endarch solenostele enclosing a large pith and with internal as well as external phloem. The structure of the root zone, however, agrees with that of other species of the genus. Compressions of the *Caulopteris* type are found occasionally in the lower Pennsylvanian of North America and in beds of similar age at other places.

Numerous silicified trunks of *Psaronius* have been found in the late Pennsylvanian and early Permian in southeastern Ohio mainly in the vicinity of Athens. Recent studies have revealed the presence of about five species but detailed descriptions are not yet available. Some of the trunks are as much as 3 feet in diameter, and when cut and polished show an attractively colored ornamental pattern. Years ago large numbers of these were collected and sold as inexpensive gem stones.

POST-PALEOZOIC FERNS

No account of the ferns of the past is complete without reference to some of the better known Mesozoic and Cenozoic forms. The Coenopteridales do not appear to have left any descendants, but by late Permian time other ferns had come into existence, which were able to adapt themselves more satisfactorily to the unfavorable surroundings.

Fossil ferns attributed to the Matoniaceae show that this family, which at present is restricted to the East Indies, once had a wide range. One of the Mesozoic members of this family is *Phlebopteris* (often called *Laccopteris*) which ranges from the Triassic to the Lower Cretaceous. The pedate fronds of this fern resemble those of *Matonia pectinata*. The pinnules are linear, are attached by a broad base, have a distinct midrib with anastomosing veins, and the circular sori are arranged in two rows, one on each side of the midrib. The sori contain 5 to 15 sporangia, each with an oblique annulus. In the compressed condition it is sometimes difficult to recognize the annulus and the sorus is sometimes mistaken for a synangium. The best known American species is

P. Smithii (Figs. 91 and 92), from the Chinle formation (Upper Triassic) of Arizona.

FIG. 91.—*Phlebopteris Smithii.* Sterile frond. Chinle formation, Triassic. Petrified Forest National Monument, Arizona. Slightly reduced.

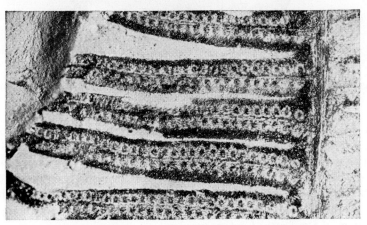

FIG. 92.—*Phlebopteris Smithii.* Fertile frond. × about 5.

The Osmundaceae are abundant in the Mesozoic. *Todites Williamsonii* is world-wide in its distribution. Petrified stems are assigned to *Osmundites* and *Paradoxopteris*. *Osmundites skidegatensis*, a fern in which the stem resembles that of *Osmunda regalis*, is from the Lower

Cretaceous rocks of Queen Charlotte Island off the coast of British Columbia. Similar stems have been found in the Tertiary of central Oregon (Fig. 93). *Osmunda* does not exist in the Recent flora of the western part of North America. One of the most common of the Mesozoic Osmundaceae is *Cladophlebis*. In North America alone nearly 40 species have been indentified from strata ranging from the Triassic to the Lower Cretaceous. It is probably most abundant in the Jurassic although it is well represented in the Lower Cretaceous Potomac formation.

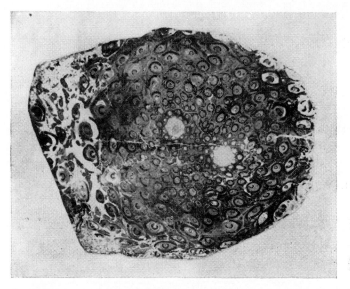

FIG. 93.—*Osmundites oregonensis*. Transverse section of stem. Tertiary. Post, Oregon. Natural size.

A Mesozoic fern genus of exceptional interest is *Tempskya*. It occurs at several horizons within the Cretaceous but in Western North America, where several species have been found, it is claimed to be restricted to the lower part of the Upper Cretaceous in beds nearly comparable to the Colorado group. It may, however, occur in older beds. In Maryland *Tempskya* is found in the Patapsco formation in the uppermost part of the Lower Cretaceous.

Tempskya occurs as silicified trunklike structures, which are sometimes several inches in diameter and as much as 9 feet long. They are usually straight or slightly curved, and are round or irregularly oval in cross section. The trunk masses are often club-shaped or conical, with the largest end representing the apex. They sometimes bear a superficial resemblance to silicified palm trunks with which they are frequently confused.

Anatomical studies of the trunks of *Tempskya* have shown them to consist of much branched stem systems in which the branches are held together by a profuse growth of adventitious roots. It appears that the base of the plant contained one stem that branched by repeated dichotomies. These branches and the adventitious roots which they produced grew in a more or less parallel direction and formed a large ropelike mass which is called the "false stem" (Fig. 94). A few living ferns produce similar organs, examples being *Todea barbara* (Osmundaceae) and *Hemitelia crenulata* (Cyatheaceae).

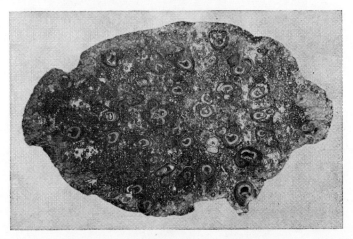

Fig. 94.—*Tempskya grandis.* Transverse section of false stem showing the individual stems embedded within the root mass. Cretaceous. Wheatland County, Montana. Natural size.

The individual stems of *Tempskya* (Fig. 95*A*) have amphiphloic siphonosteles (solenosteles), which vary in different species from 0.5 to 1 cm. in diameter. Each individual stem is complete with cortex, endodermis, pericycle, phloem, xylem, and pith. The central part of the pith is sclerotic. Of the leaves only the bases are known, and they arise from one side of the individual stems in two ranks. The orientation of the stems within the trunk with respect to the direction of departure of the leaf traces has furnished clues to the probable habit of the plant. In some species the traces depart from the stems on the side nearest the periphery of the false stem, thus indicating radial symmetry and a more or less upright habit. Examples are *T. grandis* and *T. wyomingensis*. *T. minor* and *T. Knowltoni*, on the other hand, show the traces departing mainly in one direction, indicative of a reclining or dorsiventral habit

for the plant.   It is believed that in most species the leaves were attached
only to the apex of the false stem.   However, the outermost tissues are
seldom well preserved, and whether the trunks possessed any kind of
lateral appendages is unknown.

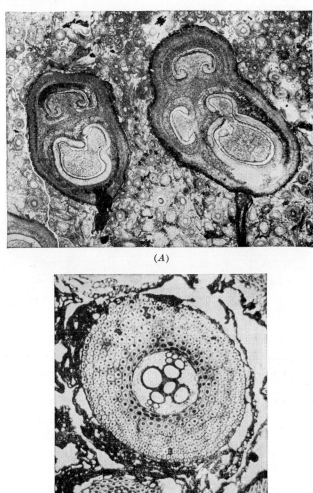

(A)

(B)

Fig. 95.—(*A*) *Tempskya wyomingensis.*   Transverse section of two stems surrounded by
roots.   The stem at the left shows two departing traces and the one at the right three.
Each shows an attached root.   Cretaceous.   Bighorn Basin, Wyoming.   × 6.   (*B*)
Tempskya Wesselii.   Transverse section of root.   Great Falls, Montana.   × about 40.

The roots (Fig. 95*B*) are smaller and more numerous than the stems.
They arise adventitiously and extend backward toward the base of the
trunk, but their course outside is unknown.   It is possible that in the

upright species the roots took the form of props for the support of the trunks. The xylem strand in all species is diarch. Two to four very large tracheids are usually present, which are arranged side by side in a line at right angles to the plane of the protoxylem poles. The epidermis sometimes bears root hairs.

The relationships of *Tempskya* to any of the modern ferns are obscure although certain points of resemblance to the Schizaeaceae, the Loxomiaceae, and the Gleicheniaceae have been noted. It has been tentatively placed within a family of its own, the Tempskyaceae.

Schizaeopsis is a genus founded upon material from the lowermost Cretaceous of Maryland originally assigned to the Ginkgoales. The fan-shaped frond, which is about 11 cm. long, is divided into ribbonlike segments by a series of dichotomies of different depths. Each segment bears at the tip several spindle-shaped sporangia about 4 mm. long. Its affinities appear to be with the Schizaeaceae.

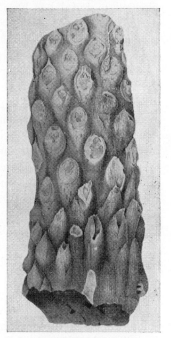

The Cyatheaceae, the family embracing the majority of the living tree ferns, was in existence during the middle part of the Mesozoic era. *Cyathocaulis* and *Cebotiocaulis* are structurally preserved cyatheaceous types from the Jurassic and early Cretaceous of Japan. *Protopteris* (Fig. 96) is a Lower Cretaceous tree fern which was originally described from Bohemia and which may be related to the Recent *Dicksonia*. The trunk bears persistent leaf bases, which reveal on the exposed surface a horseshoe-shaped trace strand with inwardly curled ends and indented sides. The vascular system is similar to the Recent genera *Alsophila* and *Cyathea* in

Fig. 96.—*Protopteris Sternbergii.* (*After Corda.*)

that it consists of a large cylindrical stele broken by leaf gaps, but the continuous bundle trace is more suggestive of *Dicksonia*.

The ferns of the Cenozoic belong almost exclusively to modern families, and although many species have been recorded from the rocks of this era, they are a subordinate element of the flora. They do, however, reveal some interesting information concerning the distribution of certain ones during the recent past. For example, the genus *Lygodium*, which is now mainly tropical, was until recently of world-wide distribu-

tion.   The fossil record has shown the retreat of other forms from northerly localities to the tropics within comparatively recent times.   It is
significant to note that the most widely spread of living ferns, *Pteridium
aquilinum*, is not represented in the fossil record.

Compressions of the floating leaves of the aquatic fern *Salvinia* have
been found in the Tertiary of all continents except Australia.   Nearly
half of the recorded occurrences are from Europe but there are a few
from North America.   The best known American fossil species are
Salvinia elliptica from the Eocene of Washington, and *S. preauriculata*
from the Wilcox group of Tennessee and the Bridger formation of
Wyoming.   *Azolla* is rare in the Tertiary but some very interesting
silicified material has been found in the Intertrappean beds of India.
Azolla intertrappea consists of megaspores with the floats attached, and
fastened to the fibrils of the megaspores are microspore massulae with
the very characteristic anchor-tipped glochidia.

Silicified tree ferns belonging to the Cyatheaceae have been found in
the Tertiary of East Africa and Texas.   The East African plant, *Dendrop-
teridium cyatheoides*, is represented by a decorticated stem about 3
inches in diameter, which shows the typical cyathean dictyostelic
structure.   There is a single series of nine peripheral meristeles, which
turn outward at the margins.   There are numerous medullary bundles in
the pith which are apparently associated with the leaf traces which
also consist of separate bundles.   Although the affinities of the fossil
are obviously with the Cyatheaceae, it cannot be identified with any
modern genus.

Cyathodendron texanum is a cyatheaceous tree fern from the Eocene
of Texas.   The stems are as much as 3 inches in diameter, and each
contains a large central pithlike tissue containing concentric bundles
and surrounded by a solenostele that appears continuous in transverse
section.   Leaf gaps, however, are present, but they appear to be closed
by a peculiar downward extension of vascular tissue, which leaves an
oblique opening through the stele.   Through this opening some of the
medullary bundles pass to contribute to the leaf-trace supply.   This
plant is believed to be a generalized type combining certain characteristics of the three tribes of the Recent Cyatheaceae.

References

ARNOLD, C.A.: Structure and relationships of some Middle Devonian plants from
     western New York, *Am. Jour. Botany*, **27**, 1940.
———: Silicified plant remains from the Mesozoic and Tertiary of Western North
     America, I. Ferns, *Michigan Acad. Sci. Papers*, **30**, 1945.
BERTRAND, P.: "Études sur la fronde des Zygoptéridées," Lille, 1909.

————: Observations sur la classification des fougères anciennes du Dévonien et du Carbonifère, *Soc. Bot. France Bull.*, **80**, 1933.

————: Contribution des Cladoxylees de Saalfeld, *Palaeontographica*, **80** (Abt. **B**), 1935.

Corsin, P.: "Contribution a l'étude des fougères anciennes du groupe des Inversicatenales," Lille, 1937.

Daugherty, L.H.: The Upper Triassic flora of Arizona, *Carnegie Inst. Washington, Pub.* 526, 1941.

Gillette, N.J.: Morphology of some American species of *Psaronius*, *Bot. Gazette*, **99**, 1937.

Gordon, W.T.: On the structure and affinities of *Metaclepsydropsis duplex*, *Royal Soc. Edinburgh Trans.*, **48**, 1911.

Hirmer, M.: "Handbuch der Paläobotanik," Munich, 1927.

————, and L. Hoerhammer: Morphologie, Systematik, und geographische Verbreitung der fossilen und recenten Matoniaceen, *Palaeontographica*, **80** (Abt. **B**), 1936.

Høeg, O.A.: The Downtonian and Devonian flora of Spitzbergen, *Skr. Norges Sval. Ish.*, **83**, 1942.

Holden, H.S.: On the structure and affinities of *Ankyropteris corrugata*, *Royal Soc. London Philos. Trans.*, **B**, **218**, 1930.

Hoskins, J.H.: Contributions to the Coal Measures flora of Illinois, *Am. Midland Naturalist*, **12**, 1930.

Radforth, N.W.: An analysis and comparison of the structural features of *Dactylotheca plumosa* Artis sp. and *Senftenbergia ophiodermatica* Göppert sp., *Royal Soc. Edinburgh Trans.*, **59**, 1938.

Read, C.B., and R.W. Brown: American Cretaceous ferns of the genus *Tempskya*, *U. S. Geol. Survey Prof. Paper* **186 E**, 1937.

Read, C. B.: A new fern from the Johns Valley shale of Oklahoma, *Am. Jour. Botany*, **25**, 1938.

————: The evolution of habit in *Tempskya*, *Lloydia*, **2**, (1), 1939.

Renault, B.: Bassin houiller et permien d' Autun et d'Epinac. Flore Fossile, *Étude des Gîtes Minéraux de la France*, 1896.

Sahni, B.: On the structure of *Zygopteris primaria* (Cotta) and on the relations between the genera *Zygopteris*, *Etapteris*, and *Botrychioxylon*, *Royal Soc. London Philos. Trans.*, **B 222**, 1932.

————: On a Paleozoic tree-fern, *Grammatopteris Baldaufi* (Beck) Hirmer, a link between the Zygopteridaceae and the Osmundaceae, *Annals of Botany*, **46**, 1932.

Scott, D.H.: A Paleozoic fern, the *Zygopteris Grayi* of Williamson, *Annals of Botany*, **26**, 1912.

————: "Studies in Fossil Botany," 3d ed., Vol. 1, London, 1920.

CHAPTER IX

THE PTERIDOSPERMS

The pteridosperms, or seed-ferns, may be tersely defined as plants with fernlike foliage, which bore seeds. Some of them were trees, but others were smaller plants of reclining or sprawling habit. A few petrified stems attributed to pteridosperms have been found in the Devonian, but the group did not become important until the Carboniferous during which time it also reached its developmental climax. The plants were still abundant at the onset of the Permian ice age, and a few persisted in modified form as late as mid-Jurassic. The discovery of this remarkable group of seed plants soon after the turn of the present century has been one of the greatest in the history of paleobotany, and its importance has been far-reaching. In the first place it brought to light a heretofore unknown group of seed plants, and secondly it placed the ferns in a more subordinate position in the Carboniferous flora. Morphologically the pteridosperms stand between the ferns and the cycads.

The history of the discovery of the pteridosperms includes a series of events dating back to 1877. In that year Grand'Eury mentioned the structural similarity between the petioles of *Alethopteris, Neuropteris,* and *Odontopteris,* and certain other petioles which he at the time called *Aulacopteris* but later identified as *Myeloxylon.* Then in 1883 Stur commented upon the persistent sterile condition of many Carboniferous leaf types, and suggested that these forms may not be ferns as they were always supposed. In 1887 Williamson, after studying *Heterangium tiliaeoides* and *Kaloxylon Hookeri,* concluded that these plants were different from any then living, and that they combined the characters of ferns and cycads. Schenk in 1889, and Weber and Sterzel in 1896, reported that the *Myeloxylon* type of petiole was borne on medullosan stems. Thus, the stems, petioles, and leaves of *Medullosa* were assembled, and on this basis Potonié in 1899 proposed the class name Cycadofilices for *Medullosa* with the object of designating a transition group between the ferns and higher seed plants. It should be noted that up to this time nothing was known of the seeds borne by these Cycadofilices, and the only criteria for the establishment of this group were anatomical. However, additional evidence was accumulating, and in 1903 Oliver and

204

Scott announced their epoch-making discovery of the identity of the seed *Lagenostoma Lomaxi* and the stem known as *Lyginopteris oldhamia*, and proposed the name Pteridospermae "for those Cycadofilices that bore seeds." This connection was founded upon the similarity of the capitate glands borne upon the seed cupules, the stem, and the rachis of the fronds of *Sphenopteris Hoeninghausi*, although it was not until 1930 that the actual attachment of the seed and leaf was discovered. In 1904 Kidston described specimens of *Neuropteris heterophylla* with seeds attached, thus verifying the previous discovery of Oliver and Scott concerning the seed-bearing habit of the Pteridospermae. Since then, seeds have been discovered attached to several other fronds, and the Pteridospermae as a plant group are firmly established. Although the name Cycadofilices was proposed first, Pteridospermae has become widely adopted as being more nearly descriptive of the class.

The pteridosperms are a large and diversified plant group, and as with most fossil plants, their remains are usually fragmentary. The problem of correctly assigning fertile parts of the plant to the appropriate vegetative parts is one of the greatest with which paleobotanists are confronted. The taxonomic limits of the group have never been precisely determined, and there exist numerous detached organs, both vegetative and reproductive, whose assignment to the pteridosperms is tentative pending proof of connection with other parts known to belong to the group. However, by disregarding for the present the Mesozoic members, and considering only the better known Paleozoic forms, the seed-ferns show certain well-defined characteristics, which may be listed as follows:

1. The leaves are large and frondlike, belonging to various form genera such as *Alethopteris*, *Neuropteris*, *Sphenopteris*, etc.

2. The seeds, which are not produced in cones or inflorescences of any kind, are borne upon modified or unmodified foliage, but if modified, the differentiation is relatively slight.

3. The seeds bear considerable resemblance to those of Recent cycads.

4. The leaf traces are relatively large, consisting of a single strand or of several strands.

5. Secondary wood and phloem was formed.

6. The primary wood is usually mesarch and is present in fairly large amounts.

7. The tracheids of the secondary wood bear bordered pits which are confined mostly to the radial walls.

8. The pollen-producing organs are (as far as known) exannulate sporangia sometimes grouped into synangia that are difficult to distinguish from the fructifications of ferns.

9. The leaves in all forms so far investigated bear a resistant cuticle that does not disintegrate when subjected to oxidative maceration followed by alkali treatment.

Position in the Plant Kingdom

Taxonomically the pteridosperms are classified along with the other naked-seeded Pteropsida in the Gymnospermae. In many respects they may seem to stand well apart from the other members of this class, especially the Cordaitales, Coniferales, and Gnetales, but there is ample evidence of a considerable degree of kinship with the cycads and ginkgos. Affinities with these groups are expressed in numerous anatomical, foliar, and seed characteristics. It is also believed that the pteridosperms produced motile male gametes as do the Recent cycads and ginkgos. The cycads are readily separable from the pteridosperms by their definitely organized strobili, but were it not for the fact that in the genus *Cycas* the megasporophylls bear considerable resemblance to foliage leaves, this difference might be sufficient to distinguish them entirely. As it is, however, the situation in *Cycas* (*C. revoluta* in particular) points to the derivation of the cycadaceous ovulate strobilus from the seed-bearing frond of the pteridosperm type. It is well within the limits of possibility that the pteridosperms, the cycads and cycadeoids, and possibly the ginkgos also, constitute a unified phyletic line that not only has long been distinct from the other gymnosperms but may even have had a separate origin.

Although the pteridosperms are in many respects transitional between the cycadophytes and the ferns, the recent trends in morphological thought have been to stress their gymnospermous traits as being more indicative of their true position in the plant kingdom. The fernlike appearance of the foliage, while probably retained from ancient times, is in some respects more apparent than real. The cuticles of all pteridosperms examined are resistant to oxidative maceration followed by alkali in which respect they resemble gymnosperms. Fern cuticles dissolve when subjected to such treatment. Too little is known of the habit of the Paleozoic pteridosperms to permit much generalizing, but in many pteridosperms the fronds branched into two equal parts, as in *Sphenopteris Hoeninghausi, Diplothmema furcata, Diplopteridium teilianum,* etc., and they were arranged in spirals on the main trunks, which in many cases were rather strongly reinforced with secondary wood. The type of frond division shown by the zygopterid ferns is not known among the pteridosperms. Probably the most difficult organs to distinguish are the fructifications of the Paleozoic ferns and the microsporangiate organs of the contemporary pteridosperms. It is generally believed that the tufted type such as occurs in *Telangium* and *Crossotheca*, and in which the clusters are situated at the tips of naked pedicels, is pteridospermous, and that the pinnule-borne synangium reminiscent of the Recent Marattiaceae is filicinean. However, there are many examples of exannulate singly borne fossil sporangia, which in the present state of our knowledge cannot be assigned with confidence.

Origin.—The problem of the origin of the pteridosperms is similar to that of all other Paleozoic plant groups in that the limiting factor is the recognition of the earliest preserved members. The oldest plants that have been assigned to this class are certain Upper Devonian members of the Calamopityaceae in which only the stems are preserved. The foliar and reproductive organs are unknown, and until we know whether these plants bore seeds, their inclusion within the Pteridospermae must remain tentative. It is probable that the pteridosperms originated at some time during the Devonian period from fernlike stock, which had not advanced far beyond the psilophytic stage. It is not necessary to assume that all pteridosperms arose from the same source, and it is possible that the different groups had their origin separately from different psilophytic ferns which, while all were progressing in the direction of seed development, differed among themselves in vegetative characters. A close connection between the original pteridosperms and the Psilophytales is suggested by the fact that the oldest known seeds were borne singly or in pairs at the tips of bifurcated stalks in a manner strongly reminiscent of the terminal sporangium of the Psilophytales. Leaf-borne seeds seem to have developed later. The main obstacle in deriving the pteridosperms from cryptogamic ancestors had always been the lack of any known heterosporous fern which could have given rise to the seed habit, but it appears now that the obvious heterosporous condition in *Archaeopteris latifolia* may partly span this break in continuity.

F I G. 9 7.—*Lageno-spermum* (*Calymmatotheca*) *imparirameum*. Pair of seed cupules. Price formation, Mississippian. Virginia. × 2½.

The only evidence we have of the existence of pteridosperms in the Devonian consists of a few petrified stems. No organs positively identified as seeds have been found below the Lower Carboniferous. In the Pocono and Price formations, which lie at the base of the Mississippian and overlie the Devonian in the eastern part of North America, we find at several places cupulate seeds of the *Lagenostoma* or *Calymmatotheca* type (Fig. 97). These fructifications consist of small cupules, about 3 mm. wide and 10 mm. long, situated in pairs upon the tips of leafless, slightly unequally bifurcated stalks. Each cupule consists of about five acicular bracts fused together at the base for about two-thirds of their length with the ends free. Within and attached at the base of each oval cup is a small seed. Similar cupulate organs also occur in the Upper Devonian but no seeds have been observed within them so, consequently what they produced is unknown. It is often difficult to distinguish between early seed cupules and tufts of slender microsporangia borne at the ends of slender stalks. The Middle Devonian *Eospermatopteris* is often cited as the oldest seed plant, even though it has been known for some time that seed production by this plant is extremely doubtful. Some of the seedlike bodies attributed to it contain spores, and moreover, the vegetative parts from the type locality in the Catskill Mountains are indistinguishable from *Aneurophyton*. Some authors place such plants as *Cephalopteris*, *Aneurophyton*, and *Sphenopteridium*, which are wholly or partly Devonian, in the pteridosperms on evidence which is not conclusive.

SUBDIVISIONS OF THE PALEOZOIC PTERIDOSPERMS

Although any division of the Pteridospermae into smaller groups is tentative, it is, nevertheless, desirable to assign some of the better known types to groups that are subject to such modification as may become necessary as our knowledge of them accumulates. The best known groups, here given family ranking, are the following:

1. *Lyginopteridaceae*. This family includes forms with monostelic stems and petioles ordinarily with a single vascular strand. The seeds are mostly small. The family is typified by the stems *Lyginopteris* and *Heterangium*, and *Sphenopteris* or *Pecopteris* foliage.

2. *Medullosaceae*. This family is characterized by plants with polystelic stems and large petioles containing a number of scattered bundles. The seeds are often large. It contains the stem types *Medullosa* and *Sutcliffia*. The foliage consists mainly of *Alethopteris* and *Neuropteris*.

3. *Calamopityaceae*. This family is a somewhat diversified assemblage of monostelic stems some of which bore petioles containing a number of bundles. It is confined, as far as we know, to the Upper Devonian and Lower Carboniferous. As at present constituted it is the largest family of the Pteridospermae.

THE LYGINOPTERIDACEAE

Calymmatotheca Hoeninghausi.—This name is used for the plant with stems known as *Lyginopteris oldhamia*. It occurs in the Lower and middle Coal Measures of England and in beds of similar age in continental Europe and North America. The bulk of our information on its structure and morphology has been obtained from coal-balls found in Lancashire and Yorkshire.

The history of our knowledge of *Calymmatotheca Hoeninghausi* dates from 1828 when Brongniart described the frond *Sphenopteris Hoeninghausi*, which was later found to belong to it. Binney in 1866 described the stem under the name of *Dadoxylon oldhamium*. This stem was later assigned to *Lyginodendron*, a genus based upon cortical impressions, but when it was found that the type specimens of *Lyginodendron* belonged to plants of another group, Potonié substituted the name *Lyginopteris* in 1899. In 1903, Oliver and Scott showed that the seed *Lagenostoma Lomaxi* belonged to *Lyginopteris oldhamia*, and in 1929 Jongmans demonstrated the identity of the gland-studded cupulate envelope enclosing the seed and of *Calymmatotheca*, a structure named by Stur in 1877. Accordingly, therefore, *Calymmatotheca* is the correct generic name and *Hoeninghausi* the oldest specific name, hence the combination of the two constitutes the valid binomial for the plant. *Rachiopteris aspera* and *Kaloxylon Hookeri* represent the petioles and roots, respectively. For the detached parts the several names originally proposed are still used. *Calymmatotheca Hoeninghausi* is not only an example of a complex nomenclatorial situation but it also shows how a fossil plant may become reconstructed by the gradual piecing together of separate parts. The one remaining part of *C. Hoeninghausi* to be identified is the pollen-bearing organ.

The stem, *Lyginopteris oldhamia* (Fig. 98), bore spirally arranged fronds. No very large stems have been found, the largest being only 4 cm. in diameter and the smallest about 2 mm. It is probable that larger stems existed. The structure is very similar in stems of all sizes, the main differences being in the relative amounts of some of the tissues. The center is occupied by a large pith that is parenchymatous except for scattered masses of stone cells, the so-called "sclerotic nests" (Fig. 99). Around the pith is a ring of separate mesarch primary wood strands in which the protoxylem is located nearer the edge toward the outside of the stem. On the side of the protoxylem toward the center of the stem is a small strand of parenchyma. Surrounding the protoxylem and parenchyma is the metaxylem, which is wider on the pith side.

The primary xylem strands constitute a part of the leaf-trace system. The strands as seen in a single transverse section are about five in number, and when a trace forms the protoxylem first appears as two masses placed side by side. The strand then divides with one branch continuing upward as the reparatory strand and the other bending slightly to one

side to constitute the trace.   The trace then bends outward and passes
through the secondary wood which, except where it is interrupted by the

Fig. 98.—*Calymmatotheca Hoeninghausi* (*Lyginopteris oldhamia*).   Cross section of small
stem: *a*, schlerenchyma bands in outer cortex; *b*, inner cortex and phloem; *c*, double leaf
trace; *d*, secondary xylem; *e*, primary xylem; *f*, pith.   Coal Measures.   Great Britain.
× 5.

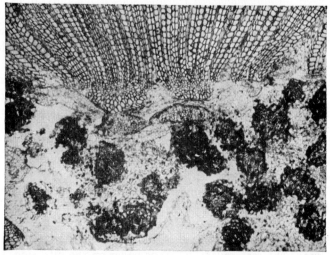

Fig. 99.—*Calymmatotheca Hoeninghausi* (*Lyginopteris oldhamia*).   Transverse section
showing two mesarch primary xylem strands, two small strands of anomalously developed
internal secondary xylem and masses of sclerotic cells in the pith.   × 12.

outward passage of the traces, is continuous around the ring of circummedullary primary strands.   The trace is accompanied during its
passage through the secondary wood by a wedge-shaped arc of secondary

wood of its own, which persists beyond the pericycle where it gradually disappears. Outside the secondary wood zone the trace divides to form a double strand, but the two unite again in the base of the petiole.

The amount of secondary xylem varies according to the size of the stems, being thicker in the larger ones. It consists of rather large tracheids with multiseriate bordered pits on the radial walls and numerous parenchymatous rays some of which are quite wide. As mentioned, the continuity of the secondary wood is interrupted by the outward passage of the traces, and in some stems these interruptions have the effect of

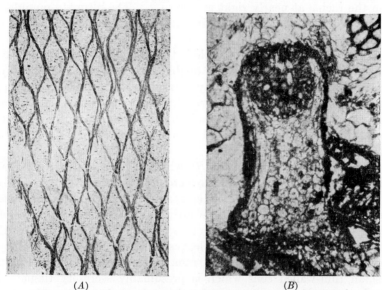

(A)   (B)

FIG. 100.—*Calymmatotheca Hoeninghausi (Lyginopteris oldhamia).*
(A) Vertical (tangential) section through outer cortex showing network of sclerenchyma strands. × 3⅓. (B) Capitate gland showing secretory cells near tip. × about 60.

breaking the woody ring into about four well-separated but unequal masses. The parenchyma separating these masses is that which accompanies the trace on its adaxial side. This tissue extends upward for a short distance and above it the two flanking wood masses join. Morphologically, the spaces which interrupt the continuity of the wood are leaf gaps which are prolonged through the secondary cylinder.

The secondary xylem is surrounded by a cambium and a band of phloem that is sometimes preserved. The pericycle is a well-marked tissue of short cells with embedded "sclerotic nests," and in the outer part there is a band of internal periderm. The inner cortex is parenchymatous but the outer cortex contains very characteristic radially broadened fibrous strands which form a vertical network (Fig. 100A).

Different specimens of *Lyginopteris oldhamia* vary considerably in structure, but the variations are not of specific magnitude. In some small stems the primary xylem strands are confluent and form a wholly or partly continuous tissue around the pith. Sometimes secondary xylem is formed to the inside of the primary xylem and in one recorded instance a complete cylinder of medullary xylem exists.

The frond of *Calymmatothea Hoeninghausi* is highly compound with a gland-studded rachis that divides by an equal dichotomy above the lowermost pinnae. The pinnae are oppositely arranged, and stand out from the rachis at nearly right angles to it. The ultimate pinnules are small-lobed cuneate leaflets borne alternately on the smallest subdivisions of the rachis.

The xylem strand of the petiole is trough-shaped with the concavity above. Sometimes the strand is double with two grooves on the upper surface. The xylem is completely surrounded with phloem and on the lower (abaxial) side there are several protoxylem groups. The fibrous outer cortex of the stem continues into the petiole where the fibers take the form of more or less rounded hypodermal strands. The inner cortex contains plates of stone cells.

The capitate glands (Fig. 100*B*), which have supplied the criteria for the connection between the aboveground parts of the plant, cover all parts except the roots. The spines are flask-shaped, sometimes as much as 3 mm. high, but have no vascular supply. They therefore belong to the category of "emergences" and are of a higher order than epidermal hairs but lower than leaves. Each gland bears a globular tip which ranges from 0.12 to 0.40 mm. in diameter, and inside is a mass of small thin-walled cells which is presumably secretory tissue. The nature of the substance secreted by these glands is unknown but it was probably oily or waxy material that protected the growing parts from fungous attacks.

Concerning the habit of *Lyginopteris oldhamia*, the radial symmetry of the stem indicates that the plant grew upright, although it seems impossible that such a small stem could have supported a large crown of leaves without additional support. It is assumed that the plant reclined somewhat against other plants, or against steep cliffs, and grew in thickets and junglelike associations. The roots were produced adventitiously and grew from among the leaves as well as on the older stem portions from which the leaves had fallen.

Lagenostoma Lomaxi (Fig. 101*A*), the seed of *Lyginopteris oidhamia*, was terminal on the ultimate and naked ramifications of the frond. The seed is a small barrel-shaped organ measuring approximately 4.25 mm. in diameter by 5.5 mm. in length. It is orthotropous and symmetrical, and the single integument is adherent to the nucellus except in the apical region beneath the canopy. The canopy, that portion of the integument which surrounds and overhangs the pollen chamber, is divided into nine loculi with a vascular strand extending into each. This loculed structure is supposed to be a mechanism for supplying the apex of the seed with water. The pollen chamber is complex, and in the apical cavity between the free portion of the integument and the nucellus there arises from the latter a flask-shaped prolongation that

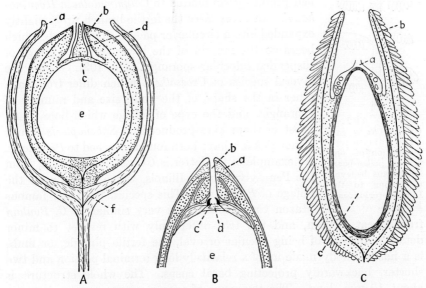

FIG. 101.—Seeds of the Lyginopteridaceae.

(*A*) *Lagenostoma Lomaxi.* Diagrammatic longitudinal section: *a*, cupule; *b*, canopy of integument; *c*, central column of pollen chamber; *d*, pollen chamber; *e*, female gametophyte; *f*, vascular supply. (*After Oliver.*) Greatly enlarged.

(*B*) *Conostoma oblongum:* *a*, integument; *b*, micropyle; *c*, lagenostome; *d*, plinth; *e*, female gametophyte. (*After Oliver and Salisbury.*) Greatly enlarged.

(*C*) *Physostoma elegans:* *a*, pollen chamber; *b*, integument with quill-like hairs; *c*, female gametophyte. (*After Oliver.*) Greatly enlarged.

extends upward through the micropyle. Arising inside this is a column of nucellar tissue that leaves only a narrow cylindrical space for the reception of the pollen. This flask-shaped structure, called the "lagenostome," is situated within the micropyle and suggested the name *Lagenostoma.* In common with other Paleozoic seeds, it appears to have had no embryo. The seed was situated within a cupule of gland-bearing husks, which completely enveloped it before maturity. It was by means of these glands that the connection between the seed, leaf, and stem was first established. Recently, the seeds have been observed in actual connection with the fronds.

The pollen-bearing organ of *Calymmatotheca Hoeninghausi* was once thought to be a fructification of the type known as *Crossotheca*, but recent investigations have cast doubt upon this. Although never found

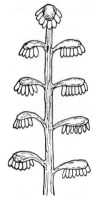

in actual connection, it is now believed that *Telangium* is the microsporangiate structure. Were it not for the uncertainty concerning the identity of this organ, the entire plant would be known.

Crossotheca is believed to belong to some unidentified pteridosperm related to *Calymmatotheca Hoeninghausi*. In *Crossotheca* the fertile branch tip is slightly expanded into a circular or paddle-shaped limb, which bears on the margin of the lower surface a fringe of suspended bilocular sporangia (Fig. 102). There are several species of *Crossotheca* which differ from each other in the shape of the limb, size and number of sporangia, and the type of foliage which bore them. Most of them were produced on *Sphenopteris* fronds although a few have been found attached to *Pecopteris*. An example of the latter is *Crossotheca sagittata*, from the Pennsylvanian of Illinois, which is borne on the foliage of *Pecopteris*. This species, from the famous Mazon Creek locality, is very similar to *C. Boulayi* from Great Britain, and the two differ only with respect to minor details. Instead of being circular or oval, the fertile pinnule, or limb, is a hastate body made up of a relatively long terminal portion and two shorter, backwardly projecting basal cusps. The whole structure is about 10 mm. long. The two rows of pendulous sporangia completely cover the lower surface. In *C. sagittata* each limb bears 30 or more sporangia, which contain smooth spores measuring 43 to 58 microns in diameter (Fig. 103).

FIG. 102.— *Crossotheca sp.* Diagrammatic representation of one of the pollen-bearing organs believed to belong to the Lyginopteridaceae. Enlarged.

Telangium is often found in intimate association with certain species of *Sphenopteris*. This fructification consists of 6 to 25 small fusiform sporangia attached to a small disc borne on the tip of an ultimate division of a frond

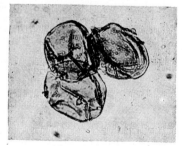

FIG. 103.—*Crossotheca sagittata.* Microspores. Pennsylvanian. Mazon Creek, Illinois. × about 300.

ramification. In some species these terminal discs are small and consist mostly of the coalesced bases of the sporangia, but in others the disc is a quadrilobate structure about 3 mm. in diameter. The forward portions of the sporangia are free, and sections through petrified specimens have shown that each sporangium consists of two compartments separated by a thin partition extending lengthwise. The spores are numerous and small, obviously microspores. The genus *Telangium* was originally founded by Benson on petrified coal-ball material, but it has often been recognized in the compression state. It is frequently associated with *Sphenopteris affinis* and *S. bifida.* In these species the frond rachis dichotomizes to produce two equal portions, and two subopposite pinnae are borne on the rachis below the fork. Each of the two branches of the main rachis forks once or twice more, and the branches thus formed bear alternately arranged secondary and tertiary pinnae. The pinnules are rhomboidal or fan-shaped, and are split into three or four cuneate segments. The fertile pinnae are believed to have been borne at or just below the main fork of the frond.

Seed cupules of the *Calymmatotheca* type sometimes occur with *Telangium.* The seeds, however, have never been found in any cupules that happened to be associated with *Telangium* although it is believed that they belong to the Lagenostomales group which was borne by *Lyginopteris* and similar stems.

Telangium differs from *Crossotheca* in having sporangia that stand upright on the terminal disc and continue the line of the rachis whereas in *Crossotheca* they form a fringe around the margin of the flattened limb.

Diplopteridium teilianum, from the Lower Carboniferous of Wales, is a sphenopterid type with fructifications resembling *Telangium.* The frond is forked in a manner similar to that of *Sphenopteris affinis,* but the fertile branch is situated upright in the axil between the two main branches. This branch, which is shorter than the two side branches, is dichotomously forked several times and the sporangial clusters are borne at the tips of the ultimate ramifications.

A frond that is divided nearly to the base into two equal divisions has been observed in several other pteridosperms in addition to those just described. It is present in *Sphenopteris Hoeninghausi* (the frond attached to *Lyginopteris oldhamia*) and has also been observed in *Alethopteris* and *Neuropteris.* Some other frond genera believed to be pteridosperms and exhibiting this habit are *Adiantites, Aneimites, Callipteris, Diplotmema, Eremopteris, Mariopteris, Palmatopteris,* and *Sphenopteridium.* However, this habit was not universal among pteridosperms because in some the fructifications were borne either among the vegetative leaves or on the upper portions of divisions bearing vegetative foliage.

The application of the name *Calymmatotheca* is by no means limited to complete or nearly complete plants such as *C. Hoeninghausi,* but is extended generally to seed cupules borne at the extremities of frond subdivisions on which no vegetative foliage occurs. The *Calymmatotheca* cupule enclosed the seed before maturity, but upon ripening the structure split lengthwise into four to six coriaceous segments (Fig. 97). The seed was attached at the base of the cup.

Fructifications of the *Calymmatotheca* type illustrate one of the technicalities of paleobotanical nomenclature. This name is used for seed cups, but if the seed happens to be present within, the name *Lagenospermum* may be used instead. If the seed is petrified, it may be referred to *Lagenostoma* or some similar genus of petrified seeds. If conditions of preservation are such that it is impossible to tell whether the object is a seed cupule or a cluster of elongate microsporangia, it may be called *Pterispermostrobus* or *Calathiops*. However, it is impossible to avoid such multiplicity of names when dealing with plants that are preserved in so many ways.

A very interesting cupule from the lower Carboniferous Oil Shale Group of Scotland has recently been described by Andrews as *Megatheca Thomasii*. Only one specimen has been found. It is a tulip-shaped object 6.2 cm. long and 2.3 cm. wide in the lower portion. It is the largest cupulate fructification known. The six leathery lobes into which the cupule is split are joined for a short distance above the base. The surface of the lower part of the lobes is featured by a network of sclerenchyma strands, which disappear about halfway to the apex. No seeds were present within the cupule, and whether it bore a single large seed or a group of smaller ones is unknown. It is intimately associated with fronds of *Sphenopteris affinis*. From its general appearance it is probably the seed husk of a lyginopterid pteridosperm, but except for this its affinities are unknown.

Tetrastichia bupatides, also from the Oil Shale Group of Scotland, is a petrified petiolate axis. The small stem, which is 1 cm. or less in diameter, is elliptical in cross section because of the decurrent petioles. The petioles are in four rows made up of subopposite pairs. The primary xylem is a cruciform structure of scalariform and reticulate tracheids, and in some parts of the plant it is surrounded by a layer of secondary wood which conforms to the outline of the primary portion. The petiole base is swollen into a pulvinus, and at about 12 cm. beyond the base, the petiole forks by an equal dichotomy. The xylem strand of the petiole below the fork has somewhat the shape of a butterfly. It is U-shaped with two lobes on the lower side. Beyond the fork the configuration is simpler. The cortex of the main stem is divisible into an inner, a middle, and an outer zone. The middle part contains sclerotic nests and the outer layer contains a longitudinal network of hypodermal sclerenchyma.

Although *Tetrastichia* is regarded as a primitive member of the Lyginopteridaceae, it shows a number of fern characteristics, the main ones being the general shape and structure of the axis, the position of the protoxylem, the scalariform and reticulate pitting of the tracheids and the manner of departure, and the form of the petiole strand. The petiole strand, however, is also quite similar in shape to that of *Lygino-rachis*, and this, combined with the structure of the cortex, is considered sufficient to identify the plant with the pteridosperms. Furthermore, the structure of the cortex appears to be such that it could produce a compression pattern similar to that displayed by *Sphenopteris affinis*.

Heterangium.—This stem genus contains about one dozen species of which the best known is *Heterangium Grievii* (Fig. 104) from the Lower Carboniferous at Pettycur, Scotland.

Stems of *Heterangium Grievii* are small, normally being about 1.5 to 2 cm. in diameter. Most of them are straight, slender, and unbranched, and until quite recently no branched specimens had been observed. In cross section the stems are somewhat angular because of the large decurrent leaf bases. The foliage is believed to be *Sphenopteris elegans*. Although the fronds have never been found attached to the stem, the compressions of the leaf stalks of *S. elegans*, both large and small, often show characteristic cross marks in the cortical portion, which are believed to be caused by transverse sclerotic plates such as occur in the inner cortex of petrified *Heterangium* stems.

Heterangium Grievii contains a single fairly large protostele. The tracheids of the primary xylem extend to the center of the stem, but instead of constituting a solid core, they are separated by intervening parenchyma into groups. The inner tracheids are rather large and bear multiseriate bordered pits on all their walls. Toward the outside of the primary xylem core, the tracheids become somewhat smaller and are arranged into mesarch strands. There are normally about 20 of these small strands, which form a nearly continuous layer. The protoxylem is not centrally located but is situated nearer the outer edge.

The primary xylem is surrounded by a thin zone of secondary wood that is seldom more than a few millimeters thick. The secondary wood consists of tracheids bearing multiseriate bordered pits on their radial walls and of rather large rays. All these rays extend to the primary wood and some of them merge with the parenchymatous tissue distributed among the primary tracheids.

The leaf trace is a single strand. When it first departs from the primary cylinder, it has a single protoxylem point that soon splits into two. Along the lower part of its course the strand is collateral, but it becomes concentric by the time the petiole base is reached.

As previously stated, one of the characteristic features of *Heterangium Grievii* is the arrangement of the sclerotic tissues in the cortex, and it is often possible to identify the stems even when the tissues are completely carbonized. Just beneath the epidermis there is a narrow zone containing vertical sclerenchyma strands, which lie parallel to each other. To the inside of this zone, and occupying the greater part of the width of the cortex, is a layer containing transverse plates of stone cells. These plates succeed one another in regular vertical series and are separated by parenchyma.

The other species of *Heterangium* are all from the Upper Carboniferous. Scott has divided the genus into three subgenera. In two of them,

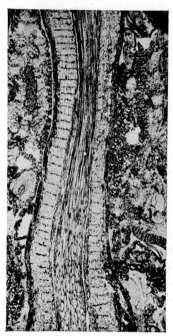

Fig. 104.—*Heterangium Grievii.* Longitudinal section of stem showing transverse sclerotic plates in the inner cortex. Calciferous Sandstone. Pettycur, Scotland. × 6.

Eu-Heterangium and *Polyangium*, there is no evidence of further evolutionary development or connection with higher groups, but the third group, *Lyginangium*, is transitional with *Lyginopteris*. In *Eu-Heterangium*, typified by *Heterangium Grievii*, the trace strand is single at its point of origin, whereas in *Polyangium* (an example being *H. tiliaeoides*) the trace originates as a pair of bundles. The range in anatomical structure revealed by the different species of *Heterangium* reveals a noteworthy transition from fernlike to more distinctly pteridospermous characteristics. The primary xylem of *Eu-Heterangium* bears a great resemblance to that of the living genus *Gleichenia*, which is probably the most familiar example of a protostelic fern. In this genus the xylem core consists of an admixture of tracheids and parenchyma, and the protoxylem occupies the mesarch position near the periphery. In the subgenus *Lyginangium* the xylem extends to the center of the core as in other species of the genus, but it is small in amount and the peripheral strands are larger and reduced in number. The resemblance to *Lyginopteris oldhamia* is therefore close.

Sphaerostoma ovale, the probable seed of *Heterangium Grievii*, is similar to *Lagenostoma*, but lacks the long projecting tubelike pollen chamber. A conspicuous feature of the seed is the "frill," which consists of lobes of the integument around the micropyle. The epidermis of the seed bears numerous short papillae, which probably secreted mucilaginous material. The seed was surrounded by a glandless cupule.

The seeds of the Lyginopteridaceae belong to the so-called "Lagenostomales," an artificial assemblage of Paleozoic seeds mostly of small size, and in which the nucellus is joined to the integument nearly to the archegonial region. The upper free part of the integument, known as the "canopy," consists of lobes that are joined in some seeds (*Lagenostoma*) but separate in others (*Sphaerostoma*). These canopy lobes are upward extensions of the ribs on the body of the seed. The vascular system supplies the integument only (in contrast to those of the Trigonocarpales, described later), but the strands extend into the canopy, which is often chambered and is believed to serve for water storage. Another structure believed to characterize the Lagenostomales is a tapetal layer around the megaspore. Some of the seeds of this group (*Lagenostoma* and *Sphaerostoma*) are borne within a cupule of bracts (*Calymmatotheca*). *Conostoma* (Fig. 101*B*) and *Physostoma* (Fig. 101*C*) are two additional members of this group of seeds that have never been found attached to foliage or stems.

Some other sphenopterids in addition to *Sphenopteris Hoeninghausi* are known to have borne seeds. One of these is *S. alata* from Great Britain. Its seeds are small oval bodies 3 or 4 mm. long, hexagonal in cross section, and pointed at the apex. As with the other lyginopterids, a cupule was present.

THE MEDULLOSACEAE

The Medullosaceae is a family of pteridosperms that was abundant and widely spread during Carboniferous times. Ancient members of the family were probably in existence during the Mississippian or Lower Carboniferous, but they did not become numerous until the beginning of the great coal age. Remains of the family are present in quantity in the early and Middle Coal Measures when it probably reached the zenith of its development, but it persisted throughout this epoch and extended into the Permian where several forms are known.

Petrified stems, petioles, and roots of *Medullosa* are frequently encountered in coal-balls, but the best evidence of the important place occupied by the Medullosaceae in the Carboniferous flora is the abundance of such leaf types as *Alethopteris* (Fig. 105*A*) and *Neuropteris* (Figs. 105*A* and 106), which, on the basis of intimacy of association and anatomical structure, are regarded as belonging in part at least to this family. As mentioned previously, Grand'Eury as early as 1877 noted a structural similarity between the petioles of these foliage genera and

FIG. 105. (A) *Alethopteris Serlii.* Mazon Creek, Illinois. Natural size. (B) *Neuropteris Schlehani.* St. Charles, Michigan. × 2.

Myeloxylon, which is now known to be the petiole borne on *Medullosa* stems. Other foliage forms also probably belonging to the Medullosaceae are *Linopteris, Lonchopteris,* and *Odontopteris.* The seeds usually attributed to the Medullosaceae are *Trigonocarpus* and other members of the Trigonocarpales. The pollen-bearing organs are believed to be complex synangia of which *Whittleseya, Aulacotheca,* and *Codonotheca*

(A) (B)

Fig. 106.—*Neuropteris tenuifolia.* (A) Vegetative pinna. Mazon Creek, Illinois. Natural size. (B) Pinna bearing a terminal seed. Yorkian of Great Britain. (*W. Hemingway.*) Slightly enlarged.

are examples. In addition to *Medullosa,* the stems of *Sutcliffia* and *Colpoxylon* have been assigned to this family.

Medullosa.—From the anatomical standpoint *Medullosa* belongs to the category of anomalous stems in which the cambium, although behaving in a normal manner, occupies abnormal positions. The organization of the stems of some species is among the most peculiar in the plant kingdom. Of the considerable number of species that have been described, the best known is probably *Medullosa anglica* from the Lower and Middle Coal Measures of Great Britain, and it will serve adequately as a typical member of the genus.

The stem of *Medullosa anglica* (Fig. 107A) is large, 7 to 8 cm. in diameter, and is unequally three angled due to the presence of very large, decurrent, spirally arranged leaf bases. The stem contains three separate steles which are somewhat irregular in shape, and each is pro-

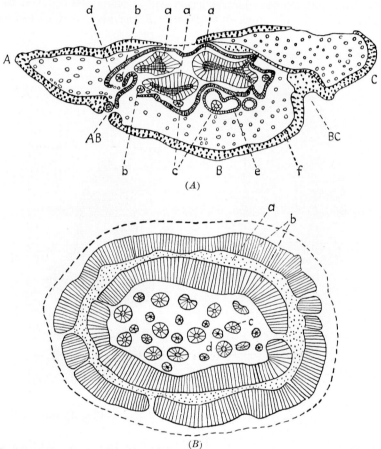

FIG. 107.—(A) *Medullosa anglica.* Cross section of stem. A, B, C, leaf bases; AB, BC, positions of next two leaf bases above: *a*, the three steles; *b*, leaf traces; *c*, accessory vascular strands; *d*, periderm; *e*, isolated periderm ring; *f*, outer cortex containing sclerenchyma strands. (*After Scott.*)

(B) *Medullosa stellata.* *a*, partial pith of peripheral solenostele; *b*, secondary wood of peripheral solenostele; *c*, "star ring"; *d*, partial pith. (*From Steidtmann after Weber and Sterzel.*)

vided with its own zone of secondary wood and phloem which completely surrounds the primary xylem core (Fig. 108). In this respect *Medullosa* resembles *Cladoxylon* described in the chapter on the ancient ferns. There is, however, no reason to suspect any close relation between the two. As the individual steles extend along the stem, they join and redivide at long intervals although the tripartite construction is generally present throughout. The spaces between the steles are not comparable in any way to the leaf gaps in other stems with separate bundles.

The individual steles are similar structurally to the single stele in the stem of *Heterangium*. The central part consists of a core of large tracheids and parenchyma without a pith, and with the protoxylem in a mesarch position. The secondary wood consists of multiseriate pitted tracheids and large rays. The stem as a whole, however, is very different

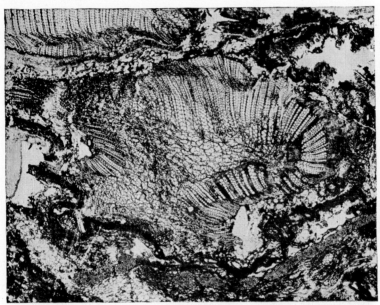

Fig. 108.—*Medullosa anglica*. Single stele showing central core of primary xylem surrounded by secondary xylem. Coal Measures. Great Britain. × 6.

from *Heterangium*, because as it grew in diameter, the increase was accomplished by the enlargement of each of the three steles, which possessed a complete cambium of its own. Often the greatest amount of secondary accretion was on the side of the stele toward the interior of the stem.

Each stele gave off leaf traces from the side facing the outside of the stem. When the trace bundle first became free from the stele, it was a concentric strand with one or more protoxylem points and surrounded by secondary wood. As it traversed the cortex, the secondary tissues disappeared, and the bundle finally split into many small collateral bundles with the protoxylem in the exarch position. The individual leaf then received the strand complex of several traces.

The cortex consists of short-celled parenchyma and numerous gum canals. The most conspicuous feature of the cortex is the presence within it of a layer of internal periderm that encloses the stelar region and separates the cortex into an inner and an outer zone. The periderm is locally interrupted where the traces and roots pass out. The tissue seems to be of the nature of phellem formed on the outside of a phellogen, but whether it was suberized is unknown. It is a remarkable tissue in that the cells are seldom crushed even though their walls appear very thin. The cortex outside the periderm contains numerous simple leaf-

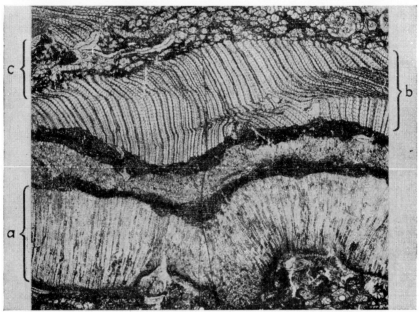

Fig. 109.—*Medullosa Noei*. Portion of stem showing: *a*, periderm; *b*, secondary xylem; *c*, primary xylem. McLeansboro formation. Illinois. × 6.

trace strands identical with those found in the petioles. Scattered throughout the outer cortex are also occasional accessory vascular strands, probably of secondary origin, which recall similar cortical bundles in some of the Recent cycads.

A recently described relative of *Medullosa anglica* is *M. distelica*, from the McLeansboro formation of Illinois, so named because of the presence of the two steles seen in a cross section of the stem. The stem is small, measuring about 4 by 8 cm., with the individual steles measuring about 7 by 18 mm. The secondary wood is thickest toward the center of the stem, and the two steles are enclosed by a narrow band of continuous internal periderm. The internodes are long, and the stem is surrounded by three decurrent leaf bases. In addition to the presence of two steles instead of three, *M. distelica* differs further from *M. anglica* in the extreme asymmetry of the steles. The secondary wood on the side toward the cortex is very thin, and there is little or no secondary tissue associated with the outgoing leaf traces. The petioles are similar in structure to those of other species.

Another *Medullosa* from the McLeansboro formation is *M. Noei* (Fig. 109), a species also similar to *M. anglica* but differing from it in several respects. The original specimen is incomplete, but the stem appears to have been at least 20 cm. in diameter. One complete stele and portions of two others, all very similar to those of *M. anglica*, are present. The assumption is that there were three steles in the complete stem. The stelar portion is enclosed by the usual layer of internal periderm and the outer cortex contains the expected complex of numerous trace bundles. In the interior of the stem between the steles there are scattered structures of unknown origin or function called "periderm strands." These consist of small masses of thin-walled cells surrounded by a sheath of peridermlike tissue. They bear some resemblance to the accessory bundles in *M. anglica* and the so-called "star rings" in certain Permian species of the genus. There is an additional pecularity in the primary wood in that it contains in addition to the normal tracheids and conjunctive parenchyma "tracheid bundles," which are groups of tracheids surrounded by tissue which appears identical in structure with the periderm tissue found in the cortex. The stem is surrounded by three large leaf bases but the angularity of the stem is less pronounced than in *M. anglica*.

Another close relative of *Medullosa anglica* is *M. Thompsonii*, from the Des Moines series of Iowa. The stem is about 3.6 cm. in diameter, has three steles in which the secondary wood is rather weakly developed on the outside, and a cortex with few secretory canals. The outstanding differences between this species and *M. anglica* are the sparcity of secretory canals and the greater amount of parenchyma in the middle cortex.

Two English species related to *Medullosa anglica* are *M. pusilla* and *M. centrofilis;* both have small stems. *M. centrofilis* is distinguished by the presence of a fourth stele in the center of the stem, which is a small concentric strand situated between the three larger ones. *M. pusilla* is a small species with a stem about 2 cm. in diameter. It may possibly represent a small atypical axis of *M. anglica*. Both *M. pusilla* and *M. centrofilis* are from the Lower Coal Measures.

Schopf has recently given *Medullosa anglica* and the related species formal taxonomic status by segregating them into the subgenus *Anglorota*. The plants of this group were relatively small or of medium size and had weak stems without massive accessory strands augmenting the main steles. The secondary wood is endocentric (thickest on the side toward the center of the stem). Internal periderm is relatively abundant and the whole stem is embraced by three large decurrent leaf bases. In these respects the Anglorota medullosans stand out more or less in contrast to the Permian species, a few of which will be summarily described. *M. Noei* is, in certain respects, transitional between the Anglorota and the later forms.

The Permian species are distinctive in showing greater structural complexity. The steles are more numerous than in the older forms and are often highly differentiated among themselves. A feature possessed in common by most of them is that there is an outer set of peripheral

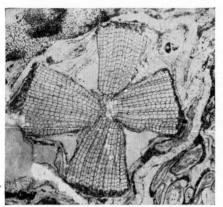

FIG. 110.—*Medullosa Noei.* Transverse section of small root. × 5.

steles surrounding or partly enclosing an inner set of smaller ones. In some species, *M. stellata* for example (Fig. 107*B*), the peripheral steles form a series of concentric rings separated by parenchyma, whereas in others, as in *M. Leuckarti,* there may be two or three large separate steles placed in a ring around the central portion. In the center of the stems of both species are the smaller steles called star rings. The star rings are uniformly thickened all the way around and vary in number from few to many. In the largest stems the whole stelar complex is sometimes reinforced with successive layers of extrafascicular wood and periderm. A number of variations of the forms mentioned above are known, but all species agree in petiole structure and the polystelic organization of the main stems.

The roots of *Medullosa* developed adventitiously. In *M. anglica* the triarch primary wood core is surrounded by considerable secondary wood which is not continuous, but opposite each protoxylem point a broad ray divides the wood into wedge-shaped segments. The roots of *M. Noei* are similar except that most of them are tetrarch (Fig. 110).

Detached medullosan petioles are placed in the organ genus *Myeloxylon.* Petioles are usually present in large numbers in coal-balls containing the stems, and associated with them are leaves of the *Alethopteris* type. The petioles are round or oval in cross section, and vary in diameter from several centimeters at the base to a few millimeters in the higher parts of the fronds. The structure of the petioles is almost identical with that of the large leaf bases found attached to the stems, a fact of great value in assigning detached leafstalks to their respective stems.

Several species of *Myeloxylon* have been described, but one showing excellent tissue preservation is *M. missouriensis* from the Kansas City group of the Missouri series (Fig. 111). Just beneath the epidermis is a layer of partly coalescent sclerenchyma strands, which are rounded or

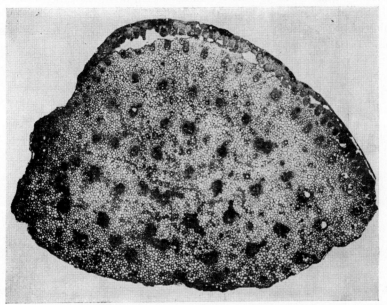

FIG. 111.—*Myeloxylon missouriensis.* Chanute shale, Middle Pennsylvanian. Kansas City, Kansas. × 4.

slightly elongated radially. The interior is filled with ground tissue containing scattered numerous vascular bundles and gum canals.

The species of *Myeloxylon* differ with respect to the form and distribution of the subepidermal sclerenchyma and in the abundance and distribution of the gum canals or secretory ducts. In *Myeloxylon zonatum*, the petiole of *Medullosa Noei*, the sclerotic strands are irregular with a distinct ring of vascular bundles bounding the inner surface, and immediately within this is a row of large gum canals. In the petiole of *Medullosa Thompsonii*, there are few canals, a feature possessed in common with the stem. The medullosan petiole bears considerable resemblance to that of a cycad, which it was supposed to be before its attachment was discovered. Some investigators have been prone to draw comparisons between medullosan petioles and monocotyledonous stems on the basis of similarities that are obviously superficial upon careful examination.

The individual *Myeloxylon* bundle consists of a compact mass of tracheids ranging from small to large, and an accompanying strand of phloem separated from the xylem by a thin parenchyma layer (Fig. 112*A*). The smallest xylem elements are toward the phloem. Nothing unusual is presented by the structure of the phloem. The tracheids have closely spaced spiral or scalariform thickenings. In most species the bundle is enclosed wholly or in part by a bundle sheath of thick-walled cells.

The gum canals are often filled with black material. The canal, which may be as much as a quarter of a millimeter in diameter, is surrounded by a layer of thin-walled secretory cells.

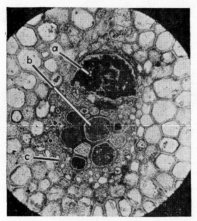

(A)   (B)

Fig. 112.—(*A*) *Myeloxylon missouriensis.* Vascular bundle: *a* phloem; *b*, xylem; *c*, bundle sheath. × about 50.
(*B*) *Aulacotheca Campbelli.* New River series. West Virginia. × 3.

Colpoxylon.—This genus, from the Permian of France, differs from *Medullosa* in having, throughout the greater length of the stem, a single, large, irregular stele. At places the stele divides to form several. This combination of monostely and polystely within the same stem is its chief feature of interest. The petioles are of the *Myeloxylon* type.

Sutcliffia.—This is a Coal Measures genus of two species characterized by the presence in the stem of a large central stele of the *Medullosa* type, but surrounded by numerous subsidiary steles, which branch from it. An additional difference is that the petiole bundles are concentric. There are also numerous extrafascicular bundles forming a network outside the normal stele system which possess secondary wood. A band of periderm is present within the cortex that is believed to have caused exfoliation of the outermost tissues.

Pollen-bearing Organs Attributed to the Medullosaceae.—With the Medullosaceae as with the Lyginopteridaceae there has been less certainty concerning the indentity of the pollen-bearing organs than of most of the other parts of the plant. However, there are several fructifications now believed to be medullosan although the evidence seldom consists of more than association with *Neuropteris* and other types of foliage. One of the most familiar of these is *Whittleseya* (Fig. 113*C*), first discovered many years ago by Newberry above the Sharon coal in Ohio. In the compressed condition it resembles a longitudinally ribbed leaf with a serrate distal margin, and for a long time it was referred to the Ginkgoales. Later, spores were discovered in the carbonized matrix of the compressions, and this, combined with the fact that it is usually inti-

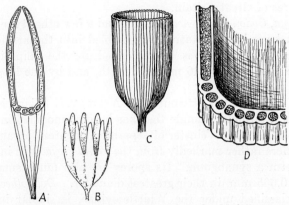

Fig. 113.—Microsporangiate organs attributed to the Medullosaceae. (*A*) *Aulacotheca.* × 3. (*B*) *Codonotheca.* × 2. (*C*) *Whittleseya.* × 1. (*D*) *Whittleseya* in sectional view. × 4. *All after Halle.*

mately associated with *Alethopteris*, led certain investigators to suspect that it might be a pteridosperm fructification. The structure of *Whittleseya* was recently revealed by Halle who succeeded in preparing microtome sections of the carbonized organs. These sections show that the organ is a campanulate synangium consisting of a series of lengthwise tubular pollen cavities around an empty center (Fig. 113*D*). The synangium is quite large, being as much as 5 cm. long and 3 cm. or more broad. The surface is marked with parallel lengthwise striations, which terminate in the small teeth at the truncate apex. The base is rounded and is usually prolonged into a short attachment stalk. In Ohio, where *Whittleseya* was first discovered, it was not found attached, but attached specimens have been reported from other places.

Codonotheca (Fig. 113*B*) is similar to *Whittleseya* except that it is more slender and the rim of the cup is deeply cut between the tubular pollen cavities so that they project as semifree segments attached to a campanulate base. This fructification occurs in the plant-bearing nodules at the Mazon Creek locality in Illinois. It has never been found attached, but is associated with several species of *Neuropteris*.

Aulacotheca (Figs. 112*B* and 113*A*) is an oval elongated organ of similar construction to *Whittleseya* except that it is more slender and tapers to a point at the apex. The interior apparently was hollow. It usually occurs in intimate association with *Alethopteris*, although in one instance it was found apparently attached to the frond of *Neuropteris Schlehani*. *Aulacotheca* was thought to be a seed before its structure became known. It occurs in the Pottsville of Eastern North America and in the Lower Coal Measures of Great Britain.

Whittleseya, Codonotheca, Aulacotheca, and a few other types of similar construction have been tentatively assembled into the artificial group, the Whittleseyinae, which is characterized by the ellipsoidal spores which range from 0.1 to 0.46 mm. in length, and by the long, tubular, concrescent sporangia.

Another fructification attributed to *Neuropteris* foliage is *Potoniea*. It consists of a terminal pinnule that has become modified into a broad shallow cup filled with a cluster of separate elongated sporangia. This structure differs rather markedly from the *Whittleseya* type in which the sporangia form a synangium. Its spores are also much smaller, being but 0.05 to 0.065 mm. in their greatest diameter. *Dolerotheca*, which is sometimes classified under the Whittleseyinae, is similar in shape to *Potoniea* but the sporangia are joined laterally by parenchymatous tissue.

Seeds Attributed to the Medullosaceae.—The seeds thought to belong to the medullosans are *Trigonocarpus, Hexapterospermum, Stephanospermum, Rotodontiospermum,* and other members of the so-called "*Trigonocarpales.*" In these seeds the nucellus is free from the integument except for a basal attachment. There is a double vascular system that supplies both the integument and the nucellus. The seed coat is differentiated into a fleshy outer layer (the sarcotesta) and an inner sclerotic layer (the sclerotesta.) The sclerotic layer is frequently ribbed in multiples of three. The sarcotesta decayed readily and is seldom preserved. The inner part exposed after disappearance of the fleshy part is often preserved as a sandstone or mineral cast. Such casts formed from *Trigonocarpus* are among the most common forms of fossil seeds.

Except for the free integument and double vascular system, the Trigonocarpales show many resemblances to seeds of recent cycads. There is also some evidence that fertilization was accomplished by swimming sperms. In common with other Paleozoic seeds, no embryos have been found in them.

Trigonocarpus Parkinsoni (Fig. 114), the seed supposed to belong to *Medullosa anglica*, is large, sometimes as much as 5 cm. long and 2 cm. in width. Almost half of the length is occupied by the long micropyle. In cross section the seed is circular at the mid-portion but becomes flattened along the micropylar tube. The testa consists of an outer fleshy layer, or sarcotesta, and a hard inner sclerotesta. The latter bears three prominent longitudinal ridges, which are the prominent features on the casts, and nine lesser ridges. The nucellus, which is surrounded by an epidermis, is free from the integument except for the attachment at the base. The pollen chamber is in a dome-shaped extension of the nucellus with an apical beak that extends upward into the micropylar opening. The vascular system of the seed is double (one of the diagnostic features of the Trigonocarpales) with one set of bundles supplying the nucellus and the other the integument. The latter consists of six strands that traverse the length of the seed opposite alternate sclerotestal ridges.

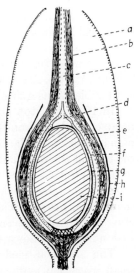

FIG. 114.—*Trigonocarpus Parkinsoni*. Diagrammatic representation of seed in longitudinal section: *a*, sarcotesta; *b*, sclerotesta; *c*, micropyle; *d*, pollen chamber; *e*, "inner flesh"; *f*, nucellus; *g*, inner vascular system; *h*, outer vascular system; *i*, female gametophyte. (*After Scott.*) Enlarged.

Sarcospermum ovale, from the Pennsylvanian of Illinois, is a probable member of the Medullosaceae, although poor preservation has obscured many details. The seed is radially symmetrical, without ribs or sutures, and has a three-layered testa. The nucellus appears to be free from the integument and has about 20 vascular strands. The whole seed measures about 3 by 1.5 cm., and is enlarged basally but pointed apically.

Another recently described medullosan seed from Illinois is *Rotodontiospermum illinoiense*. This fairly large seed is round in cross section and has a two-layered testa. The inner layer, instead of being three-angled, as in *Trigonocarpus*, bears about 20 longitudinal ribs of equal height. This ribbed layer appears fibrous rather than sclerotic and consists of a system of intertwined threads. A vascular strand traverses the length of each furrow between the ribs. The micropylar region of this seed is similar to that of *Trigonocarpus* and medullosan affinities seem further expressed by the nucellus, which is attached only at the base, and by the double vascular supply. In some respects *Rotodontio-spermum* stands somewhat apart from other members of the Trigono-carpales, but it is intimately associated with the roots, stems, and petioles of *Medullosa Noei* to which it probably belongs.

No seed has been assigned to a medullosan stem with the certainty with which *Lagenostoma* has been linked to *Lyginopteris*. A few instances are known, however, of seeds attached to *Alethopteris* and *Neuropteris*, but the assumed attribution to medullosan stem types is based upon structural resemblances and the less dependable evidence of association. There remains no doubt, however, in the minds of paleobotanists, that most of the Trigonocarpales, along with *Alethopteris* and *Neuropteris*, belong with the *Medullosa* type of stem, but little is known with respect to possible connections between species.

In 1905, Kidston announced the interesting discovery of the attach-ment of seeds to fronds of *Neuropteris heterophylla*. In this species the pinna rachis with the terminal seed bears ordinary pinnules right up to the base of the seed, and is not in any way modified or different from the seedless pinnae. The seeds are ovate and broadest at the middle, but taper to an apical beak or snout. They are about 3 cm. long and more than 1 cm. broad. The surface is longitudinally furrowed, which prob-ably indicates the existence of fibrous strands in the testa, but other-wise nothing is known of the internal structure.

The attached seeds of *Neuropteris heterophylla* were discovered in small ironstone nodules from the middle Coal Measures near Dudley, England. Kidston's photographs show on each of the two halves of the split nodules, a seed compression with a few characteristic pinnules of *N heterophylla* attached to a very short length of the pinna rachis at the base. Because of the fibrous surface Kidston referred the seeds to the genus *Rhabdocarpus*, but later authors, realizing the radiospermic character, have created for them the new genus *Neurospermum*. Seward has since identified them with Brongniart's genus *Neuropterocarpus*.

The alleged seeds of *Neuropteris hollandica* (originally identified as *N. obliqua*) from the "Coal Measures" of the Netherlands are similar to those of *N. heterophylla* but are somewhat larger, and are terminal on dichotomously branched pinnae with ordinary foliage pinnules up to the point of attachment.

In his recent textbook, "Gymnosperms: Structure and Evolution," Chamberlain has figured a restoration of *Neuropteris decipiens* with seeds attached (page 17, Fig. 11). Seed attachment to *N. decipiens*, however, has not been observed, and the concept for the restoration was no doubt derived from published figures of *N. heterophylla*.

Alethopteris Norinii, the first member of that foliage genus to be observed with seeds attached, came from the Permian of China. The longitudinally striated seeds, which somewhat resemble those of *N. heterophylla* except for the lack of the attenuated beak, are oblong-lanceolate bodies measuring about 12 by 40 mm., borne among the ordinary foliage pinnules, each apparently replacing a single pinnule.

SOME OTHER PALEOZOIC PTERIDOSPERMS

Several examples of seed attachment are known where the material cannot be referred to either the Lyginopteridaceae or the Medullosaceae. A familiar example is *Pecopteris Pluckeneti*. Attachment is at the ends of the lateral veins at the margins of the pinnules, which are sometimes slightly modified. The seeds, known in the isolated condition as *Leptotesta Grand'Euryi*, show some interesting structural details that may indicate a primitive condition. The nucellus is entirely free from the integument, even at the base, but is suspended from the arch of the seminal cavity by an elastic ring around the upper part. This ring is interpreted as erectile tissue that served to push the nucellus up against the micropyle at the time of pollination. The integument is differentiated into sarcotestal and sclerotestal layers, the former being broadened laterally into a narrow wing, thus identifying the seed as a member of the Cardiocarpales. On the other hand, the vascular system extends only into the integument, a feature in common with the Lagenostomales, and the free integument is a feature characteristic of the Trigonocarpales. In some ways this seed appears to be a composite type, and some authors consider these peculiarities sufficient to warrant placing it in a separate group, the Leptotestales.

Just previous to the discovery of the seeds of *Pecopteris Pluckeneti*, David White announced the attachment of seeds to *Aneimites fertilis* from the Pottsville (lower Pennsylvanian) of West Virginia. The small rhomboidal bilateral seeds, which were borne at the tips of reduced pinnules, average 4.5 mm. in length, and they appear to possess a narrow wing. The internal structure is unknown.

The attachment of seeds to several pteridosperms from the Permian of China has recently been announced. In addition to *Alethopteris Norinii*, already mentioned, they include *Sphenopteris tenuis*, *Pecopteris Wongii*, *Emplectopteris triangularis*, and *Nystroemia pectiniformis*. *Sphenopteris tenuis*, which is probably not a member of the Lyginopteridaceae, has small oval seeds measuring 4 to 5 mm. long produced singly at the bases of the lower surfaces of the pinnules. Little is known of the internal structure of these seeds. The position is in marked contrast to that of other species of *Sphenopteris*. In *Pecopteris Wongii* an ovoid seed about 7 mm. long appears attached by a curved stalk of about its own length to the main rachis but the relation to the vegetative pinnae is not clear. In *Emplectopteris triangularis* the narrowly winged and platyspermic seeds appear to be borne singly on the surface of the pinnules, which show no reduction or modification in any way. This method of attachment is similar to that shown by *S. tenuis*, although in the latter the attachment appears to be either to the rachis or at the extreme base of the pinnule. This attachment to unmodified pinnules is in marked contrast to that shown by earlier Paleozoic seeds, which are terminal on pinnulate or naked frond ramifications. *Nystroemia pectiniformis* is a peculiar form in which the lateral pinna rachises bear on their adaxial surfaces numerous stalklike pinnules which divide by unequal dichotomy, with each ultimate branch producing a terminal seed. The seeds are small, narrowly obovoid, platyspermic, and bicornate at the tips. *Nystrocmia* is an isolated pteridosperm of obscure affinities.

Similar to and probably congeneric with *Emplectopteris*, is *Lescuropteris* from the upper Conemaugh (middle Pennsylvanian) of western Pennsylvania, of which two species are known to bear seeds.

The Calamopityaceae

The Calamopityaceae is an assemblage of Upper Devonian and Lower Carboniferous plants, which are grouped with the Pteridospermae entirely on the basis of the anatomical structure of the stems. Foliage or reproductive organs are unknown for any of them, although a few have been found with the basal parts of the petioles attached. The family is to a large extent an artificial one, and near affinities with either of the previously discussed families are unlikely. However, in being monostelic they recall to some extent the Lyginopteridaceae, although the large petioles with numerous bundles found on a few of the stems constitute at least a superficial resemblance to the Medullosaceae.

The Calamopityaceae embrace 7 genera and about 16 species. The genera have been divided into two groups on the structure of the secondary wood. In the "monoxylic" group the wood contains rather large tracheids and broad rays, thus imparting to it a soft or spongy texture. It consists of three genera, *Stenomyelon*, *Calamopitys*, and *Diichnia*. The "pycnoxylic" group contains the four genera *Eristophyton*, *Endoxylon*, *Bilignea*, and *Sphenoxylon*. In these genera the wood is compact with small tracheids and narrow rays, and therefore resembles that of the Cordaitales and the Coniferales. The monoxylic Calamopityaceae, in contrast, have wood more like that of the Lyginopteridaceae and the Medullosaceae.

Stenomyelon.—This genus of three species is considered the most primitive of the family. The small stem in cross section is three angled with rounded corners and lacks a distinct pith. The leaf traces depart from the three lobes of the central primary xylem mass, and the protoxylem is not distinct except where it partakes of trace formation. Two species are known. One is *Stenomyelon muratum* from the late Devonian New Albany shale of Kentucky, in which the tracheids of the primary xylem are intermixed with parenchyma, thus producing a mixed pith. The leaf trace originates as a single mesarch strand that soon divides into two. In the cortex several more divisions occur so that the trace supply of the petiole consists of a number of circular or slightly oval

Fig. 115.—*Calamopitys americana.* Cross section of stem showing the mixed pith and the broad rays of the secondary wood. Section of the type specimen of Scott and Jeffrey. Upper part of the New Albany shale. Junction City, Kentucky. × about 4.

strands arranged in a horseshoe-shaped pattern. In the other species, *S. tuedianum*, from the Lower Carboniferous Calciferous Sandstone of Scotland, the mixed pith is not present, but the three lobes of solid tracheids are separated by three narrow partitions of parenchyma, which meet at the center of the stem. The secondary wood of both species is about as wide as the diameter of the primary core, and it contains numerous multiseriate rays which become narrowly fan-shaped as they approach the inner edge. The cortex is mostly homogeneous and parenchymatous except for a network of hypodermal strands in the outer part.

Calamopitys.—This genus of six species is the largest of the family. The species have been handled differently by different authors. Scott included within it the three species of *Eristophyton* (a name proposed by the Russian paleobotanist Zalessky) and retained it as a subgenus of *Calamopitys*. Since all three species of *Eristophyton* agree with respect to important characteristics of the secondary wood, it is retained here as a separate genus.

The most thoroughly investigated species of *Calamopitys* is *C. americana* (Fig. 115), from the upper part of the New Albany shale. In the center of the stem there is a relatively large pith which in large stems is as much as 13 mm. in diameter, and which contains tracheids intermixed with parenchyma. Surrounding the mixed pith is a ring of partly confluent mesarch primary xylem strands in which the protoxylem lies nearer the outer than the inner margin. A leaf trace originates as a branch from one of the circummedullary strands, and the reparatory strand continues upward without interruption and without the formation of a leaf gap. The trace divides into two before it reaches the outside of the woody cylinder, and several bundles are present in the petiole base. The secondary wood is a loosely arranged tissue of large tracheids and broad rays. There is some general resemblance between the woody axis of *C. americana* and that of *Heterangium*, but the leaf-trace supply constitutes an important difference. The woody cylinder is surrounded by a thin phloem zone and a thick cortex.

Calamopitys annularis, from beds which may range from the Upper Devonian into the Lower Carboniferous, is similar to *C. americana* in all its essential features. *C. annularis* came from Thuringia, and although wide geographical separation is no specific criterion, the name has, on this account been retained. *C. Saturni*, also from Thuringia, is different. It has a small pith only a few millimeters in diameter which is entirely parenchymatous, and which lacks medullary tracheids. Around the pith are six separate centrally mesarch primary xylem strands. *C. Saturni* is considered a more advanced species than either *C. americana* or *C. annularis*.

Diichnia.—This genus with one species, *D. kentuckiensis* (Fig. 116*A*) is from the Upper Devonian New Albany shale. The stems, which are somewhat variable in size, possess a five-angled "mixed pith" surrounded by a narrow zone of secondary wood with broad rays. The primary xylem consists mainly of small mesarch strands near the extremities of the pith angles, but separated from the secondary wood by a small amount of parenchyma. A characteristic feature of the genus is the double leaf trace, which originates as a pair of strands from adjacent pith angles. These strands pass out through the cortex and enlarge somewhat. They finally divide one or more times, and the entire set of bundles enters the petiole. The cortex is chiefly parenchymatous except for a network of hypodermal strands.

The name *Kalymma* has been assigned to detached petioles of *Calamopitys* and related genera. *K. lirata* is the petiole of *Calamopitys americana*, and *K. resinosa* is believed to belong to *Diichnia kentuckiensis*.

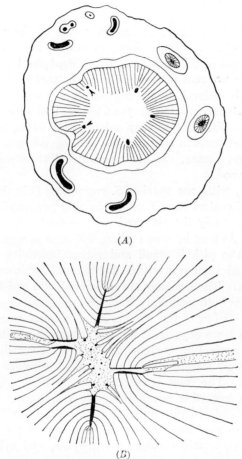

(A)

(B)

Fig. 116.—(A) *Diichnia kentuckiensis.* Cross section of stem showing the 5-angled pith and the small primary xylem strands in the angles. The two strands at the left have divided preparatory to trace emission. (*Drawn from figure by Read.*) ✕ about 1½.

(B) *Sphenoxylon eupunctata.* Cross section showing the irregular pith, the small primary xylem bundles, and the four large inwardly projecting secondary wood wedges each containing a departing trace. Pith and trace parenchyma stippled; primary xylem in solid black. (*Drawn from figure by Dale Thomas.*) ✕ about 2.

These petioles are characterized by a ring of bundles embedded in a parenchymatous ground mass and a hypodermal layer of sclerotic strands. The bundles may form a bilaterally symmetrical ring or a horseshoe pattern. They may be radially or tangentially elongated or circular. Although nothing is known of the foliage borne by these petioles, there is evidence that they supported dichotomously forked fronds. It also appears that the small branches stood out at right angles to the parent axis, which is suggestive of some of the fronds of the Lyginopteridaceae.

Pycnoxylic forms.—The best known genus of the pycnoxylic Calamopityaceae is *Eristophyton*, which, as stated, is considered by some authors a subgenus of *Calamopitys*. In *Eristophyton* the primary xylem strands are very unequal in size. The lower part of a strand is small and often entirely endarch, but as it progresses upward it becomes larger and centrally mesarch. In *E. fascicularis* the smaller strands are embedded in the pith, but as they enlarge they swing toward the edge of this tissue and come in contact with the secondary wood just before departure to form a trace. In *E. Beinertiana* all the strands lie adjacent the secondary wood.

Sphenoxylon, with one species, *S. eupunctata* (Fig. 116*B*), from the basal Portage of central New York, is the only pycnoxylic type found in North America. Only the pith and the primary and secondary xylem are preserved. As seen in cross section, the stem presents some striking peculiarities. The pith is small and rather indefinite as to its outer limits. It is quadrangular in outline due to four large inwardly projecting wedges of xylem. These wedges are associated with leaf-trace departure. The pith contains both medullary tracheids and small mesarch medullary xylem strands. A few of the strands are in contact with the secondary wood. When a trace departs, it passes through the center of one of the large inwardly projecting wedges, and within the wedge the tracheid rows curve strongly toward a small radial band of primary xylem within the tip portion. The trace is single at its point of origin but divides farther out. A few other peculiarities are also shown. The secondary tracheids bear bordered pits on their tangential as well as their radial walls. There are also distinct growth rings (structures not common in Paleozoic woods), terminal xylem parenchyma, and xylem rays which are mostly uniseriate.

Endoxylon, from the Lower Carboniferous limestone of Ayrshire, is the most advanced member of the Calamopityaceae. It contains one species, *E. zonatum,* in which large mesarch primary xylem strands are distantly spaced around the outer margin of the pith. No pith tracheids are present. A feature possessed in common with *Sphenoxylon* is a series of well-defined growth rings of which there are 14 in a cross section having a radius of 2.5 cm.

Relationships.—Very little can be said concerning the origin and relationships of the Calamopityaceae. Investigators have noted, however, that the three-angled primary xylem strand of the Devonian genus *Aneurophyton* bears considerable resemblance to that of *Stenomyelon*. *Aneurophyton* may be an early member of the calamopityean alliance which then suggests the possibility of the Calamopityaceae having developed from psilophytic ancestors in which the leaves and stems were indistinctly differentiated. The affinities of *Aneurophyton*, however, are obscure, and it is believed that this plant reproduced by spores. Although there is some similarity in the structure of the xylem of *Heterangium* and *Calamopitys*, they are distinguished on the structure of the petioles, and whatever similarities exist, are probably the result of parallel evolution. A common ancestry, if such existed, was remote.

PTERIDOSPERMS OF THE LATE PALEOZOIC AND THE MESOZOIC

The pteridosperms suffered a great decline during the Permian. They were abundant at the beginning, but with the onset of widespread

Fig. 117.—*Thinnfeldia sp.* Foliage of a probable Mesozoic pteridosperm. Upper Karroo. South Africa.

aridity which was followed in some parts of the earth by extensive glaciation, many of the genera became totally extinct. The few that survived to the end of the period did so in greatly modified form.

The early Permian pteridosperms do not differ markedly from those of the preceding Coal Measures. In the Permian rocks of Eastern North America (the Dunkard series) we find the same leaf genera as in the uppermost Pennsylvanian in that region. In the central part of the continent, however, the early Permian rocks (the so-called "Red Beds") show evidences of increasing aridity. The rocks, characterized by their red color, show remains of sun cracks, casts of salt crystals and other dessication features, and the foliage of the plants tends to be more leathery and more adapted to dry surroundings. A well-known plant of the Permian in Texas and Oklahoma is *Gigantopteris americana* (Fig. 181*A*). It has large ribbonlike fronds, which are sometimes forked. Each frond possesses a strong midrib from which straight lateral veins extend at an upward angle to the margin. The fine veinlets form a network between them. Small flattened seeds have been found in association with the leaves.

The genus *Callipteris* is a widely spread Permian leaf type believed to be a pteridosperm. *Callipteris* has thick pecopteroid or sphenopteroid pinnules somewhat decurrent at the lower margin and often connected at the base after the manner of *Alethopteris*. *Callipteris conferta* (Fig. 181*B*) is an "index fossil" of the Permian throughout the Northern Hemisphere.

The existence of pteridosperms in the Mesozoic was suspected for some time before evidence based upon fructifications established their presence in rocks of that era. *Thinnfeldia* (Fig. 117), a leaf type similar to *Supaia*, was suspected as early as 1867 as being intermediate between the ferns and the gymnosperms. More recently such genera as *Lomatopteris*, *Cycadopteris*, and *Callipteris* have been provisionally classified with the pteridosperms but without the support of the fertile parts.

The Peltaspermaceae

This family name was proposed after the discovery of the identity of certain seeds and microsporophylls with leaf fragments belonging to the form genus *Lepidopteris*. At present two species, *Lepidopteris ottonis* and *L. natalensis*, are embraced within the family.

The associated leaves, seeds, and microsporophylls of *Lepidopteris ottonis* (Fig. 118*B*) have been found in the Rhaetic (uppermost Triassic) beds of eastern Greenland. The leaf is a small bipinnate frond closely resembling that of *Callipteris* and certain other Paleozoic pteridosperms, but it is distinguished by small blisterlike swellings on the rachis. The pinnules are attached by a broad base, and the venation is of the simple open type. The cuticle is thick, and stomata occur on both surfaces of the pinnules, although they are more numerous on the surface believed to be the lower one. The seeds are suspended in a circle around the central stalk on the lower surface of a peltate cupulate disc, which is about 1.5 cm. in diameter. The seeds themselves are oval bodies about 7 mm. long and 4 mm. in diameter, and have curved micropylar beaks. In this species the seed-bearing discs have not been found on the inflorescences that bore them, but in *L. natalensis* from the Karroo formation of South Africa the inflorescence is a slender axis bearing the discs in a spiral order. In this species it appears that only one seed per disc reached maturity.

The microsporophyll of *L. ottonis* is a branched structure with microsporangia suspended from the side toward the base of the main axis. The pollen sacs are about 2 mm. long and 1 mm. wide. Dehiscence was by a lengthwise slit. The pollen grains are small, oval, bilaterally symmetrical bodies with smooth walls.

The Corystospermaceae

This family was recently instituted by Thomas to include seed- and pollen-bearing inflorescences and foliage that, although not found in attachment, are believed to be related parts. They were derived from the Molteno beds of the upper part of the Karroo formation of Natal, probably of Middle Jurassic age. The plants were probably small and bore variable fernlike fronds. The inflorescences are unisexual, consisting of a main axis with lateral branches. In the seed-bearing inflorescence the branches are axillary in small bracts, and each branch bears two or more stalked, recurved, helmet-shaped, campanulate or bivalved cupules, which in turn bear a single seed each. The seeds are ovoid or elliptical, measure 3.5 to 7 mm. in length, and have curved, bifid micropyles, which project from the cupules. The pollen-bearing inflorescences lack the subtending bracts. The tips of the ultimate branches are enlarged, and on these are a number of pendulous sporangia which contain winged pollen grains.

This family was founded upon three seed-bearing inflorescences described as *Umkomasia*, *Pilophorosperma*, and *Spermatocodon*. A fourth genus, *Zuberia*, from Argentina, has recently been added. In *Umkomasia* (Fig. 118*D* and *E*), which has two species, the branching is in one plane and the cupule is deeply divided into two lobes. *Pilophorosperma* is similar except that the helmet-shaped cupule has but one cleft, and the inner surface is covered with hairs. It contains eight species. In *Spermatocodon* some of the branches are spirally arranged in contrast to *Umkomasia* and *Pilophorosperma* in which the branching is in one plane. It has one species. The pollen-bearing organs associated with all of these are assigned to *Pteruchus* (Fig. 118*G*). Eight species have been named. The clustered sporangia are lanceolate, apparently lack a dehiscence line, and probably bore the pollen in two locules. The pollen grains have a central cell and two symmetrically placed lateral wings. The inflorescence is very similar to *Crossotheca* and would be difficult to distinguish from it except for the winged pollen grains. It may represent the persistence of the *Crossotheca* or *Telangium* type of pollen-bearing organ into the Mesozoic era.

The foliage of the Corystospermaceae is believed to be represented by several frond genera, such as *Dicroidium*, *Thinnfeldia*, *Stenopteris*, and *Pachypteris*.   Although not found actually attached these leaf types are intimately associated with the inflorescences, and the structure of the

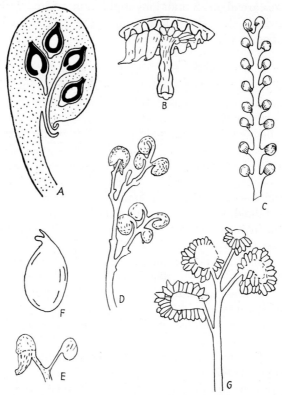

Fɪɢ. 118.—Fructifications and seeds of Rhaetic and Middle Jurassic pteridosperms. (A) *Caytonia Thomasi*. Longitudinal section of fruit showing the orthotropous ovules. (*After Harris.*)   Greatly enlarged; (B) *Lepidopteris ottonis*. Restoration of cupulate disc from which all seeds except three have fallen. (*After Harris.*)   × about 1½.   (C) *Caytonia Nathorsti*. Ovulate inflorescence showing the pinnately attached fruits. (*After Thomas.*)   Slightly reduced.   (D) *Umkomasia MacLeani*. Ovulate inflorescence. (*After Thomas.*)   Natural size. (E) *Umkomasia sp.* Pair of seed cupules. (*After Thomas.*) × 2.   (F) seed belonging to the Corystospermaceae. (*After Thomas.*)   × 3; (G) *Pteruchus africanus*. Microsporangiate inflorescence. (*After Thomas.*)   × 1½.

epodermis of both the foliage and the inflorescences is sufficiently similar to indicate relationships.

Tʜᴇ Cᴀʏᴛᴏɴɪᴀʟᴇs

This group of plants was named in 1925 from remains discovered in the Middle Jurassic rocks of the Yorkshire coast. They consist of fruit-bearing stalks (*Caytonia*), microsporophylls (*Caytonanthus*), and foliage (*Sagenopteris*). The two Yorkshire species of *Caytonia, C. Sewardi* and *C. Nathorsti*, have been identified with the foliage forms *Sagenopteris colpodes* and *S. Phillipsi*, respectively. The microsporophyll *Caytonanthus Arberi* belongs to one of these also. A third species, *Caytonia Thomasi* (leaf *S. Nilssoniana* and microsporophyll *Caytonanthus Kochi*) has been found in slightly older rocks in Eastern Greenland.

When first described *Caytonia* was compared with the angiosperms, and the genus has been referred to by numerous writers as a Jurassic representative of this class of plants. Later investigations, however, have shown that the genus is essentially gymnospermous and most closely related to the Pteridospermae.

The fruit-bearing stalk, or the megasporophyll, is a dorsiventral structure, 5 cm. or more in length, which bears shortly stalked fruits laterally in subopposite pairs (Fig. 118C). The fruit is a rounded, sac-like body, 5 mm. or less in diameter, with a "lip" situated close to the stalk. Between the "lip" and the stalk is a narrow "mouth." In *Caytonia Thomasi* (Fig. 118A), and probably in the other species also, the lip opened into the interior of the fruit, where small orthotropous seeds are situated in a row on the curved inner surface. The micropyles point toward the mouth. It appears that the mouth opened when pollination took place, but closed soon afterward, and that the seeds matured within the closed fruit. Pollen grains have been observed within the micropyles of the seeds, thus showing that the pollen actually reached the ovule and did not germinate on the lip as was formerly supposed. Angiospermy, therefore, was only partly attained in *Caytonia.*

The small seeds vary from 8 to 30 per fruit, depending somewhat upon the species. The integument is single and free from the nucellus to the base. The seeds lack the micropylar beaks present in the Peltaspermaceae and the Corystospermaceae.

The genus *Gristhorpia*, also described from Yorkshire, has been united with *Caytonia.*

Caytonanthus, the microsporangiate fructification, resembles *Caytonia* somewhat in form, but the tips of the pedicels are often subdivided, with each subdivision bearing a cluster of quadrilocular microsynangia. The pollen grains are small and doubly winged like those of pine, except that the wings are directly opposite each other.

The leaf, *Sagenopteris*, is palmately compound with three to six lanceolate leaflets, ranging from 2 to 6 cm. in length, borne on the tip of a petiole. The venation is reticulate. The identity of the leaves and fertile parts has been established by the similarities in the structure of the stomata and epidermal cells of these organs.

MORPHOLOGIC AND PHYLOGENETIC CONSIDERATIONS

Because of the complexity of the problems pertaining to the morphology and relationships of the pteridosperms, only a few outstanding features will be mentioned. The essential conclusions are listed in summary fashion as follows:

1. *The Origin of the Pteridosperms.* The pteridosperms probably developed either directly from psilophytalean stock or from some intermediate fernlike type previous to the Carboniferous. The latter course seems more likely in view of the recent discovery of heterospory in *Archaeopteris.*

2. *Relation to the Ferns.* The question may very reasonably be asked whether the pteridosperms are ferns with seeds or gymnosperms with fernlike foliage. Some botanists argue that if the seed habit is insufficient to set the so-called Spermatophyta apart as a separate division of the plant kingdom, the mere presence of seeds in the pteridosperms should not exclude this group from the ferns. The pteridosperms, they maintain, are simply seed-bearing ferns just as the lepidocarps are seed-bearing lycopods. The proponents of this view also refer to numerous fructifications, having evident affinity with the Marattiaceae, found on *Pecopteris* and *Sphenopteris* foliage, which, moreover, have never been shown not to belong to seed plants. They regard some of these as possible pollen-bearing organs of pteridosperms.

Unless a fructification of proved filicinean affinity is found attached to a plant known to be seed-bearing, or its connection is established by other means, it is useless to try to establish a connection between the ferns and seed-ferns on the basis of the fructifications. It has been emphasized before that the identity of the microsporangiate organs of the pteridosperms is based mostly on inference and association, and that those fructifications thought most likely to be such are quite different from the supposed fossil marattiaceous ferns.

Filicinean affinities among the pteridosperms are suggested by certain anatomical features as well as by the form of the foliage, but the fernlike aspect of the foliage is more apparent than real. It should be recalled that Potonié first proposed the name Cycadofilices for stems showing transitional characters between the ferns and higher seed plants. Williamson had previously noted a similar combination of structures in *Heterangium.*

3. *Relation to the Cycadophytes.* It is just as logical to call the pteridosperms cycads without strobili as ferns with seeds. Although the primary stele of *Heterangium* or *Medullosa* is on the approximate level of the protosteles of some ferns, the stem when viewed in its entirety

is decidedly cycadlike. The connection between the Medullosaceae and the cycadophytes seems especially close when all features of the structure of the stems, petioles, and seeds is taken into account.

4. *Relation to the Higher Gymnosperms.* Epidermal studies show that the foliage of the pteridosperms is typical of that of the higher gymnosperms in some respects. There is very little evidence of derivation of the Cordaitales and the Coniferales from pteridospermous ancestors, although there is a suggested connection between some of the Calamopityaceae and the Pityaceae. However, the true limits of neither of these latter groups are perfectly understood. The ginkgoalean leaf may have been derived from the wedge-shaped pinnule of some Paleozoic pteridosperm.

5. *Relation among Themselves.* Early in their history the pteridosperms became diversified into several groups, and some might have had an independent origin. The group as a whole dominated the flora of the late Paleozoic previous to the Permian ice age. Only a few pteridosperms survived this climatic barrier, and those that persisted did so in modified form as late as the Middle Jurassic.

Aside from questions of affinity, which apply to all plant groups to more or less the same extent, the main problem pertaining to the pteridosperms is finding a satisfactory means of establishing connections between detached organs. It is very seldom that seeds are found in actual attachment to vegetative parts, and it is rarer still that the male inflorescence can be correctly assigned to a seed-bearing frond. A sufficient number of attached seeds have been found to give a fairly accurate notion of which of the leaf genera belong to pteridosperms. However, some species of *Sphenopteris* and also of *Pecopteris* are known to bear fern fructifications, whereas others produce seeds. Epidermal studies may in some cases furnish valuable clues for separation where seed attachment or preserved stems are lacking.

POSITION OF THE SEED ON THE PLANT

As stated at the beginning of the chapter, none of the pteridosperms is known to have produced strobili or organs resembling them. It is true that the number of known pteridosperms, with seeds attached, probably does not exceed a dozen, but the absence of detached cones or conelike organs in the fossil record that might be attributed to pteridosperms justifies the assumption that the seeds were borne singly throughout the group.

In a recent account of some pteridosperms from China, Halle has assembled pertinent data on the position of the seed on the plant. The following examples are given:

1. *Lyginopteris oldhamia.* Cupulate seeds are terminal on naked frond ramifications. The cupules may represent modified pinnules, but this is uncertain.

2. *Neuropteris heterophylla.* The seed is terminal on the pinna rachis, which bears ordinary pinnules up to the seed.

3. *Neuropteris hollandica.* The situation here is similar to that of *N. heterophylla,* but in the single observed instance the pinna rachis is forked just below the seed with pinnules extending beyond the bifurcation.

4. *Aneimites fertilis.* The small seeds are terminal on narrow naked pedicels, which correspond to reduced frond segments.

5. *Pecopteris Pluckeneti.* The seeds are attached at the ends of the small veins at the margins of the ordinary pinnules. The pinnules are not at all, or only very slightly, modified.

6. *Alethopteris Norinii.* Attachment is on the rachis of an ultimate pinna, with unmodified pinnules below and beyond the seed. The seed therefore may replace a pinnule.

7. *Sphenopteris tenuis.* Seed borne on the lower side of the rachis of an ultimate pinna or on the extreme bases of unmodified pinnules.

8. *Emplectopteris (Lescuropteris) triangularis.* Each unmodified fertile pinnule bears a single seed on its upper surface.

Too few examples of seed attachment among pteridosperms are known to support definite conclusions concerning the evolution of the position of the seed, but those given present some interesting facts. *Lyginopteris oldhamia* and *Aneimites fertilis* are the oldest to which seeds have been found attached, and they bear their seeds terminally on strongly reduced naked frond segments. In *Neuropteris heterophylla* and *N. hollandica* from the middle Upper Carboniferous the seeds are still terminal but on unmodified pinnae. In the Upper Carboniferous *Pecopteris Pluckeneti,* the seeds have taken a marginal position on normal or slightly modified pinnules, but in the Permian species, *Alethopteris Norinii, Sphenopteris tenuis,* and *Emplectopteris triangularis,* the seeds are on the surface of the rachis or the pinnules. There is a suggestion, therefore, that there has been a transition from a terminal position on a naked rachis to a position on the surface of unmodified foliar organs. This situation is paralleled somewhat in the ferns. In the earlier ferns the sporangia are often terminal on reduced stalklike segments (as in *Archaeopteris*), but those with sporangia on the pinnule surface, such as *Asterotheca* and *Ptychocarpus,* become more frequent at a later age. Although there are probably many exceptions, and generalizations should be made only with caution, there does seem to be a parallelism between the seed of the pteridosperms and the fern sporangium with regard to position on the plant, and this may be the result of their common origin from psilophytalean ancestral types with terminal spore-bearing organs.

References

ANDREWS, H.N., JR.: On the stelar anatomy of the pteridosperms with particular reference to the secondary wood, *Annals Missouri Bot. Gardens,* **27**, 1940.

———: A new cupule from the Lower Carboniferous of Scotland, *Torrey Bot. Club Bull.* **67**, 1940.

ARNOLD, C.A.: Paleozoic seeds, *Bot. Rev.,* **4**, 1938.

BENSON, M.: *Telangium Scotti,* a new species of *Telangium (Calymmatotheca)* showing structure, *Annals of Botany,* **18**, 1904.

CHAMBERLAIN, C.J.: "Gymnosperms; structure and evolution," Chicago, 1935.

CROOKALL, R.: *Crossotheca* and *Lyginopteris oldhamia, Annals of Botany,* **44**, 1930.

GORDON, W.T.: On *Tetrastichia bupatides:* a Carboniferous pteridosperm from East Lothian, *Royal Soc. Edinburgh Trans.,* **59**, 1938.

GRAND'EURY, C.: Sur les graines trouvées attachées au *Pecopteris Pluckeneti* Schl., *Acad. Sci. Comptes rendus,* **140**, 1905.

HALLE, T.G.: Some seed-bearing pteridosperms from the Permian of China, *K. svenska vetensk. akad. Handl.,* III, **6**, 1929.

———: The structure of certain fossil spore-bearing organs believed to belong to pteridosperms, *K. svenska vetensk. akad. Handl.,* III, **12**, 1933.

HARRIS, T.M.: *Caytonia. Annals of Botany, New series,* **4**, 1940.

———: A new member of the Caytoniales, *New Phytologist,* **22**, 1933.

JONGMANS, W.J.: On the fructifications of *Sphenopteris Hoeninghausi* and its relation with *Lyginodendron oldhamium* and *Crossotheca schatzlarensis, Proc. Fifth Intern. Bot. Congr.,* 1930, Cambridge, 1931.

KIDSTON, R.: On the fructification of *Neuropteris heterophylla* Brongniart, *Royal Soc. London Philos. Trans.,* B **197**, 1905.

———: Fossil plants of the Carboniferous rocks of Great Britain, *Great Britain Geol. Survey Mem.,* **2** (1), 1923; *ibid.,* **2** (2), 1923–25.

OLIVER, F.W., and D.H. SCOTT: On the structure of the Paleozoic seed *Lagenostoma Lomaxi,* with a statement of the evidence upon which it is referred to *Lyginodendron, Royal Soc. London Philos. Trans.,* B **197**, 1905.

READ, C.B.: The flora of the New Albany shale, pt. 2, The Calamopityeae and their relationships, *U. S. Geol. Survey Prof. Paper,* 186-E, 1937.

SALISBURY, E.J.: On the structure and relationships of *Trigonocarpus shorensis* sp. nov., a new seed from the Paleozoic rocks, *Annals of Botany,* **28**, 1914.

SCHOPF, J.M.: *Medullosa distelica,* a new species of the Anglica group of Medullosa, *Am. Jour. Botany,* **26**, 1939.

SCOTT, D.H., and A.J. MASLEN: The structure of the Paleozoic seeds *Trigonocarpus Parkinsoni* Brongniart, and *P. Oliveri,* sp. nov., pt I, *Annals of Botany,* **21**, 1907.

———: On the structure and affinities of fossil plants from the Paleozoic rocks, III, On *Medullosa anglica,* a new representative of the Cycadofilicales. *Royal Soc. London Philos. Trans.,* B **191**, 1899.

———: "Studies in Fossil Botany," 3d ed., Vol. II, London, 1923.

SEWARD, A.C.: "Fossil Plants," Vol. 3, Cambridge, 1917.

STEIDTMANN, W.E.: The anatomy and affinities of *Medullosa Noei* Steidtmann, and associated foliage, roots, and seeds, *Michigan Univ., Mus. Paleontology, Contr.,* **6** (7), 1944.

THOMAS, H.H.: On some pteridosperms from the Mesozoic rocks of South Africa, *Royal Soc. London Philos. Trans.,* B **222**, 1933.

WHITE, D.: The seeds of *Aneimites, Smithsonian Misc. Coll.,* **47**, 1904.

CHAPTER X

CYCADOPHYTES AND GINKGOS

Although the Mesozoic has been called the "Age of Cycads," the appellation "Age of Cycadophytes" would be more appropriate. This is because of the existence at that time of the cycadeoids, plants related to cycads and in common parlance often referred to as such, but different from them.

The term "cycadophyte" as employed here embodies the *Cycadales*, or true cycads, both living and extinct, and the *Cycadeoidales*, a group of cycadlike plants that is entirely extinct. There are some fundamental differences between these groups. In the Cycadeoidales the fructifications are flowerlike organs with a whorl of stamens surrounding a central ovuliferous receptacle. In contrast the relatively simple cones of the Cycadales are constructed more like those of other gymnosperms. Another basic difference is found in the epidermis of the leaves. In the Cycadeoidales the guard cells of the stomata are flanked by a pair of subsidiary cells that originated from the same mother cell as the guard cells. Furthermore, the walls of the epidermal cells are usually wavy instead of straight. In the Cycadales, on the other hand, the subsidiary cells did not arise from the same mother cells as the guard cells, and the epidermal cell walls are usually straight. Other differences between the two groups will be subsequently mentioned. The main features possessed in common between these two groups are (1) the fact that they are all naked-seeded plants, and (2) the prevailingly pinnate plan of the foliar organs.

The earliest cycadophytes probably arose from pteridospermous ancestors during the late Paleozoic. We have little exact knowledge of Paleozoic cycadophytes, although it is possible that certain leaf forms such as *Eremopteris* were early members of this line. We do not know which of the two cycadophytic groups is older, although the cycadeoids did attain a much higher degree of specialization and are more prominently displayed in the fossil record. The history of the true cycads has been more obscure, and only recently have we begun to realize that they existed in considerable numbers as early as the late Triassic. Studies of leaf cuticles and fructifications have been mainly instrumental in the gradual unfolding of the history of this group.

248

The decipherable record of the cycadophytes begins with the Triassic. The group probably reached its developmental climax during the Middle or Late Jurassic, although it still remained an important element of the flora throughout the early Cretaceous. By late Cretaceous times it had declined to a position of subordination, and today the only survivors are the nine living genera of the Cycadales.

Cycadophytes are preserved in various ways. The abundant sclerenchyma and the thick layer of cutin covering the leaves are admirable adaptations for preservation in the form of compressions. All parts of the plant, with the possible exception of the roots, have been found in the petrified condition, and the perfection with which the morphology of the vegetative and reproductive structures of some members has been analyzed probably rivals that of any other group of fossil plants. Petrified foliage, however, is known mainly from young leaves contained within unopened buds at the apices of the trunks. Casts of the trunks and the large globular floral buds are occasionally found.

THE CYCADEOIDALES

Historical

Evidence that the cycadeoids attracted attention during ancient times is furnished by the discovery in 1867 of a silicified trunk in an Etruscan tomb where it had been placed more than four thousand years before. The specimen has subsequently been described under the name *Cycadeoidea etrusca*. The years immediately preceding and following the finding of this trunk mark the beginning of important cycadeoid studies in both Europe and North America. Many of the specimens described during this period had been discovered earlier, and had in the meantime reposed either in museums or in the possession of private collectors. The first cycadeoids described from Great Britain were sandstone casts of Wealden age, which were assigned to the genus *Bucklandia* in 1825. In 1875 the name *Cycadeoidea* was proposed for petrified trunks from the Isle of Portland. Several years later similar trunks were found on the Isle of Wight. This island has become the most important European locality for silicified specimens.

In 1868 Williamson published his well-known account of *Zamia gigas* from the Jurassic of the Yorkshire coast. By piecing together various fragments of the trunks, inflorescences, and foliage, he was able to reconstruct the entire plant. This species was later made the type of the genus *Williamsonia*. At about the same time Carruthers described *Bennettites Gibsonianus*, from the Lower Greensand (Lower Cretaceous) of the Isle of Wight. In contrast to Williamson's material from Yorkshire, the Isle of Wight cycadeoids are silicified and Carruthers was able to include in his description of *Bennettites* something of the internal structure of the trunks. *B. Gibsonianus* was therefore the first cycadeoid to be studied from thin sections. Soon afterward the previously described cycadeoids from the Isle of Portland were sectioned, and it was discovered that no essential differences existed between the genera *Cycadeoidea* and *Bennettites*. *Cycadeoidea*, having priority, is therefore the valid name for the genus, although *Bennettites* is still frequently used for the species from the Isle of Wight.

The first cycadeoids to receive scientific attention in America were reported by Phillip Tyson from the iron-ore beds of the Potomac formation between Baltimore and Washington in 1860. About 10 trunks were obtained at this time, but 33 years elapsed before any further studies were made. In the meantime, many had been gathered by the local inhabitants of the region who, in ignorance of their identity, called them "beehives," "wasps' nests," "corals," "mushrooms," "beef-maw stones," and various other names. The first species to be described from the collection was *Cycadeoidea marylandica*, of which the external features were figured by Fontaine in 1889. In 1897 Lester F. Ward described 6 additional species based upon a total of 70 or more trunks.

What has turned out ultimately to be the most important American locality for cycadeoids is the Lower Cretaceous rim of the Black Hills of South Dakota. Specimens have been found at a number of places along the rim, but at two localities, one near Hermosa and the other near Minnekahta between Edgemont and Hot Springs, silicified trunks have been collected in large numbers. The trunks were first noted by settlers in the Black Hills in 1878, but no scientific collections were made until 1893. The first specimens collected were sent to Lester F. Ward. The region was subsequently visited independently by Ward and by Thomas H. Macbride, and in October, 1893, Macbride described *Cycadeoidea dacotensis*. In 1899 and 1900 Ward described additional species, which made a total of 28 for the Black Hills region. At about this time George R. Wieland commenced his studies of the cycadeoids, which resulted in important discoveries concerning the floral and foliar characteristics of the group. These studies were possible mainly because of the excellent preservation of the Black Hills trunks. Altogether more than 1,000 specimens have been collected and studied.

A third locality of high ranking importance is the Freezeout Mountain region of central Wyoming where cycadeoids were discovered by O.C. Marsh in 1898. The fossiliferous beds have been assigned to the Morrison formation and are presumably of uppermost Jurassic age, although they are sometimes included in the Lower Cretaceous. These cycadeoids were smaller and less well preserved than those from the Black Hills, although in 1900 Ward described 20 species for which he created the new genus *Cycadella*. All these species, however, have since been transferred to *Cycadeoidea*.

Silicified cycadeoid trunks have been found in smaller numbers in Kansas, Colorado, California, Texas, New Mexico, and elsewhere, but nowhere have they approached the quantities secured from the three principal regions cited. Unfortunately, the once famous localities are now almost entirely depleted of material, and the visitor is fortunate to find even a mere fragment.

The Cycadeoidales consist of two families, the *Williamsoniaceae*, the older and more primitive, and the *Cycadeoidaceae*, the younger and more advanced.

The Williamsoniaceae

This family is characterized chiefly by the slender, often considerably branched stems or trunks, as contrasted with the short, stout trunks of the Cycadeoidaceae. The inflorescences of the Williamsoniaceae are either sessile or stalked, depending upon the genus, but they are never embedded in the thick ramentum of leaf bases and scales as in the other family.

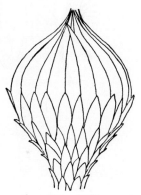

Williamsonia is a large and important genus embracing a few trunks, and a large number of leaves and fructifications. Our knowledge of it has long centered mainly around *W. gigas* from the Lower Oölite (Jurassic) of Yorkshire, and it was of this species that the various parts of the plant were first assembled. The plant, as it was originally restored by Williamson before its real affinities were known, possessed a slender trunk 2 m.

Fig. 119.—*Williamsonia gigas.* Floral bud. (*After Williamson.*)

or more tall, which was ornamented with closely set spirally arranged rhomboidal leaf scars. A dense crown of large, *Zamia*-like, simply pinnate leaves was borne at the top. The fructifications, often called "flowers," were produced among the leaves on long, scaly peduncles. Before the fructification opened it was a globular bud about 12 cm. in diameter (Fig. 119). The outer part consisted of a layer of overlapping, involucral bracts, which enclosed and protected the essential floral parts. No petrified specimens of *W. gigas* are known, and consequently the structure of the "flower" is not as thoroughly understood as one would desire. For more detailed information on the structure of the Williamsonian fructification other species are cited.

The most completely known species of *Williamsonia* is *W. sewardiana* (Fig. 120), from the Upper Gondwana (Jurassic) of the Rajmahal Hills of India. The reconstruction of the plant by Sahni shows a small tree with a sturdy columnar trunk not more than 2 m. high, and bearing at

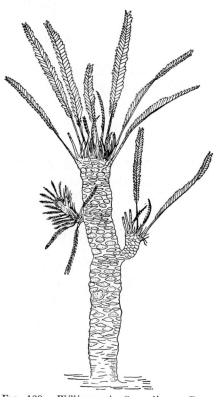

the top a crown of *Ptilophyllum* leaves. It resembles a small specimen of a living *Cycas*. Among the leaves were long pointed scales that formed small scars on the trunk surface among the larger scars of the leaves themselves. The leaf-base armor formed a thick layer around the outside, and projecting through this at places were lateral branches, some of which bore "flowers." These lateral shoots were constricted at the base, which suggests that they broke away and served as propagative shoots. These side shoots bore a few vegetative leaves, but for the most part the covering consisted of bracts and scales. The bracts on the lower regions of these shoots were simple like those on the main stem, but toward the tip those which served as bud scales were finely cut in a pinnate manner. The "flower" consisted of a central receptacle covered with long-stalked ovules and club-shaped interseminal scales. As in other Williamsonian fructifications, the "flowers" of *W. sewardiana* were unisexual, and for this species only the seed-bearing organs are known.

Fig. 120.—*Williamsonia Sewardiana.* Restoration. (*After Sahni.*)

Partial tissue preservation in the flowering shoots of *W. sewardiana* have shown some details of the internal anatomy. There is a central pith about 1 cm. in diameter surrounded by a thin ring of xylem. The rays are numerous and mostly uniseriate, although triseriate and biseriate rays are common. The xylem between the rays is compact, imparting a coniferous aspect to the wood. The tracheids are all scalariform, a feature in agreement with *Cycadeoidea*, although in *Bucklandia indica*, a preserved trunk believed to be the same as that of *W. sewardiana*, multiseriate bordered pits exist. The protoxylem is endarch. The pith and cortex are similar in structure and contain irregularly arranged secretory sacs.

Two other Williamsonian fructifications that have not been found attached are *Williamsonia spectabilis* and *W. scotica*. The former, from the Oölite of Yorkshire, is a staminate "flower" with a whorl of basally fused microsporophylls. These organs become free a short distance upward and curve inward. These free portions bear lateral pinnae that in turn bear synangia. No basal bracts appear to be present. *W. scotica*, from the Jurassic of Scotland, is another of the very few examples of a silicified Williamsonian fructification. It is a seed-bearing "flower" similar in general structural plan to the "flower" of *W. sewardiana*. At the top of the hairy peduncle is the receptacle, which is covered with closely crowded ovules and interseminal scales. Each ovule is shortly stalked and the nucellus is surrounded by an integument which is prolonged into a micropylar beak which projects through the rather firm surface layer of the "fruit" produced by the enlarged tips of the interseminal scales. Some other examples of Williamsonian "flowers," which differ only in detail from those described, a er *W. mexicana*, from southern Mexico, and *W. whitbiensis*, from Yorkshire.

Williamsoniella, also originally described from the Middle Jurassic of Yorkshire, is a plant of more spreading habit than *Williamsonia*. The best known species is *W. coronata*. The stems fork dichotomously, and the inflorescences were borne in an upright position in the angles of the dichotomies. These "flowers" are bisexual, are slightly smaller than those of *Williamsonia*, and lack the basal bracts. When the "flower" opened, the outermost parts consisted of the whorl of 12 to 16 fleshy stamens that bore two rows of pollen sacs on the upper surface, one row on either side of a median ridge. The ovulate receptacle extended upward into a sterile summit against which the tips of the infolded stamens pressed before the "flower" opened. The seeds were sessile and attached directly to the sides of the receptacle between numerous interseminal scales. The "flowers" were produced on smooth slender stalks about 3.5 cm. in length.

The foliage associated with *Williamsoniella* is known as *Nilssoniopteris*. The leaves are linear with a continuous lamina bordering a strong midrib.

Wielandiella, from the Rhaetic of Sweden, resembles *Williamsoniella* somewhat in its dichotomous mode of branching, but its foliage is different and the axillary fructifications are sessile instead of pedunculate.

The foliage is of the *Anomozamites* types, which resembles *Nilssoniopteris* except for the tendency of the lamina to become divided into small unequal segments.   The leaves were borne in dense clusters at the forks of the branches, often surrounding the inflorescences.   The "flowers" were sessile and quite similar to those of *Williamsonia* although they are believed to have been bisexual.

THE CYCADEOIDACEAE

The principal genus of the Cycadeoidaceae is *Cycadeoidea* although many authors continue to use the name *Bennettites* for certain of the

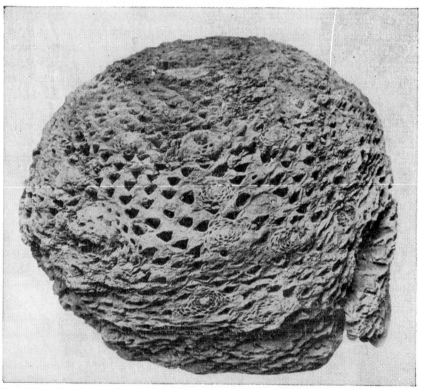

FIG. 121.—*Cycadeoidea dacotensis*.   Lateral view of unbranched trunk showing several "flower" buds.   (*After Wieland.*)   ⅕ natural size.

European species.   However, since these European species show no characteristics which distinguish them generically from those included in *Cycadeoidea*, the latter name is used in this book for all.   Other names have been proposed at times for certain specimens.   The silicified trunks from Freezeout Mountain were named *Cycadella*, and a large one from Germany, now the type of *Cycadeoidea Reichenbachiana*, was

originally called *Raumeria*. These names now have historical significance only, as *Cycadeoidea* has unquestioned priority in both instances.
Navajoia and *Monanthesia* have been proposed for cycadeoid trunks
showing certain distinctive characters in the production of the inflorescences. These, however, have not been satisfactorily set forth as
taxonomic groups, so for the time being they should best be regarded as
no more than subgenera of *Cycadeoidea*.

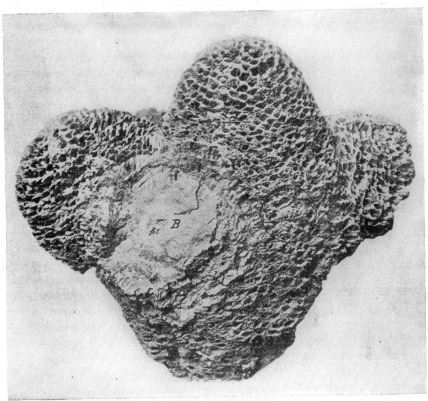

Fig. 122.—*Cycadeoidea Marshiana*.   Lateral view of branched specimen.   (*After Wieland.*)
⅕ natural size.

The members of this family have ovoid or short-columnar trunks
thickly invested by a covering of long persistent leaf bases and multicellular hairs. The "flowers" are lateral on the trunks, and are partially
sunken within the leaf base layer (Figs. 122, 123, 125, 126, and 127).
In contrast to the Williamsoniaceae, the majority of which are preserved
as compressions, most of the known Cycadeoidaceae are silicified, and
as a result, the morphology of both the vegetative and the reproductive

structures have been worked out to a degree of fullness rare among fossil plants.

Full descriptions of the anatomy of the Cycadeoidaceae have been given elsewhere (Wieland, "American Fossil Cycads;" Scott, "Studies in Fossil Botany," Vol. 2) so only a general summary will be given here. Anatomically the various species are much alike and differ mainly in detail. The trunks are subspherical, oval, or columnar, and vary from a few inches to 2 feet in diameter (Fig. 121). Most of them are under 3

Fig. 123.—*Cycadeoidea ingens.* Cross section of trunk. (*After Wieland.*) ⅕ natural size.

feet in height, although some of the columnar forms might have grown to a height of 10 to 12 feet. The short stout trunk stands in marked contrast to the slender often forked axis of the Williamsonians. Some specimens were branched into three or four short trunks all attached in a close cluster to the same basal stalk (Fig. 122).

A cross section of a cycadeoid trunk (Fig. 123) shows in the center a large pith surrounded by a thin zone of wood, which is interrupted by large raylike extensions of pith tissue. At the innermost edge of the wood is the endarch protoxylem. Vertically, the wood strands form a network, with a leaf trace arising at the lower angle of each large ray.

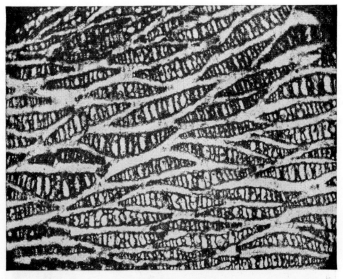

Fig. 124.—*Cycadeoidea sp.* Section of ramentum showing the multicellular hairs. Morrison formation. Bighorn Basin, Wyoming. × about 25.

Fig. 125.—*Cycadeoidea sp.* Surface of trunk showing "flower" buds and leaf bases. Lower Cretaceous. Texas. Slightly reduced.

The secondary xylem consists of scalariform tracheids and small rays, which are uniseriate or biseriate in part. The rays extend across the cambium line into the phloem and thus divide this tissue into alternating masses of parenchyma and vertical conducting tissue. The cortex is parenchymatous and contains gum canals and leaf traces. The trace is single at the point of origin, but in the cortex it divides into several mesarch strands, which become arranged in a horseshoe-shaped curve with the opening at the top. The trace supply in each petiole is derived from a single strand arising from the xylem of the stem. The course of

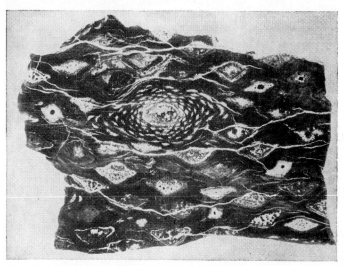

Fig. 126.—*Cycadeoidea sp.* Tangential section through leaf base mantle showing numerous leaf bases and an included "flower." Lower Cretaceous. Black Hills. ⅔ natural size.

the trace is a direct upward slope, there being no curvature around half of the circumference of the stem as in Recent cycads.

Forming a heavy layer on the outside of the trunk is a thick mantle of persistent leaf bases, and packed tightly between them is a ramentum of flattened scalelike hairs. Cut transversely, these hairs appear one cell thick at the margin, but increase to several cells in the central portion (Fig. 124).

Mature leaves have not been found attached to the silicified trunks, but in a number of specimens the young fronds have been observed folded in the terminal bud. In form, venation, and structure these leaves closely resemble those of the recent genera *Macrozamia, Encephalartos,* and *Bowenia.* Their size when mature is unknown, but the frond of *Cycadeoidea ingens* is estimated to have been 10 feet long.

The fructifications of all members of the Cycadeoidaceae were borne on short axillary shoots.   In the subgenus *Monanthesia* there is a fertile shoot in each leaf-base axil.   This subgenus contains at least three species, *Cycadeoidea blanca*, *C. magnifica*, and *C. aequalis*, all from the Mesa Verde formation of New Mexico.   In most species of *Cycadeoidea* only part of the leaves subtend fertile shoots, although in some of the large trunks more than one hundred "flowers" have been counted on a single specimen.   In at least six species all the fructifications on each plant are at the same stage of maturity.   These species are assumed to be monocarpic, that is, they fruited once during a lifetime, and then died. The name *Navajoia* has been proposed by Wieland for these monocarpic species.

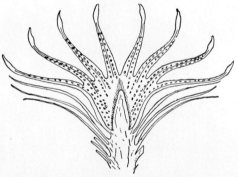

Fig. 127.—*Cycadeoidea sp.* (*C. ingens* type).   Median section of "flower" showing the central ovuliferous cone surrounded by the stamen whorl.   Position of the individual stamens is shown by the small circular scars.   (*After Wieland.*)   Slightly reduced.

The fertile shoot is a stout branchlet that extends laterally from the main trunk into the armor of leaf bases.   These shoots bear a large number (sometimes as many as 100) of spirally arranged, pinnate bracts, which closed over the fructification before maturity.   Similar structures exist in *Williamsonia*, except that in that genus the bracts are simple pointed structures without the ornate construction present in *Cycadeoidea*. Upon maturity of the "flower," the bracts spread and formed a broad saucer-shaped "perianth" surrounding the androecium and gynoecium in the center (Fig. 127).

The apex of the fertile shoot is expanded into the ovuliferous receptacle.   In *Cycadeoidea* (*Bennettites*) *Gibsonianus*, from the Isle of Wight, this receptacle is a convex cushion bearing a cluster of stalked ovules and interseminal scales (*C. Wielandi*, Fig. 128).   In most of the American forms, such as *C. dacotensis* and *C. ingens*, this cushion is more elongated or broadly cone-shaped, and the seeds have shorter stalks and are borne at right angles to the surface (Fig. 129*B*).   Also, in the latter species, the summit and base of the receptacle is sterile, with seeds borne only along the mid-portion.   The orthotropous ovules have long micropylar beaks that project outward beyond the surface of the "fruit" for some distance.   The seeds are small oval bodies only a few millimeters long, which contain embryos with two cotyledons.

The interseminal scales are about as numerous as the ovules.   The tips of the scales are enlarged into clubs that fit into the spaces between

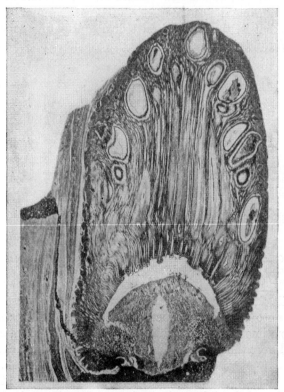

Fig. 128.—*Cycadeoidea Wielandi* (?).   Longitudinal section through ovulate receptacle. (*After Wieland.*)   X about 4.

the seeds.   These tips become fused between the ovules in such a manner as to form a continuous surface layer with openings through which the micropylar beaks project.   The ovuliferous receptacle with its stalked ovules and interseminal scales matured into an oval or elongate "fruit" an inch or two long.   The fused tips of the interseminal scales constituted a sort of "pericarp" that protected the seeds until ripe (Fig. 130*A*).

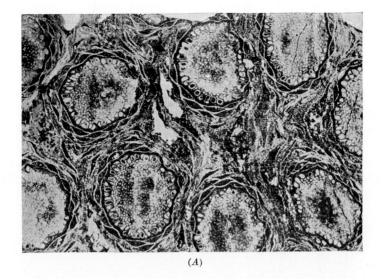

(A)

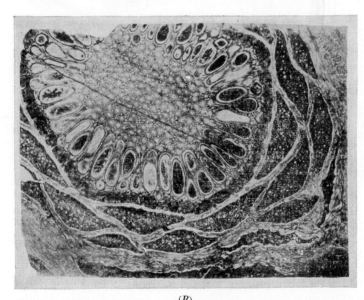

(B)

FIG. 129.—*Cycadeoidea Dartoni.* (A) Tangential section through the leaf base mantle showing several unexpanded "flowers." (*After Wieland.*) Natural size. (B) Transverse section through an unexpanded "flower" showing details of the seed cone. (*After Wieland.*) × about 4

The androecium or staminate disc surrounds the fertile shoot just below the ovuliferous receptacle.   The stamens range in number from 8 to 18 or more, depending upon the species, and are joined basally into a sheath.   The individual stamens are 10 to 12 cm. long when expanded and are pinnate organs bearing 10 or more pairs of pinnae (Fig. 130B). The pinnae at the extreme base and tip are sterile, and the stamen itself terminates in a sterile point.   The lateral pinnae bear two rows of synangia in a sublateral position.   The synangia are shortly stalked, are broader than long, and contain 20 or 30 loculi in two rows.   The synangium wall is made up of palisadelike cells, and dehiscence was by a longitudinal apical slit that caused the organ to open into two valves.

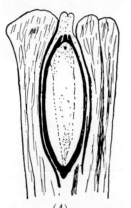
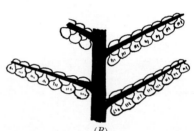

| (A) | (B) |

Fig. 130.—*Cycadeoidea dacotensis.*   (*A*) Longitudinal section of seed flanked by two interseminal scales.   (*After Wieland.*)   Enlarged.   (*B*) Portion of stamen bearing microsporangia.   (*After Wieland.*)   Enlarged.

A slit also developed in the wall of each locule.   As far as we know, the "flowers" of all species of *Cycadeoidea* are bisexual.

FOLIAGE OF THE CYCADEOIDALES

The most abundant remains of the Mesozoic cycadophytes are compressions of the foliage.   Cycadophytic foliage occurs in intimate association with remains of ferns, conifers, scouring rushes, and other plants, often in great quantities.   In Western Europe two localities famous for the abundance of cycadophytic leaf remains are in Scania in southern Sweden, in rocks of Rhaetic age, and in the Jurassic formations of the Yorkshire coast.   Important discoveries have also been made in the Liasso-Rhaetic of eastern Greenland.   On the continent of North America large quantities of leaves have been found in the Triassic of southeastern Pennsylvania and of the Richmond coal field of the Atlantic coastal plain, and in the Chinle formation (Upper Triassic) of Arizona.

Other localities are in the Liassic (Lower Jurassic) of the state of Oaxaca in southern Mexico, in the Jurassic near Roseburg, Oregon, and in the Jurassic at Cape Lisburne in northwestern Alaska. In South Africa cycadophytic fronds occur in the Karroo Series (Jurassic), and in India and Australia in various formations within the Mesozoic.

The problem of interpretation of detached cycadophytic foliage is similar in many respects to that of the fernlike leaf forms of the Paleozoic. Because of the fragmentary condition of the leaves and fructifications, it has been necessary to establish a number of artificial genera for their reception.

Features used in generic separation include shape of the leaf, whether or not the lamina is cut into separate pinnules, the venation pattern, attachment of the lamina to the rachis, serration of the lamina, the outline of the epidermal cells, and the arrangement of the subsidiary cells around the guard cells of the stomatal apparatus. It has been found that epidermal and stomatal characteristics provide clues to relationships not revealed otherwise.

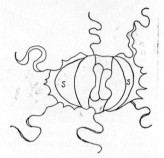

The resemblance between the cycadlike foliage of the Mesozoic and that of modern genera was noted long ago, and at first the fossil types were referred to those Recent genera which they most closely resemble. An outstanding example is the cycadophyte discovered many years ago in Yorkshire and named *Zamia gigas*, but which was ultimately found to belong to the genus *Williamsonia* and not related to any modern member.

FIG. 131.—*Williamsonia pecten*. Stomatal apparatus: *s*, subsidiary cell. (*After Thomas and Bancroft*.)

The fancied or real resemblance between extinct and Recent cycadophytic types is often expressed by such names as *Zamites*, *Diönites*, and *Cycadites*.

The cutinized remains of the epidermis of the fossil cycadophytes have been extensively investigated. Bornemann recorded observations on the structure of the cuticles as early as 1856, and in the years following similar work was done by Schenk and others. Several years ago Nathorst investigated the abundant and well-preserved remains in Scania and described a large number of types. In 1913, Thomas and Bancroft completed an epoch-making study of the material from Yorkshire, and more recently Harris has published a series of reports in which are described numerous cycadophytic types from Eastern Greenland. Florin has embraced in similar studies the entire gymnosperm complex and some of the pteridosperms.

A classification of fossil cycadophyte leaves was proposed by Thomas and Bancroft. The two groups coincide to some extent with the two natural orders of the cycadophytes, and this classification has been adopted with slight modification by subsequent authors.

The two foliage groups proposed by Thomas and Bancroft are the Bennettitales and the Nilssoniales. These so closely parallel the two natural cycadophyte groups that they can be appropriately discussed in connection with them.

For complete descriptions of the leaf genera of the Mesozoic cycadophytes, more extensive works such as Seward's "Fossil Plants," Volume 3, and the recent papers by Harris and by Florin should be consulted. Nevertheless a few of the best-known forms will be briefly discussed here.

FIG. 132.—*Williamsonia (Ptilophyllum) pecten.* Eskearine Series. England.

The constant characteristics of the Bennettitales are (1) the presence on each side of the pair of guard cells of a lateral subsidiary cell formed from the same mother cell as the guard cells, and (2) the heavy cutinization of the outer and dorsal walls of the guard cells. Other features that occur less constantly are (1) the transverse orientation of the stomata at right angles to the veins, (2) the arrangement of the epidermal cells in distinct rows, and (3) the wavy outline of the cell walls (Fig. 131). The most familiar genera are *Ptilophyllum* (Fig. 132), *Zamites*, *Otozamites* (Fig. 133), *Dictyozamites*, *Pterophyllum*, *Anomozamites*, *Nilssoniopteris* (*Taeniopteris*), and *Pseudocycas*.

Ptilophyllum.—This variable leaf type has linear, straight or slightly falcate pinnae with parallel or subparallel veins. The pinnae are characterized by their attachment to the upper surface of the rachis which they almost cover. The pinna base is rounded or slightly arcuate with a broad attachment, although the upper angle is free. One well-known species, *Ptilophyllum pecten* (Fig. 132), is attributed to *Williamsonia*.

Zamites.—This genus does not resemble *Zamia* as the name implies but (depending upon how it is defined) bears a closer resemblance to some of the species of *Encephalartos*, *Ceratozamia*, and *Macrozamia*. However, a close relationship to any of these Recent cycads is not to be assumed. *Zamites* resembles *Ptilophyllum* in possessing linear or lanceolate pinnae attached to the upper surface of the rachis, and with parallel or slightly divergent veins. The main difference is the constricted pinna base. The foliage of *Williamsonia gigas* is of this type.

Otozamites.—This form is separated from *Ptilophyllum* and *Zamites* by the arcuate bases of the pinnae and by the spreading veins. In some forms with long slender pinnae the veins may be nearly parallel, thus making it difficult in some instances to distinguish it from *Ptilophyllum*. *Otozamites* is believed to belong to the Williamsoniaceae. *O. powelli* (Fig. 133) occurs in abundance at some places in the Shinarump and Chinle formations (Upper Triassic) of Arizona, New Mexico, and Utah.

Dictyozamites.—This leaf type resembles *Otozamites* and differs mainly in the anastomosing veins. It is characteristic of the Middle Jurassic, and probably belongs to the Williamsoniaceae.

Pterophyllum.—In this genus the pinnae are linear with almost parallel margins, and the attachment is lateral on the rachis by the full width of the base. *Pterophyllum* resembles *Nilssonia* and distinction between the two is not always possible if epidermal characters are not observed. One peculiarity is that the epidermal cell walls of *Pterophyllum* are usually straight, and not wavy after the manner characteristic of most Bennettitalean leaves. *Pterophyllum* is a widely distributed genus and probably represents the foliage of a number of Mesozoic cycadeoids.

FIG. 133.—*Otozamites Powelli.* Chinle formation, Triassic. Petrified Forest National Monument, Arizona.

Anomozamites.—This leaf is very similar to *Pterophyllum* and is combined with it by some authors. In *Anomozamites* the segments of the lamina are usually less than twice as long as broad, whereas in *Pterophyllum* they are usually more than twice as long. Also, the leaf segments of *Anomozamites* are sometimes of irregular length. The best distinction, however, is in the epidermis. The epidermal cell walls of

Anomozamites are usually wavy and not straight, and the stomata are irregularly disposed. *Anomozamites Nilssoni* is thought to be the leaf of *Wielandiella*.

Pseudocycas (*Cycadites*).—The pinnae of this leaf form possess a single midrib but no lateral veins. Thus they bear a superficial resemblance to the recent *Cycas*. The epidermal cells and stomata, however, are typically Bennettitalean.

Nilssoniopteris.—This name has been applied to leaves of the *Taeniopteris* group, which show Bennettitalean stomata and epidermal cells. (The non-Bennettitalean segregate of *Taeniopteris* is *Doratophyllum*. For leaves in which the epidermis is not preserved, the name *Taeniopteris* is retained.) The leaf is slender with a strong midrib, an undivided lamina, and forked or simple veins that pass to the margin at right angles to the midrib or at a slight upward slant. *Nilssoniopteris vittata* is probably the leaf of *Williamsoniella coronata*.

MORPHOLOGICAL CONSIDERATIONS

The cycadeoids are often called "the Flowering Plants of the Mesozoic," a very appropriate appellation if a close relationship is not implied. Wieland believes that the group did give rise to some of the modern flowering plant types. If this assumption could be proved, it would partly solve the old and perplexing problem of angiosperm origin.

The cycadeoid inflorescence is a strikingly flowerlike structure that is similar in several respects to the flower produced by the modern *Magnolia*. The spiral series of bracts beneath the stamen ring of the cycadeoid may be compared to the calyx and corolla of *Magnolia* in which the number of parts is also large and in spiral sequence. Then the comparatively large number of stamens in the staminate disc finds a counterpart in the numerous stamens of *Magnolia*. Also, the ovuliferous receptacle of *Magnolia* bears a cluster of separate fruits on a cone-shaped receptacle. These floral similarities are accompanied by vegetative structures which are also similar. In *Magnolia* the vessels and tracheids bear scalariform pits, a feature present in all of the Cycadeoidaceae and to some extent at least in the Williamsoniaceae.

These resemblances, as prominent as they are, can be considered as nothing more than analogies brought about during a long course of parallel development. Here we see in the cycadeoid line, which probably had its ancestry among the Paleozoic pteridosperms, a remarkable example of a high degree of specialization of the reproductive organs, a situation which was somewhat analogous to the development of the lepidocarp line in the lycopods. The cycadeoid ovule is strictly gymnospermous, and it lacks the carpel of the *Magnolia*. The so-called "pericarp" of the cycadeoid "fruit" consists of nothing other than the fused heads of the interseminal scales, which give to the mature object a hard outer surface. This is simply a device for the protection of the immature seeds, and is not homologous with the individual pericarps of the numerous fruits of the *Magnolia* ovuliferous receptacle. Furthermore, the cycadeoid stamen is quite frondlike, probably reflecting its filicinean or pteridospermous origin, and it bears little resemblance of any kind to the stamens of a true flower. Then, too, the foliar characteristics and the general habit of the plant supply additional evidence that the cycadeoids and the true flowering plants are not members of the same developmental line. The two groups furnish one of the best examples of parallel development known in the plant kingdom.

Comparisons with Recent Cycads and Ferns.—It is impossible within this chapter to discuss in detail the intricate and far-reaching questions concerning the possible relationships of the cycadeoids and other plants. Brief mention, however, will be made of some of the more obvious points of resemblance between the Mesozoic types and the recent Cycadales. For further details, books devoted more exclusively to plant morphology should be consulted. (Chamberlain, "Gymnosperms: Structure and Evolution).

Mention has already been made of the resemblance between the enfolded fronds of *Cycadeoidea ingens* and those of *Encephalartos, Macrozamia,* and *Bowenia.* Foliage attributed to the entire group of cycadeoids, however, consists of many types which agree more or less in epidermal structure but differ from that of the Cycadales. As to habit, almost as great a range of variation exists throughout one group as in the other. The cycads range from the small species of *Zamia* with their underground stems to *Cycas media,* which is reported to reach a height of 23 m. No known cycadeoid trunks ever reached this size, although some gradation exists between the short stout and the slender branched forms.

In internal structure the trunks of the cycadeoids and those of the Recent cycads show few fundamental differences and many points of genuine resemblance. The greatest single difference is in the course of the leaf trace through the cortex. In the Recent cycads the trace bundle half encircles the stem before it enters the leaf, but in the cycadeoids it follows a direct course. None of the cycadeoid trunks have been found with medullary bundles, such as occasionally occur in the Recent genera, although layers of internal periderm occur sporadically in both groups. The leaf-trace armor is persistent in some genera of the Cycadales, but in others it is pushed off by the periderm that develops beneath.

Were it not for the more fundamental differences depicted by the fructifications and by the epidermal cells and stomata, the slight anatomical differences that may be noted between the two cycadophyte groups would hardly be sufficient to place them in different orders or even in separate families. The differences in the fructifications are too obvious to require amplification, and to attempt to derive the Recent cycad cone from the "flower" of the Mesozoic cycadeoid would be almost as far-fetched as trying to draw a direct homology between the flower of the angiosperms and the cone of the Coniferae. The inflorescence, probably more than anything else, shows that the cycads and the cycadeoids are separate branches of the phylogenetic tree. The closest resemblance is shown by the seeds, which are essentially alike in both groups.

The entire cycadophyte complex shows evidence of pteridospermous ancestry. This is revealed by the general architectural plan of the foliage, the mesarch leaf-trace bundles, the loosely arranged secondary wood with large tracheids and numerous rays, the large pith (with medullary bundles in the cycads), and especially the structure of the seed. Also, it is quite probable that swimming sperms were produced by both pteridosperms and cycadeoids, just as in the Recent cycad order.

THE CYCADALES

As explained earlier in the present chapter, there is ample evidence that true cycads existed contemporaneously with the cycadeoids during the Mesozoic. They are not represented by the fine series of silicified trunks so magnificently displayed by the cycadeoids, but the remains are convincing nevertheless. The evidence of their existence is furnished by fructifications definitely cycadean and by epidermal fragments.

Foliage of the Fossil Cycadales

The foliage of the Mesozoic Cycadales is embraced within the Nilssoniales of Thomas and Bancroft.

The Nilssoniales are distinguished from the Bennettitales by a set of contrasting characters, which, however, they possess in common with many other gymnosperms. These are (1) a ring of several subsidiary cells that originated independently of the guard cells, and (2) no heavy thickenings in the cuticle of the guard cells. Less constantly occurring features are (1) the irregular orientation of the stomata, (2) the epidermal cells not in well-marked rows, and (3) the epideral cell walls, which are usually straight. *Nilssonia* (Fig. 134) is the only genus that conforms strictly to this definition but by allowing certain exceptions in particular instances the group embraces *Ctenis, Pseudoctenis, Doratophyllum,* and *Macrotaeniopteris.*

The same type of epidermal cell and stoma that characterizes the Bennettitales has been found in several instances on fructifications attributed to the Cycadeoidales, and in like manner there is a distinct relation between the Nilssonian frond and other remains undoubtedly cycadalean.

Nilssonia.—This genus is similar to *Pterophyllum* but the lamina is attached to the upper surface of the rachis, and the epidermal structure is of the simple type agreeing with the Cycadales, Ginkgoales, and Coniferales. In some forms the lamina is entire, but in others it is cut into pinnae. On some specimens the leaf is dissected only in the upper portion with the lower part entire.

Nilssonia appears to be the leaf of a cycadaceous plant that bore staminate cones of the *Androstrobus* type and carpellate cones known as *Beania*. Among living cycads the nearest relatives seem to be among the Zamioideae.

Ctenis.—The leaves assigned to this genus resemble those of Recent cycads in that there is a strong rachis bearing broadly attached pinnae with parallel or nearly parallel veins. Several veins enter the base of the pinna from the rachis, and a characteristic feature is the occurrence of occasional cross connections between them. In some species the walls of the epidermal cells are wavy.

Pseudoctenis.—This leaf form resembles *Ctenis* but differs in the absence of vein connections. It agrees macroscopically with *Pterophyllum*. The stomata are of the Nilssonian type although in *Pseudoctenis spectabilis* they are often transversely oriented.

Fig. 134.—*Nilssonia compta.* Jurassic. Cayton Bay. Yorkshire.

Doratophyllum.—This genus has been segregated from *Taeniopteris* on account of the cycadean type of stomata present. The leaf is oblong-lanceolate, and the lamina arises from the sides of the midrib, distinguishing it from *Nilssonia*. The veins fork occasionally but never anastomose. *Doratophyllum* is known only from Sweden and Eastern Greenland in beds of Liasso-Rhaetic age.

Macrotaeniopteris.—Aside from the fact that this genus possesses non-Bennettitalean epidermal cells and stomata, its main characteristic is its size.   The leaf is long-obovate with an entire margin.   Some leaves are as much as 33 cm. broad, but complete leaves are seldom if ever found.   Restorations of *Macrotaeniopteris magnifolia* from the Triassic of Pennsylvania show a group of simple leaves arising from the ground, presumably attached to a rhizome.   Some attain a height of approximately 1 m.   It is more probable, however, that the leaves were borne on upright trunks.

FRUCTIFICATIONS

There are several Mesozoic cone types that have been attributed to the Cycadales but on evidence not always satisfactory or conclusive.

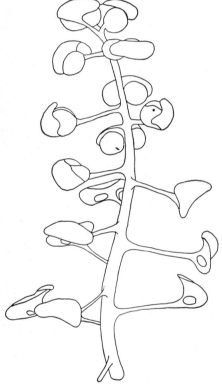

FIG. 135.—*Beania gracilis.*   Restoration of ovulate inflorescence.   The lower mega-sporophylls have shed their seeds.   (*After Harris.*)   Slightly reduced.

Some of them have probably been confused with cones of the araucarian conifers.

Androstrobus.—This cone genus had long been suspected as being the staminate fructification of some unidentified Mesozoic cycad, but satisfactory proof was lacking until some additional specimens were recently discovered in the Jurassic rocks of Yorkshire. The Yorkshire material contains two species, *Androstrobus manis* and *A. wonnacotti.*

Androstrobus manis is a cone about 5 cm. long and 2 cm. in diameter. The spirally arranged microsporophylls consist of a horizontal portion which widens rapidly from the point of attachment, and a distal part which forms a broad rhomboidal scale. The fairly large, finger-shaped microsporangia, measuring about 1.2 by 0.7 mm., are attached to the lower surface, and appear to be arranged in clusters with the dehiscence apertures facing each other. The pollen grains are oval, averaging about 36 by 26 microns in length and breadth, respectively.

Androstrobus wonnacotti is similar to *A. manis,* the essential differences being its smaller size, and the square rather than rhomboidal terminal bract of the microsporophyll.

Aside from the general structural plan, two important resemblances between *Androstrobus* and the staminate cones of recent cycads are the thick walls of the microsporangia and the presence of a cuticle lining the inner cavity. These features are absent from the conifers with which *Androstrobus* might be confused. Furthermore, the stomata on the microsporophylls are characteristically cycadean.

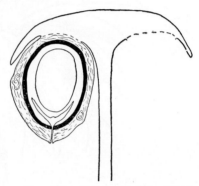

Fig. 136.—*Beania gracilis.* Longitudinal section of megasporophyll and ovule. (*After Harris.*) × about 2.

Beania.—This genus has long been in use for cones organized similarly to the seed-bearing cones of *Zamia.* The structure of *Beania* is clearly revealed in some recently discovered specimens of *B. gracilis* (Figs. 135 and 136) from the Jurassic of Yorkshire. The cone is a loose spike that sometimes attained a length of 10 cm. The seed stalks are spirally arranged at right angles to the axis. The apex is peltate, and on the surface facing the axis are two sessile seeds, one on each side of the stalk. The seeds are oval, and the largest ones measure about 13 by 16 mm. The micropyle is directed away from the base. The seed is constructed similarly to those of Recent cycads. The nucellus is surrounded by a two-layered integument, the inner one being sclerotic and the outer fleshy. A feature not in common with Recent cycads is the presence of a few resin lumps in the fleshy layer of the integument. As nearly as the innermost tissues of the seed can be interpreted, they are identical with those in Recent forms.

Epidermal structures indicate specific identity between the staminate *Androstrobus manis* and the ovulate *Beania gracilis*. It is also believed that *Androstrobus* and *Beania* are the cones of a plant that bore *Nilssonia* leaves. The reasons for this view are (1) that the cones are found in

(*A*)

Fig. 137. (*A*) *Bjuvia simplex*. Restoration. × ⅟₃₀. (*B*) *Palaeocycas integer*, the megasporophyll of *B. simplex*. × ⅟₁₆. Rhaetic of Sweden. (*After Florin.*)

intimate association with the leaves, and (2) that the structure of the epidermis and stomata on the cone scales corresponds to that of the leaves. *Androstrobus manis* is associated with *Nilssonia compta*, and *Androstrobus wonnacotti* with *Nilssonia orientalis*.

The relation of *Androstrobus* and *Beania* to any particular genus of living cycads is not as evident as the mere fact that they are both definitely cycadaceous. *Androstrobus* agrees in all essential features with the staminate cones of any of the living cycad genera. As for *Beania*, the peltate carpel suggests affinity with the *Zamia* group of recent cycads, although the loose structure of the cone is different from any modern genus.

Nothing is known of the trunks on which the above-described organs were borne.

Cycadospadix, from the late Triassic, is a small organ only a few centimeters long, which bears resemblance to the megasporophyll of *Cycas*. Although it has always been regarded as a seed-bearing organ of some kind, no attached seeds have been observed. It is a flattened object with a broadly lanceolate or triangular deeply cut limb that terminates a rather stout pedicel. The structure of the epidermis more closely resembles that of a cycadeoid rather than that of a true cycad, and instead of being a megasporophyll it may be one of the bracts which clothed the stalk of a cycadeoid inflorescence.

Another fossil that is shown by the structure of the epidermis to be definitely cycadaceous is *Paleocycas* from the Rhaetic of Sweden. It resembles *Cycadospadix* except that the broadened terminal expansion has an entire margin, and the slender attachment stalk shows short lateral projections which apparently bore seeds (Fig. 137*B*). Florin has attempted a reconstruction of the plant that bore the *Paleocycas* fructification. His figure shows a straight, unbranched, columnar trunk, 9 or 10 feet high, bearing at the top a crown of *Taeniopteris* leaves. The megasporophylls (*Paleocycas*) are spirally arranged in a dense cluster at the apex of the stem. Whether the staminate organs were produced on the same plant with the seeds is unknown. The name *Bjuvia simplex* has been given to the reconstructed plant (Fig. 137*A*).

THE GINKGOALES

One of the most important achievements of paleobotanical research has been the gradual revelation of the geologic history of the Ginkgoales, of which the Maidenhair tree, *Ginkgo biloba*, is the sole surviving member. The situation has been well stated by Seward: "Evidence furnished by fossils enables us to state with confidence that the *Ginkgo* group of trees— the Ginkgoales—was once almost world-wide in its distribution and comprised many genera and species: a few million years ago this section fell from its high estate and is now represented by a single species."

The *Ginkgo* tree is an object of interest to scientist and layman. Its dichotomously veined, fan-shaped leaves are unique among the trees of temperate climates. This feature, along with the production of motile sperms (a feature shared by the cycads), and the terminally borne seeds constitute a set of primitive characteristics that have earned for the tree the appellation of "living fossil." Anatomically *Ginkgo* resembles Recent conifers, its peculiarities being the double leaf-trace and the lysigenous resin ducts within the leaf tissue.

It is often believed that *Ginkgo biloba* no longer exists in its natural state, and that it has been saved from extinction only by having been cultivated for centuries in the monastery gardens of certain Asiatic religious orders. However, within recent years exploring parties have brought information of it growing apparently wild in remote parts of China. Whether it is genuinely native in these places, or whether it has merely survived from the abandoned culture of long ages ago is quite unknown, but the fact seems well established that the genus has survived in Eastern Asia longer than anywhere else.

The Ginkgoales are ancient plants which reached their developmental climax during the middle of the Mesozoic era. They probably developed from pteridospermous ancestors during the latter part of the Paleozoic in close proximity to the Cordaitales and cycadophytes. Points in common with the pteridosperms are revealed by the structure of the seeds, the swimming sperms, and the double leaf trace. The ovulate fructification of *Ginkgo* is a rather indefinitely organized structure, and abnormally developed specimens sometimes bear ovules on partly developed leaves that are strongly reminiscent of the leaf-borne seeds of certain pteridosperms.

Our knowledge of extinct Ginkgoales is founded almost exclusively upon leaf remains (Fig. 138). Fructifications, except for isolated seeds sometimes found with the leaves, are rare and not well preserved. Leaf characters that are used in classification include outline of the blade, extent and depth of the lobing, presence or absence of a distinct petiole, manner of forking of the veins (whether strictly dichotomous or partly sympodial), the shape of the epidermal cells, and the arrangement of the stomata. The megascopic characters are extremely variable, especially those pertaining to lobing, and accurate specific or even generic determinations are often difficult where tissues are not preserved.

The ginkgoalean leaf is a wedge-shaped or broadly obcuneate organ with an expanded apex and a pointed base. In the broadly expanded forms the inferior margins tend to be concave, but they may be straight or slightly convex in the slender types. In some genera (including *Ginkgo*) the abruptly tapering wedge-shaped base is prolonged into a distinct petiole. In others a petiole is lacking. Many of the leaves of some genera have entire apical margins, but in others a terminal notch divides the leaf into lobes of approximately equal size. The depth of this apical notch varies even on a single branch. The lateral lobes may

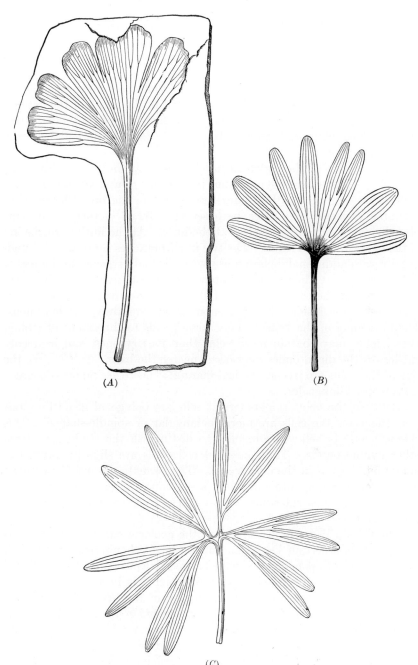

Fig. 138.—(A) *Ginkgo digitata.* Jurassic. Spitzbergen. (*After Heer.*) (B) *Ginkgo siberica.* Jurassic. Siberia. (*After Heer.*); (C) *Ginkgo lepida.* Jurassic. Siberia. (*After Heer.*)

be variously cut, following either a strictly dichotomous or more irregular plan. In the latter case, one of the lobes is usually larger, and it is separated from the adjacent ones by a deeper notch. The central notch, however, is always deepest in normal leaves. Some genera (*Baiera*) are characterized by very deep incisions that divide the lamina into narrow slender segments. The variability, however, in the depth of the notches renders generic separation on these grounds rather uncertain unless more constant characters such as are often expressed in epidermal structure are recognizable.

The venation of the ginkgoalean leaf is strictly dichotomous in some forms but approaches a sympodium in others. Simple dichotomy prevails in these genera with slender leaves, but in the broad leaves considerable variation may often be observed. In all of them, two veins placed side by side enter the leaf at the base. At the initial broadening of the blade each vein forks. The two inner branches continue (although usually with additional dichotomies) with slight divergence to the region of the leaf apex. The two lateral branches may subdivide by dichotomy, or, as is often the case, they may follow the lower margin and give off branches to the inside. Further subdivisions are always dichotomous. The branching of the veins usually corresponds to the extent of lobing. Broad lobes may contain more veins than narrow ones, and in deeply cut leaves the dichotomies may occur nearer the base. In all cases the tips of the veins are free at the leaf margin. There are no cross connections in the Ginkgoales.

Between the veins the epidermal cells are polygonal in outline, but over the veins the cells are more rectangular or spindle-shaped. This characteristic is believed adequate to distinguish the Ginkgoales from other gymnosperms. The epidermal cell walls are slightly wavy, but not as much so as in the cycadeoids. The stomata are usually without definite orientation, although they are frequently in ill-defined rows. From four to seven subsidiary cells form a circle around each slightly sunken stoma.

The Ginkgoales extend as far into the geologic past as the late Permian, although certain Devonian and Lower Carboniferous members of the form genus *Psygmophyllum* have on insufficient evidence been placed within the Ginkgoales. Another fossil which has been responsible for exaggerated ideas concerning the antiquity of the *Ginkgo* line is *Ginkgophyllum hibernicum*, from the Upper Devonian of Ireland. It has large fan-shaped leaves with a deeply cut lamina. It is not essentially different, however, from certain Devonian fernlike fossils, for example, *Archaeopteris obtusa*, which also bears a superficial resemblance to *Ginkgo*. The Carboniferous *Whittleseya*, a campanulate organ probably belonging to the Medullosaceae, was originally classified with the Ginkgoales, and although it has been retained there by some authors of modern textbooks on plant morphology, it has been known for some time not to belong to this group. *Ginkgo biloba* is one of the oldest of living plants, and may indeed be the oldest living genus of seed plants.

Ginkgo and Ginkgoites.—Many authors consistently use *Ginkgoites* for all fossil leaves resembling *Ginkgo* even though they may be indistinguishable from the leaves of the living species. Florin has proposed that *Ginkgoites* be applied to *Ginkgo*-like leaves which may be distinguished from *Ginkgo* by anatomical characteristics, or where the structure is not preserved. This is objectionable in that it makes *Ginkgoites* both a natural genus and a form genus.

Ginkgo (or *Ginkgoites*) leaves occur in rocks ranging in age from late Triassic or earliest Jurassic to the present. Distribution of the genus has been almost world-wide. *G. digitata* (Fig. 138*A*), is a widely distributed Jurassic species with leaves similar in size and shape to those of *G. biloba*. The leaf has a long petiole, and the semiorbicular or obcuneate blade may be either entire or deeply dissected. The deeply cut leaves seem to be quite prevalent, although the extent of variation is probably as great as in the living species. In North America *G. digitata* has been found in Douglas County, Oregon, and at Cape Lisburne, Alaska. Elsewhere it has been recorded in Australia, Japan, Turkestan, Yorkshire, and Franz Joseph Land.

Ginkgo adiantoides is either identical with the living species, or represents a series of Tertiary forms so like *G. biloba* as to be indistinguishable from it on the evidence available. Its leaves are often fairly large, and the lobing, while quite variable, is often somewhat suppressed. In North America it has been found in the Eocene of Alaska, Alberta, Montana, South Dakota, and Wyoming. It occurs in the Oligocene of British Columbia and in the Miocene (though seldom in abundance) in central Oregon and in eastern Washington. It persisted in Western Europe as late as the Pliocene. Since then, however, the genus seems to have been restricted more or less to its present limited habitat in Eastern Asia.

Baiera.—This genus ranges from the Rhaetic to the Lower Cretaceous. It possesses deeply cut leaves in which the central incision extends almost to the base (Fig. 139*A*). Less deep incisions divide the two lateral lobes into slender linear segments each of which contain but two to four parallel veins. The leaf is petiolate, similar to that of *Ginkgo*, but in some of the slender forms the narrowing of the lamina continues so near the point of attachment that a petiole seems almost nonexistent.

Other genera.—There are several other Mesozoic ginkgoalean genera that possess long, slender, ribbonlike, sessile leaves produced in clusters upon scaly dwarf shoots.   One of these is *Sphenobaiera,* which is the oldest of the Ginkgoales, being known from the late Permian.   Other similar ones are *Czekanowskia, Hartzia, Windwardia* (Fig. 139*B*) *Torellia,* and *Phoenicopsis.*   In the last three the leaves are undivided at the apex, and are distinguished from each other mostly upon epidermal and

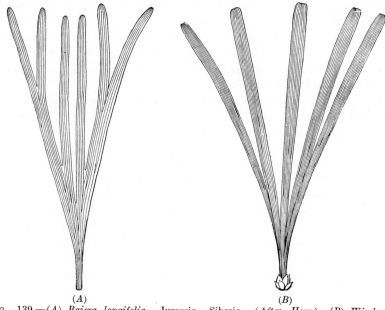

(A)   (B)

Fig. 139.—(*A*) *Baiera longifolia.* Jurassic. Siberia. (*After Heer.*) (*B*) *Windwardia Crookallii.* Lower Cretaceous. Franz Joseph Land. (*After Florin.*) ⅔ natural size.

stomatal characters.   *Torellia* is the only sessile genus to outlive the Mesozoic.   It became extinct, however, in the early Tertiary.

References

Cycadeoidales and Cycadales

CHAMBERLAIN, C.J.: "Gymnosperms: Structure and Evolution," Chicago, 1935.
FLORIN, R.: Studien über die Cycadales des Mesozoikums, *K. svenska Vetensk. akad. Handl.,* III, **12**, 1933.
HARRIS, T.M.: The fossil flora of Scoresby Sound, East Greenland, pt. 2, *Medelelser om Grönland,* **85**, 1932.
———: Cones of extinct Cycadales from the Jurassic rocks of Yorkshire, *Royal Soc. London Philos. Trans.,* **B 231** (577), 1941.
SAHNI, B.: A petrified *Williamsonia* from the Rajmahal Hills, India, *Palaeontologia Indica,* **20**, 1932.
SCOTT, D.H.: "Studies in Fossil Botany," 3d ed., Vol. II, London, 1923.
THOMAS, H.H., and N. BANCROFT: On the cuticles of some recent and fossil cycadean fronds. *Linnean Soc. London Jour.* (2d ser.), Botany, **8**(5), 1913.

WARD, L.F.: The Cretaceous formation of the Black Hills as indicated by the fossil plants, *U. S. Geol. Survey*, 19th Ann. Rept, pt. 2, 1899.

———: Status of the Mesozoic floras of the United States, *U. S. Geol. Survey, Mon.* 48, pt. 1, text, pt. 2, atlas, 1905.

WIELAND, G.R.: American fossil cycads. *Carnegie Inst. Washington Pub.* 34, 1, 1906; 2, 1916.

———: Fossil cycads, with special reference to *Raumeria Reichenbachiana* Goepp. sp. of the Zwinger of Dresden, *Palaeontographica.*, 79, Abt. B, 1934.

Ginkgoales

FLORIN, R.: Die fossilen Ginkgophyten von Franz Joseph Land, *Palaeontographica*, 81, 1936; 82, 1937.

HARRIS, T.M.: The fossil flora of Scoresby Sound, East Greenland, pt. 4, Ginkgoales, etc., *Meddelelser om Grönland*, 112 (1), 1935.

ÔISHI, G.: A study of the cuticles of some Mesozoic gymnospermous plants from China and Manchuria. *Tôhoku Imp. Univ. Sci. Repts., ser. (Geology)*, 12, 1928–32.

SEWARD, A.C.: The story of the maidenhair tree, *Sci. Progress*, 32, 1938.

CHAPTER XI

THE CORDAITALES

The Cordaitales is a group of Paleozoic arborescent gymnosperms, which, together with the seed-ferns, constituted the bulk of the seed plants of the Carboniferous coal forests. The name is here used in the ordinal sense although some authors give it class designation. It is based upon the old genus *Cordaites*, which was proposed many years ago by Unger for certain types of leaves and stems which are almost universally present in Carboniferous plant-bearing rocks. Since the genus was first described, other plant parts bearing different names have been brought together within the order, and Scott and Seward, the most outstanding of modern English paleobotanical writers, have organized them into three families, the Pityeae, the Cordaiteae, and the Poroxyleae. The original genus, *Cordaites*, together with all vegetative and reproductive parts described under different names but believed to be related to it, constitute the family Cordaiteae. The Pityeae embraces a group of genera founded entirely upon vegetative parts, mostly stems and trunks; and the Poroxyleae consists of one genus, *Poroxylon*, which shows certain distinctive features in the structure of the stem. Thus constituted the Cordaitales is an artificial order based principally on anatomical features of vegetative parts. The acquisition of information on the reproductive structures of those genera for which none are at present known might necessitate considerable reorganization within the group.

Although the Cordaitales and the pteridosperms were nearly contemporaneous throughout the later part of the Paleozoic era, they differed markedly in habit and external appearance. Also the Cordaitales show much less evidence of connection with the ferns. Their origin is obscure, and the oldest known members reveal such an advanced state of development that it is certain that they were preceded by a long evolutionary stage which is entirely unknown.

Gymnospermous characters in wood from the early Carboniferous and Devonian rocks were discovered long ago. In 1831, the English paleobotanists, Lindley and Hutton, described silicified tree trunks from the Lower Carboniferous of Scotland as *Pinites Withami*. The largest of these, from the Craigleith quarry near Edinburgh, was 5 feet in diameter and 47 feet long. Its tissue structure was well preserved and since

280

then detailed studies have been made of its anatomy. Another of the early writers on fossil woods was Henry Witham, whose "Observations on the Internal Structure of Fossil Vegetables" (published in 1833) was a classic on the anatomy of Paleozoic woods in the years that followed. Witham figured a variety of structurally preserved woods from Scotland and the north of England, some from the Lower Carboniferous rocks but others from higher series. He redescribed *Pinites Withami* of Lindley and Hutton, and for certain other trunks from two localities on the Tweed in Berwickshire proposed the generic name *Pitus*, a name later changed to *Pitys*. *Pitys* now embraces several species including *Pinites Withami*, which was subsequently transferred to it.

Another wood fragment, the first to be recognized in rocks of pre-Carboniferous age, was found by the Scottish quarryman and amateur geologist, Hugh Miller, in the Old Red Sandstone near Cromarty, Scotland. This fragment, along with others subsequently gathered from the same locality, has been fully described under the name of *Palaeopitys Milleri*. It bears less resemblance to the wood of a modern conifer than Miller believed although it may belong to some early gymnospermous plant.

THE PITYEAE

This family is made up of genera founded mainly upon anatomical characteristics of the trunks. All contain a central pith surrounded by separate strands of mesarch primary xylem, of which some are in contact with the secondary xylem while others (even within the same stem) are embedded at variable distances within the pith. The pith is mainly parenchymatous, but it may contain scattered tracheids, groups of tracheids, or even complete mesarch strands. When a leaf trace departs, one of the circummedullary strands branches, and one branch bends outward to form the trace. The reparatory strand continues upward apparently without the formation of a definite leaf gap. The secondary wood is compact and consists of tracheids and rays. The foliage is known for only one species (*Pitys Dayi*) and we are completely ignorant of the fructifications of any of the members. If more were known of the foliar and reproductive organs of the Pityeae, it is likely that they would show a sufficient number of differences to set the group apart as an order distinct from the Cordaitales, but from the anatomical characters, we conclude that they are probably closer to the Cordaiteae although relationship with the Calamopityaceae is strongly indicated in some of the forms. The retention of the Pityeae within the Cordaitales is therefore provisional pending more information on their fructifications and foliage.

The Pityeae are known only in the Devonian and the Lower Carboniferous and are therefore the oldest of the three families of the Cordaitales. They are classified under the Cordaitales solely on the basis of the structure of the wood. By virtue of the structural resemblance between the woods within the group seed production in the Pityeae is inferred but not proved, and there exists the possibility that some or all of its members might have reproduced by the more primitive cryptogamic methods. The only support that can be brought forth in favor of this is that no seeds or fructifications positively attributable to seed plants

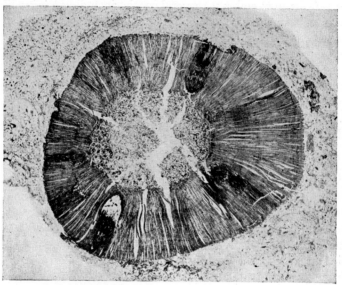

Fig. 140.—*Callixylon Zalesskyi.* Transverse section of the woody cylinder showing the large pith and several outgoing leaf traces. Genundewa limestone, Upper Devonian. Ontario County, New York. × 3.

have been found in rocks older than Lower Carboniferous. Seedlike fructifications have upon a few occasions been reported from the Devonian, but subsequent investigations have always shown them to be either spore-bearing organs or remains of an indeterminate nature. This of course amounts only to negative evidence and is not proof that seed plants were nonexistent during the Devonian. The anatomical evidence that the Pityeae are seed plants seems stronger than the negative evidence that they are not, and in this book they are considered to be such.

The most abundant and widely distributed genus of the Pityeae is *Callixylon*, which occurs throughout the entire vertical extent of the Upper Devonian in the eastern part of North America and to a more limited extent elsewhere. In North America, it occurs in greatest quantity in the New Albany shales of Indiana and in black shales of similar age in Ohio, Kentucky, Michigan, and Ontario. It is also found in the Woodford chert of Oklahoma and the Arkansas novaculite. In New York *Callixylon* is present in the more seaward phases of the great Catskill Delta deposits, ranging from the Genesee shale to the top of the Upper Devonian. It is especially abundant in the Genundewa limestone in the central and western part of the state. The genus was first reported from the Upper Devonian of the Donetz Basin in South Russia.

Callixylon was a large tree. Trunks 5 feet in diameter have been reported from Oklahoma, and numerous specimens 2 or 3 feet in diameter have been found in the New Albany shale. The trunk was tall and straight and bore a crown of slender branches at the top. The habit might have been somewhat like that of a modern *Araucaria* (see Chap. XII).

The wood of *Callixylon* is often found in an amazing state of preservation and it has been possible to study its structure in detail. The pith of a large trunk is often a centimeter or more in diameter, and is sometimes partly septate due to stretching of the tissues during growth. This is a feature shared by *Cordaites*. The innermost wood is a ring of separate mesarch primary xylem strands, which are mostly in contact with the secondary wood, although a few are slightly embedded within the pith (Fig. 141A). The principal rays of the secondary wood widen slightly as they approach the pith thereby dividing the inner part of the wood into wedge-shaped segments at the points of which the primary xylem strands are located. The leaf traces were formed by division of one of these strands. The trace strand bends outward and passes directly into the secondary wood, and the reparatory strand continues without interruption up the stem. The secondary wood grew to considerable thickness and developed weakly defined growth rings. The rays in most species are narrow and of variable height although in some, as *C. Newberryi*, for example, some of the rays may be sufficiently broadened as to appear oval in cross section. The tracheids are thick walled and compactly arranged without intervening wood parenchyma. The radial walls are heavily pitted and the tangential walls are sparingly so. The pits on the radial walls are alternately and multiseriately arranged, but the distribution on the walls is not uniform. The pits are segregated into groups with unpitted spaces between them, and these pit groups in adjacent tracheids are in a regular radial alignment, thus producing a very characteristic appearance in radial section (Fig. 141B). This feature of the pitting is sufficiently distinctive to identify the genus even in fragmentary or poorly preserved material. In no other wood, living or fossil, is this particular kind of arrangement known to exist.

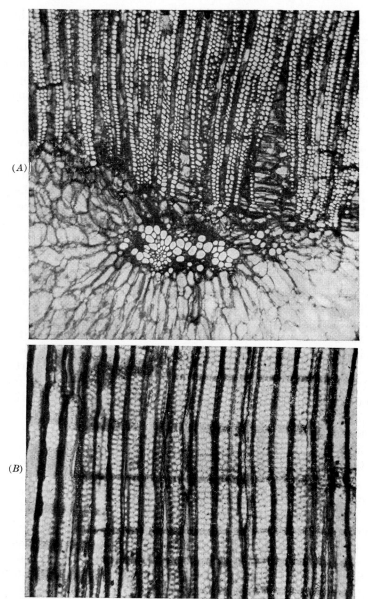

Fig. 141.—*Callixylon Newberryi.* (*A*) Transverse section showing part of the pith, a mesarch primary wood strand slightly immersed in the pith, and the inner part of the secondary wood which is divided into narrow segments by rays which broaden as they approach the pith. New Albany shale, Upper Devonian. Indiana. ✕ about 35.

(*B*) Radial section through the secondary wood showing the regular alignment of the pit groups on the radial walls of the tracheids. New Albany shale, Upper Devonian. Indiana. ✕ about 175.

The number of pits per cluster varies considerably between species and even within the same specimen although the normal variation is from 6 to 12. The pits are mostly hexagonal in outline, and the aperture is a diagonal slit.

Several species of *Callixylon*, mostly from North America, have been described on the basis of the structural differences in the secondary wood. The most thoroughly studied species is *Callixylon Newberryi* which occurs abundantly in the New Albany shale of southern Indiana and Ohio. It is distinguished from the other species of the genus principally by its broader rays which are sometimes as much as four cells wide. In most species the rays are uniseriate or only partly biseriate. *C. Zalesskyi* (Fig. 140) occurs rather abundantly in the lower part of the Upper Devonian in central and western New York, mainly in the Genundewa limestone, and mostly as small twigs only a few centimeters in diameter. Its rays are narrower than those of *C. Newberryi*, but both species contain scattered ray tracheids. *C. erianum*, similar in age to *C. Zalesskyi*, is distinguished by the regularly arranged tracheids in the rather high uniseriate and partially biseriate rays. *C. whiteanum*, from the Woodford chert of Oklahoma, closely resembles *C. Zalesskyi*, but ray tracheids appear to be lacking. *C. whiteanum* is much like *C. Trifilievi*, the Russian species upon which the genus was founded. The rays of the latter, however, are noticeably smaller than in most of the American species.

Callixylon is not known in rocks proved younger than Upper Devonian, and the genus may therefore be regarded as a fairly reliable indicator of Upper Devonian age of rocks that contain it. It appears to have been the dominant forest tree of that period and existed contemporaneously with the fern *Archaeopteris*.

A Lower Carboniferous tree closely related to *Callixylon* is *Pitys*, in which there is also a large pith and mesarch primary wood. The primary wood strands are more numerous than in *Callixylon* and some of them are submerged for a slight distance in the pith. In some species, such as *P. Dayi*, the strands are scattered throughout this tissue. The secondary wood has broad oval rays, and the pitting is continuous and not grouped as in *Callixylon*. Otherwise the wood of the two genera is similar. The presence of rather weakly developed growth rings in the secondary wood of some species of *Pitys* suggests that the trees grew on dry land rather than in swamps.

The discovery of the foliage of *Pitys Dayi* has added something to our knowledge of the affinities of the Pityeae. The so-called "leaves" are fleshy structures about 4 by 6 mm. in diameter and 50 mm. in length.

They resemble phylloclads more than true leaves because no lamina is present, and in transverse section they resemble the petioles of the Lyginopteridaceae. The epidermis bears hairs and sunken stomata. At the ends of the twigs these phylloclads are clustered into buds. They were shed regularly and remained attached for only a short distance below the stem tips. This kind of foliage is quite different from that known to belong to any of the Cordaiteae and it bears more resemblance to that of the araucarian conifers. It therefore tends to separate the Pityeae and the Cordaiteae, and suggests at least a remote connection with the Araucariaceae. The Pityeae and the Cordaiteae were in all probability independent families as early as Upper Devonian time but we know nothing of their history previous to that time.

Archaeopitys, with one species, *A. Eastmanii*, from the late Upper Devonian of Kentucky, is a close relative of *Pitys*. The stems have a large continuous pith with mesarch primary strands scattered throughout, and numerous circummedullary strands at the inner surface of the wood. Some and possibly all of the last mentioned strands divide to form leaf traces, and as the trace passes through the wood it becomes surrounded by secondary wood of its own. The reparatory strand, above the point of departure of the trace, swings inward and becomes a medullary strand. The secondary wood is dense, like that of *Callixylon* and *Cordaites*, and consists of multiseriate pitted tracheids, and rays that are both uniseriate and multiseriate. Nothing is known of the foliage of this genus.

Some of the similarities between the Pityeae and certain genera of the Calamopityaceae should be mentioned. In *Calamopitys*, *Callixylon*, *Pitys*, and *Archaeopitys* no true leaf gap is present in the primary vascular cylinder, but the reparatory strand continues upward without interruption from the point of departure of the trace. In some species of *Calamopitys* and *Pitys* there is a "mixed pith," which is a pith consisting of a mixture of xylem elements and parenchyma, and in certain species of both genera there is a tendency toward complete medullation whereby the primary xylem is reduced to small circummedullary strands. The secondary wood of the calamopityean genus *Eristophyton* is pycnoxylic (dense wooded) much like that of *Callixylon* and *Archaeopitys*: the differences relate only to details, and all are on approximately the same plane of development. The Pityeae as a group are more advanced than the Calamopityaceae, but as they are nearly contemporaneous geologically one cannot be considered the ancestral stock of the other. The term "iceloplasmic" has been proposed for forms such as those mentioned here which may represent terminal members of lateral offshoots and which may appear on structural grounds to represent a phylogenetic sequence, but which because of contemporaneity cannot be regarded as veritable ancestors or descendants. The Calamopityaceae and the Pityeae disappear from the geological sequence at about the same time and they may in some manner be connecting links between some very early stock and the araucarian conifers. This assumption, it should be stressed, is to a great extent hypothetical and requires substantiation.

The possible affinity between the Calamopityaceae and the Pityeae is further emphasized by the recent discovery of two stems, *Megalomyelon* and *Cauloxylon*, in the Reed Springs formation of Mississippian age from Missouri. Both types have paired leaf traces arising from separate bundles, which is a calamopityean feature. They differ, however, in the structure of the secondary wood; that of *Cauloxylon* is more distinctly calamopityean, but in *Megalomyelon* the wood is dense and the rays narrow. A third genus, *Pycnoxylon*, has dense wood and a single leaf trace at the place of origin, but in the secondary wood it divides into two in a vertical plane. The primary xylem strands, which are partly submarginal, are endarch in the vicinity of the nodes. All three forms from the Reed Springs formation possess a large pith and numerous bundles of which *Megalomyelon* and *Cauloxylon* have about 100 each, and *Pycnoxylon* about 45. They all show a combination of pityean and calamopityean characteristics, although *Pycnoxylon*, because of a tendency toward endarchy, is more cordaitean.

The Cordaiteae

Any account of the Cordaiteae centers principally around the genus *Cordaites*, which is known mostly from an assemblage of dismembered parts separately designated under a variety of names. The wood, for example, is known as *Cordaioxylon*, *Dadoxylon*, or *Mesoxylon*. The roots are *Amyelon*, the inflorescenses are *Cordaianthus*, the defoliated branch compressions are *Cordaicladus*, the pith casts are *Artisia*, and the seeds are variously called *Cordaicarpus*, *Cardiocarpus*, *Mitrospermum*, or *Samaropsis*, depending upon the form and method of preservation. *Cordaites* is essentially a genus of the Carboniferous. There is no satisfactory evidence of its existence in the Devonian although it is occasionally mentioned in rocks of that age in the literature. If it survived into the Mesozoic, its post-Paleozoic occurrence is restricted to a few forms. *Dadoxylon Chaneyi*, from the Upper Triassic Chinle formation of Arizona, is a stem that displays many essential cordaitean characteristics such as a large pith, partially biseriate pitting, and numerous and fairly distinct primary wood strands in contact with the secondary wood. The exact affinities, however, are somewhat problematic.

The cordaitean remains most frequently encountered are compressions of the long slender leaves (Figs. 142 and 143). These occasionally reach a length of 1 m. and taper gradually toward the apex and base. Typical examples are linear although they vary from lanceolate to spathulate in form. Some taper to a lancelike point but others are blunt or rounded at the tip. Still other cordaitean leaves are of the *Psygmophyllum* type—broad fanshaped organs split apically into several slender segments. Cordaitean leaves possess no midrib, but a series of fine parallel lines traverse the length of the blade. Attachment was either by a broad vertically compressed base or by a transversely elongated elliptical base. They were borne in spiral sequence on the smaller branches that formed the crown of the tree.

Fig. 142.—*Cordaites principalis.* Nearly complete leaf. Saginaw group, Lower Pennsylvanian. Grand Ledge, Michigan. ½ natural size.

The different species of cordaitean leaves show considerable variation in internal structure. In some (*C. lingulatus*) the palisade and spongy layers are well differentiated, but in others (*C. angulosostriatus*) the photosynthetic tissue is quite uniform. All leaves that have been investigated contain considerable sclerenchyma. In some it occurs as laterally coalescent hypodermal ribs, with the largest ribs above and below the bundles (*C. angulosostriatus*). In other species this tissue accompanies the bundles only, and the mesophyll between them abuts directly against the epidermis (*C. lingulatus*). Separate fibrous strands sometimes alternate with the bundles, and in *C. Felicis,* which is believed to belong to *Mesoxylon,* these strands extend completely across the interior of the leaf from one epidermal layer to the other. The bundles, which are either mesarch or exarch, are completely surrounded by a bundle sheath of thick-walled cells.

The epidermis of the cordaitean leaf is covered with a moderately thick cuticle which is sometimes preserved, and which occasionally reveals the pattern of the epidermal cells and the form and arrangement of the stomata (Fig. 144). In those cordaitean leaves in which the epidermis has been investigated the stomata are distributed in bands lengthwise of the leaf. In some species the stomata within a stomatiferous band are arranged in lengthwise rows, but in others the distribution

Fig. 143.—Inflorescence and foliage-bearing branches of *Cordaites*. (*After Grand 'Eury.*)
Reduced.

is irregular. The stomatiferous bands are separated by nonstomatiferous strips that probably, in most species at least, correspond to the scle-renchymatous strips inside the leaf. However, the relation between the internal and external features of cordaitean foliage has not been investigated in more than a few forms. The relative widths of the stomatiferous and nonstomatiferous bands vary in different species. The epidermal cells of the nonstomatiferous bands are rectangular in surface view, their walls are straight, and they are often arranged in regular series for long distances. The guard cells are surrounded by four to six subsidiary cells. Those lying beside the guard cells often bear a cuticular ridge that arches slightly over the stomatal opening. Cuticular ridges and papillae may be seen on some of the other epidermal cells also. Each stoma has two polar subsidiary cells, which abut on the ends of the guard cells. When the stomata are in definite longitudinal rows and close together, the polar subsidiary cell may serve to connect consecutive stomata. The epidermal cells within

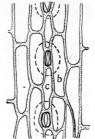

Fig. 144.—*Cordaites sp.* Epidermis: *b*, lateral subsidiary cell; *c*, polar subsidiary cell. (*After Florin.*)

a stomatal band are usually shorter in proportion to their width than those between the bands.

The size and form of the cordaitean leaves led many of the early paleobotanists to place them with the monocotyledons. Some thought they belonged to palms, and the associated circular or heart-shaped seeds were mistaken for palm fruits. Other investigators compared them with the cycads. It was not until 1877, when the French paleobotanist Grand'Eury published his epoch-making work on the French Carboniferous flora, that the question was settled and the true position of *Cordaites* in the plant kingdom was finally decided. Grand'Eury's conclusions

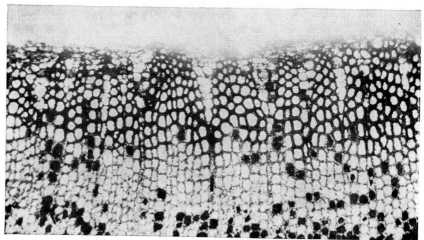

FIG. 145.—*Cordaites michiganensis.* Inner part of the wood showing the narrowing of the woody segments as they approach the pith. Lower Pennsylvanian. Grand Ledge, Michigan. × 50.

concerning the gymnospermous nature of *Cordaites* were soon confirmed by Renault's investigations of the inflorescences.

Cordaitean foliage is often present in large quantities in the shales and sandstones in close proximity to Carboniferous coal beds. It is not unusual to find the surface of large slabs nearly covered with the criss-crossed leaves, but because of their great length they are seldom found complete.

The cordaitean trunk contains a large pith, a relatively small amount of primary wood, and a thick layer of secondary wood. The primary wood, which is inconspicuous, is made up of small tracheid strands situated at the tips of narrow wood wedges formed by a slight broadening of the rays as they approach the pith (Fig. 145). The tracheids of the primary wood are annular, spiral, scalariform, reticulate, and pitted, and the transition zone, especially in the spiral and scalariform region, is relatively broad. There is no sharp demarcation between the primary and the secondary wood.

The primary wood in *Cordaites* is usually regarded as endarch, although in *Mesoxylon* the mesarch condition prevails in those bundles which give rise to the leaf traces. It is believed that the stems of a large number of typical cordaiteans is of the *Mesoxylon* type, although it is manifestly improper to assign any stem to *Mesoxylon* unless the mesarch character of the bundles is demonstrable.

The secondary wood of *Cordaites* (and *Mesoxylon*) was produced as a thick layer around the thin discontinuous primary xylem cylinder and pith. The tracheids are long and slender, and bordered pits cover the

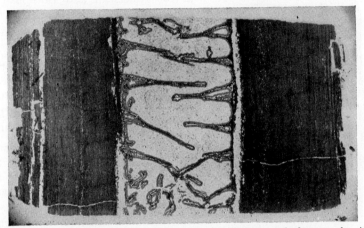

FIG. 146.—*Cordaites transversa.* Radial longitudinal section through the stem showing the pith septations. Coal Measures. England. × 2.

radial walls. The pits may be in one, two, three, or more rows. They commonly show an alternate arrangement and a hexagonal outline due to mutual pressure. The rays are simple structures commonly not more than one cell broad. The ray cells are thin walled and sometimes contain dark resinous material. Xylem parenchyma or resin cells are lacking. Growth rings are seldom present in the coal-swamp members but appear more frequently in the late Permian forms.

The pith is a prominent feature in the cordaitean stem (Fig. 146). Ordinarily it ranges from 1 to 4 cm. in diameter, although specimens have been recorded with a pith 10 cm. or more in diameter. The tissue is entirely parenchyma, and it is not known to contain tracheids or sclerotic cells. Gummy material may sometimes be present. During elongation of the stem tip the pith became transversely ruptured at close intervals resulting in the formation of a series of diaphragms separated by lacunae similar to those of the modern *Juglans*. On the pith casts of *Cordaites* the position of these septations is shown by a series of shallow ringlike constrictions (Fig. 147).

Cordaitean pith casts have long been designated by the name *Artisia*, which was given them by Sternberg in honor of his friend Artis in 1838.

FIG. 147.—*Artisia sp.* Cordaitean pith cast. Lower Pennsylvanian. Alabama. Natural size.

Artis however,, had previously created for them the genus *Sternbergia*,a name since discarded because it designates a Recent genus of flowering plants. *Artisia transversa*, which ranges up to 10 cm. in diameter, is a common form in the American and European Carboniferous. Another so-called "species" frequently mentioned in the literature is *A. approximata*. It is impossible to give satisfactory descriptions of these casts because the size and surface markings are extremely variable and, moreover, a transversely septate pith is not restricted in the plant kingdom to the Cordaiteae.

Petrified cordaitean wood is occasionally found in coal-balls and other types of petrifactions, but when the primary wood is lacking, as it usually is, criteria for satisfactory determination of species are somewhat limited. Nevertheless, many species based wholly upon microscopic details of the secondary wood have been named.

The best example of a structurally preserved cordaitean stem to be found in North America is *Cordaites michiganensis* (Fig. 145) from the Saginaw group of lower Pennsylvanian age in Michigan. The stem portion, which was 16 cm. long and 4 cm. in diameter, was infiltrated with calcium carbonate and iron pyrites. The large pith, which was about 2 cm. across, was replaced with a core of crystalline sphalerite in which no cellular structure had been retained. The septations of the pith were therefore not visible. In a cross section of this stem the secondary wood forms a layer about 1 cm. in extent and is of the type previously described for cordaitean stems in that it consists of compactly placed tracheids with multiseriate bordered pits on the radial walls and narrow rays. As the rays approach the pith, they broaden and thus separate the innermost wood into narrow, wedge-shaped segments. At the inner extremities of these wedges is located the primary wood, which is small in amount. Longitudinal sections show tracheids of the closely wound spiral type near the points of the wedges, and between these and the pitted tracheids are one or two cells showing reticulate wall thickenings. No enlarged trace-forming strands are visible in the transverse section, hence the specimen appears to be a true *Cordaites.*

Cordaites iowensis, from the Des Moines series of Iowa, differs from *C. michiganensis* in having multiseriate bordered pits in the first-formed secondary wood but uniseriately arranged pits in the later wood. Since the primary wood is not well preserved, it is impossible to determine whether this species belongs to *Cordaites* or to *Mesoxylon,* but lacking this one diagnostic feature *Cordaites* is the only appropriate genus for it.

Soon after the middle of the last century Sir William Dawson described several specimens of wood from the Eastern United States and Canada under the name *Dadoxy-lon.* Most of them are undoubt-edly cordaitean but some have been found upon more recent and thorough examination, to belong to other plant groups. The wood that Dawson named *Dadoxylon Hallii,* for example, belongs to *Aneurophyton,* and his *D. New-berryi* exhibits the characteristic pitting of *Callixylon.* An un-doubted member of the Cordaiteae is Dawson's *D. ouangondianum* from St. Johns, New Brunswick. Dawson thought its source was Middle Devonian but later in-

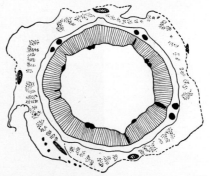

Fig. 148.—*Mesoxylon multirame.* Semi-diagrammatic transverse section of stem showing the large pith, the leaf trace bundles (solid black), the secondary wood, and the cortex with fibrous tissue (stippled). (*After Scott.*) Slightly reduced.

vestigations have shown that the *Dadoxylon* sandstone from which it came is of lower Pennsylvanian age. Our knowledge of *D. ouangon-dianum* is sketchy but Dawson figures the septate pith, which provides fairly satisfactory proof of its affinities.

Additional species of *Dadoxylon* from Pennsylvania, Kansas, Illinois, and Ohio have subsequently been described by other authors. In other continents, species of *Dadoxylon* have been described from Brazil, England, France, Saxony, South Africa, New South Wales, and elsewhere. Our knowledge of many of these woods is in a very unsatisfactory state and a list of species would serve no useful purpose. However, they do show the wide distribution, both geographically and stratigraphically, of large cordaitean forest trees during Carboniferous and Permian times.

Mesoxylon was originally described by Scott and Maslen from the Lower Coal Measures of Great Britain where some half-dozen species have been discovered. The genus has since been reported elsewhere and is probably of wide occurrence, but the obscure anatomical feature that

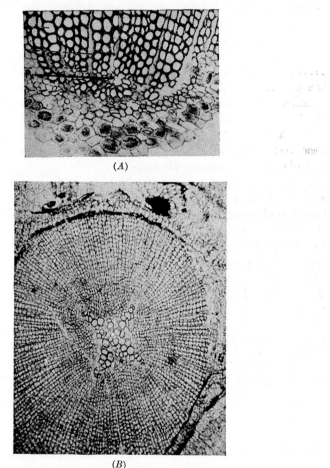

(*A*)

(*B*)

Fig. 149.—(*A*) *Mesoxylon platypodium.* Transverse section of one of a pair of leaf trace bundles at the margin of the pith. *Px*, position of the protoxylem. (*After Maslen.*) ✕ about 50. (*B*) *Amyelon radicans.* Root of *Cordaites* showing the tetrarch primary xylem strand. Coal Measures. Great Britain. ✕ about 15.

distinguishes it from *Cordaites* is probably a reason for its having been frequently overlooked. The calcified stems in which the distinctive features of *Mesoxylon* were first noticed had previously been assigned to *Cordaites* and *Poroxylon.* In the stem the pith is relatively large (as in a typical *Cordaites*) and discoid. The wood is dense, with small tracheids that bear multiseriate bordered pits on their radial walls, and narrow, mostly uniseriate rays.   The generic character lies in the leaf trace bundles.   The trace originates as a single strand along the inner surface of the wood and then divides into two strands.   At the point where these strands originate they are mesarch, thus differing from those of *Cordaites* which according to the earlier authors are "absolutely deprived of centripetal wood."   The outer cortex of the stems is strengthened by a network of sclerenchymatous strands.   The several species differ only with respect to details, some of which may be of questionable constancy.   In *M. Sutcliffii*, for instance, the leaf bases are crowded, completely covering the stem surface.   In other species they are described as less crowded or

(A)   *(B)*   *(C)*

Fig. 150.—*(A) Cordaianthus ampullaceus*. Portion of inflorescence showing the acicular bracts and the axillary scale-covered dwarf shoots. The seeds have become detached. Saginaw group. Grand Ledge, Michigan. About natural size. *(B) Cordaianthus ampullaceus*. Detached seed. Saginaw group. Grand Ledge, Michigan. × 2. *(C) Samaropsis annulatus*. A flattened winged seed with circular wing, apparently belonging to the Cordaitales. Saginaw group. Grand Ledge, Michigan. × 1½.

moderately scattered, and in *M. Lomaxii* and *M. platypodium* they are definitely scattered.   A specific distinction of more fundamental importance has to do with the separation of the trace bundles.   In *M. Sutcliffii* the twin bundles arise at a low level and extend upward as a pair through several internodes before they depart through the secondary wood.   In *M. poroxyloides* the division occurs just below the point where the trace bends outward to traverse the woody cylinder, the lower part being a single strand.   In *M. multirame* (Fig. 148) the situation is somewhat like that in *M. Sutcliffii* except that the twin bundles lie in contact with each other.   In *M. platypodium* the members of the pair are far apart along the pith margin and further subdivisions occur before they leave the wood (Fig. 149*A*).   The leaves of all species receive several trace strands (16 in *M. Sutcliffii*), and in some species axillary buds are present.

M. multirame has been found to bear axillary shoots having the structure of *Cordaianthus*, the inflorescence of the Cordaiteae.

The roots of *Cordaites* are described as *Amyelon*. The central protostele is diarch, triarch, or tetrarch and is surrounded by secondary xylem (Fig. 149*B*). A rather extensive periderm layer was produced deep within the cortex.

The fructifications of the Cordaiteae are lax inflorescenses 10 cm. or more in length borne on the stems among the leaves (Figs. 143, 150*A*, and 151*A*). The defoliated stems often show small round inflorescence scars located a slight distance above the larger and transversely broadened leaf scars. Compressed inflorescenses are frequent but petrified specimens are rare. It is from the latter, however, that most of our knowledge of their structure and morphology is obtained. The main axis of the inflorescence is a slender stalk bearing stiff taper-pointed bracts. Within the axil of each bract is a short budlike strobilus (Fig. 150*A*). In most compressions the strobili and their subtending bracts have the appearance of being arranged in two rows on opposite sides of the inflorescence axis, but petrified specimens sometimes show in cross section four strobili arranged nearly at right angles to each other. The arrangement is probably a spiral or a modified spiral.

The most recently investigated cordaitean inflorescence is *Cordaianthus Shuleri* from the middle Pennsylvanian of Iowa and Kansas. The individual strobili are short, being only slightly more than 1 cm. long and about one-half as wide. The strobilus consists of a short woody axis or dwarf shoot bearing spirally arranged overlapping appendages, some of which are fertile. The fertile appendages bear terminal ovules or pollen sacs, but never, as far as known, both within the same inflorescence. The strobili of the male inflorescences contain one to six or more fertile bracts (microsporophylls) each of which supports as many as six terminal elongated pollen sacs. Each female strobilus usually contains one to four ovuliferous appendages (megasporophylls). In some species the megasporophylls are forked and bear two terminal flattened ovules. The male and female inflorescences are always similar in appearance, and it is difficult to distinguish between them unless pollen sacs or ovules are present. In the young stages the fertile appendages are short and retain the ovules between the closely appressed sterile appendages on the dwarf shoot. In mature specimens the fertile appendages have elongated so that the ripened seeds are suspended out beyond the bracts on slender stalks.

Until recently most of our knowledge of the internal organization of the cordaitean strobilus was based upon silicified French material from the late Carboniferous. In the French forms the seed pedicels were described by Renault as being in an axillary position, but it is now believed that this original interpretation was wrong. The situation is well shown in *Cordaianthus fluitans* and the afore-mentioned *C. Shuleri*, which bear their fertile appendages in the same spiral with the sterile ones.

The pollen grains of *Cordaites* are globoid bodies usually measuring less than 100 microns in their greatest dimension. As with pollen grains and heavily walled spores generally, the wall consists of an intine and a heavily cutinized exine. The exine is inflated into a single large air sac

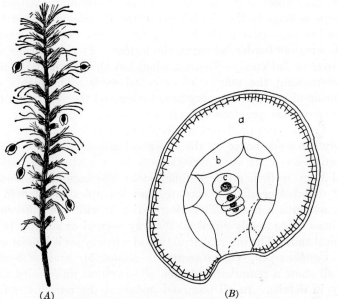

(A)   (B)

Fig. 151.—(*A*) *Cordaianthus sp.* Middle Pennsylvanian. Maryland. (*After Berry.*) Reduced. (*B*) *Cordaianthus sp.* Pollen grain: *a*, air cavity of exine; *b* and *c*, cells of the internal cavity. (*After Florin.*) × 710.

that completely surrounds the intine except for a small area at one side of the grain. This air sac rendered the pollen grain buoyant and facilitated dispersal by wind. Well-preserved fructifications sometimes show developmental stages. The cavity of the young grain is lined with a layer of prothallial cells which at first surrounded a large cell within which nuclear division took place to produce the potential spermatogenous cells. Four or five spermatogenous cells may be seen in some specimens. The pollen grain is essentially an antheridium similar to that produced by modern gymnosperms (Fig. 151*B*). It is probable that pollen tubes were produced when the pollen germinated, although fertilization was apparently carried out as in Recent cycads.

Most of the seeds of the Cordaitales are of the "Cardiocarpales" type. The cardiocarps are flattened, heart-shaped seeds in which the nucellar portion is surrounded by an integument of which the outer layer is laterally expanded into an encircling wing (Figs. 150*B* and *C*). These winged seeds are often associated in considerable numbers with cordaitean foliage and in a few instances they have been found attached to the inflorescences. It should be noted that in the Cordaiteae both the seeds and the pollen grains are winged, a fact that stresses the important role of wind in the life cycle of the plant.

The only absolutely positive means of determining whether a given seed type belongs to the Cordaiteae or the Pteridospermae is to find it attached to vegetative parts, but in general those flattened seeds with a narrow winglike border belong to the former. Pteridospermic seeds are often oval or flattened, without a wing, but this is not an infallible rule for determining the affinity of detached seeds. The highly constant association of seeds with a very broad wing and the vegetative fronds of *Megalopteris* has previously been mentioned. Furthermore, the full range of seed types produced by the Cordaiteae is unknown. Morphologically the seeds borne by the two plant groups are quite similar.

Cordaitean seeds vary considerably in size although they seldom exceed 3 cm. in their greatest dimension. The base of the seed is commonly rounded or slightly cordate, and the apical portion in some is drawn out into a beak. The wing varies in width in different forms, sometimes being nearly equal all the way around or slightly broader at the apical end. The internal structure of a few types is known, examples being *Cardiocarpon*, *Cycadinocarpus*, *Diplotesta*, and *Mitrospermum*. They all show a remarkable degree of structural uniformity and differ mainly in details. In all observed instances the nucellus is free from the integument except at the base, and the vascular supply is double, as in *Trigonocarpus*. The nucellar beak of the pollen chamber projects upward into the micropyle. It is at once evident that the resemblance between the seeds of the pteridosperms and the Cordaiteae is close. There is, however, in the Cordaiteae a tendency toward simplification, and the elaborate lobes and frills sometimes present in pteridospermic seeds are lacking. As explained in the discussion of the pteridosperms, cordaitean seeds are constructed according to the same general plan as those of the Recent cycads. There is a single integument with a more or less fleshy outer layer, a sarcotesta, and an inner firm layer, the sclerotesta. Fertilization apparently was accomplished by swimming sperms. There is no evidence of fertilization by a pollen tube in any Paleozoic seed. If a pollen tube was present, it probably functioned as an absorbative organ as in Recent cycads and the Ginkgo.

THE POROXYLEAE

The Poroxyleae contains one genus, *Poroxylon*, with two species from the Permo-Carboniferous of France. Since nothing has been added to our knowledge of this genus within recent years it will be treated briefly.

The stem of *Poroxylon* has a large pith surrounded by a ring of exarch primary xylem strands that is the most distinctive feature of the genus. These bundles are in contact with the secondary wood, which, unlike that of *Cordaites*, is made up of large cells and broad rays. In this respect it is almost identical with *Lyginopteris*. The leaves closely resemble those of *Cordaites*, and seeds known as *Rhabdocarpus* have been found in association with the leaves and stems.

AFFINITIES OF THE CORDAITEAE

The Cordaiteae represent a state of comparatively high development among Paleozoic plants, and phylogenetically they occupy a position that would be approximately intermediate between the Recent cycads and conifers. This does not imply, however, that they are the direct ancestors of the conifers or were derived from cycadaceous predecessors. They were on the same approximate level as the modern *Ginkgo* with which they show several traits in common.

Structural characteristics of the Cordaiteae, which are also shown by the cycads, are the large pith, the abundant sclerenchyma in the leaf, and the relatively simple and fairly large seeds. It is probable that they produced motile sperms. All these seem to be primitive features which both groups have retained from some early but as yet undetermined ancestor.

Points of resemblance between the Cordaiteae and the Ginkgoales are the probable motility of the cordaitean sperm and the double leaf trace. There is also some resemblance between the leaf of the two groups, and mention has already been made of the *Psygmophyllum* type believed to have been borne by some cordaitean trunks. The Recent *Ginkgo* exhibits a number of characteristics that may have been retained from some ancient form ancestral to both the Ginkgoales and the Cordaiteae. The motile sperms produced by the cycads and ginkgos apparently were shared by all Paleozoic seed plants but fertilization by means of the pollen tube was probably a later development.

Between the Cordaiteae and the Pteridospermae the similarities are not as conspicuous as the differences. The strongest bond between them is the structure of the seeds, which, as heretofore explained, are not readily separable. On the other hand they were borne differently by the two groups because nothing resembling a seed-bearing inflorescence is known for any of the seed-ferns. Then there is little in common between the foliage of the two groups, and numerous modifications would be required to produce a cordaitean leaf from the pteridospermic frond. As to the anatomy of the stems, a large pith is the rule in both groups. *Mesoxylon* has retained a certain amount of centripetal wood and the leaf trace is a double strand. *Poroxylon* probably comes nearest to linking the Pteridospermae and the Cordaitales because it has *Lyginopteris*-like secondary wood, double traces, and exarch primary wood. Except in such genera as *Eristophyton*, the secondary wood of the Pteridospermae is made up of large tracheids and rather broad numerous rays (monoxylic), which contrast rather sharply with the denser (pycnoxylic) structure of the Cordaiteae, Pityeae, and the conifers. Multiseriate bordered pits cover the tracheid walls of the Pteridospermae and most of the Cordaiteae.

The most apparent of the resemblances between the Cordaiteae and the modern conifers is the lofty arborescent habit, a feature shared by the Pityeae and possibly also by the Poroxyleae. As none of the pteridosperms are known to have become large trees, the Cordaitales are the largest of the known Paleozoic seed plants. Although the internal anatomy of a cordaitean leaf may more closely resemble that of a cycadean pinnule than a typical coniferous leaf, there is some external resemblance to the leaves of the eastern Asiatic *Podocarpus* and the araucarian genus *Agathis*. The leaves of these genera are the largest of the conifers, sometimes reaching a length of 29 cm. in *Podocarpus* and 17 cm. in *Agathis*.

The presence of multiseriate bordered pits on the tracheid walls and the absence of resin canals from the wood of the Cordaiteae has long been looked upon as prima-facie evidence that the Cordaiteae are more closely related to the araucarian conifers of the Southern Hemisphere than to the Pinaceae and the other northern families. These similarities, coupled with the fact that wood showing araucarian characteristics (*Araucarioxylon*) is the predominating form of coniferous wood throughout the Mesozoic, are mainly responsible for the prevailing views of araucarian antiquity and theories that the araucarians are cordaitean derivatives. However, the additional light that has been thrown on the morphology of the cordaitean inflorescence during the past two decades has directed attention to possible homologies between this organ and the abietinean[1] cone, homologies which receive support in a series of

---

[1] The Abietineae, or Abietaceae, depending upon whether the group is given subordinal or family status, includes those genera placed by Pilger in the Pinaceae in Engler and Prantl, "Die natürlichen Pflanzenfamilien," Vol. 13, 1926. They are *Abies, Cedrus, Keteleeria, Larix, Picea, Pinus, Pseudolarix, Pseudotsuga,* and *Tsuga*.

connecting links supplied by the Paleozoic coniferous genera *Lebachia*, *Ernestiodendron*, and *Pseudovoltzia*. These genera are described on a later page.

References

ARNOLD, C.A.: On *Callixylon Newberryi* (Dawson) Elkins et Wieland, *Michigan Univ. Mus. Paleontology, Contr.*, **3**, 1931.

————: Cordaitean wood from the Pennsylvanian of Michigan and Ohio, *Bot. Gazette*, **91**, 1931.

BERTRAND, C.E.: Le bourgeon femelle des Cordaites, d'après les préparations de Bernard Renault, *Soc. Sci. Nancy Bull.*, **58**, 1911.

CHAMBERLAIN, C.J.: "Gymnosperms: Structure and Evolution," Chicago, 1935.

CRIBBS, E.J.: *Cauloxylon ambiguum* gen. et sp. nov., a new fossil plant from the Reed Springs formation of Missouri, *Am. Jour. Botany*, **26**, 1939.

DARRAH, W.C.: The flora of Iowa coal balls. III. Cordaianthus. *Harvard Univ. Bot. Mus. Leaflets*, **8**, 1940.

FLORIN, R.: On the structure of the pollen grains in the Cordaitales, *Svensk. bot. tidskr.*, **30**, 1936.

————: The morphology of the female fructifications in *Cordaites* and conifers of Paleozoic age, *Bot. Not. Lund*, 1939.

GORDON, W.T.: The genus *Pitys*, Witham, emend., *Royal Soc. Edinburgh Trans.*, **58**, pt. 2, 1935.

GRAND'EURY, C.: Mémoire sur la flore carbonifère du département de la Loire et du centre de la France, *Mém. l'Acad' Sci.* Inst. France, **24**, Paris, 1877.

MASLEN, A.J.: The structure of *Mesoxylon Sutcliffii*. *Annals of Botany*, **25**, 1911.

————: The structure of *Mesoxylon platypodium* and *Mesoxyloides*, *Annals of Botany*, **44**, 1930.

SCHOUTE, J.C.: La nature morphologique du bourgeon féminin des *Cordaites*, *Rec. trav. bot. Néerlandais*, **22**, 1925.

SCOTT, D.H.: The structure of *Mesoxylon multirame*, *Annals of Botany*, **32**, 1918.

————: The fertile shoot of *Mesoxylon* and an allied genus, *Annals of Botany*, **33**, 1919.

————: "Studies in Fossil Botany," Vol. 2, London, 1923.

————, and A.J. MASLEN: On *Mesoxylon*, a new genus of Cordaitales, *Annals of Botany*, **24**, 1910.

SEWARD, A.C.: "Fossil Plants," Vol. 3, Cambridge, 1917.

CHAPTER XII

THE ANCIENT CONIFERS

The conifers are woody, naked-seeded plants, which often grow to a large size. In vegetative features such as habit, the entire leaves, and the extensive development of secondary wood, many of them show a decided resemblance to the Cordaitales, although differences may be noted in the structure of the reproductive organs, the manner in which the seeds are borne, and in the somewhat lesser amounts of primary wood. Nevertheless, the resemblances are sufficient to suggest some degree of relationship. In most conifers the leaves are needlelike or scalelike, and differ markedly from those of the Cycadales and Ginkgoales. An additional feature that sets the conifers apart from their more pteridospermlike contemporaries is the production of nonmotile sperms at the time of fertilization. The fructifications of most conifers are unisexual conelike inflorescences in which the seeds are borne on spirally arranged cone scales which in most genera are hard and woody. The pollen is produced in pollen sacs attached to the lower surface of stamens arranged in spiral sequence on the axis of the staminate cone.

The geological history of the conifers begins with the Upper Carboniferous (Pennsylvanian) epoch, and the order probably reached its developmental climax in the late Jurassic or early Cretaceous. During the late Mesozoic it began to decline, probably as a result of more successful competition on the part of the angiosperms whose phenomenal rise to dominance took place rather abruptly during the early and Middle Cretaceous. Since late Mesozoic times the decline of the conifers has gone on slowly but steadily. Certain genera and species, as the sequoias and the Monterey cypress, are probably making their last stand in restricted localities along the Pacific Coast, and some sudden climatic or other environmental alteration would probably bring about their complete elimination. Others, such as the pines, spruces, junipers, and firs seem well established and apparently are not facing immediate extinction from natural causes.

THE PALEOZOIC CONIFERS

The conifers of the Paleozoic are known to us mostly from compressed foliage-bearing twigs, inflorescences, seeds and pollen, all of which are conveniently although somewhat arbitrarily grouped into the family Voltziaceae. The members of this family are woody plants with small spirally arranged leaves of somewhat araucarioid appearance and rather small loosely constructed conelike inflorescences, which are essentially intermediate between the inflorescences of the Cordaitales and the cones of the Pinaceae. The group ranges from the early or middle Pennsylvanian into the Jurassic.

The most fully investigated Paleozoic coniferous genus is *Lebachia*, which was recently proposed by Florin for parts belonging to the form genera *Walchia*, *Walchiostrobus*, *Walchianthus*, *Gomphostrobus*, *Cordaicarpus*, and *Pollenites*. Leaf-bearing twigs in which stomatal characteristics are not determinable are still assigned to *Walchia*. *Gomphostrobus* is the name given to certain bifurcated leaves, which are found on both the fertile and vegetative parts, and *Walchiostrobus*, *Walchianthus*, *Cordaicarpus*, and *Pollenites* refer to seed-bearing cones, staminate cones, seeds, and pollen, respectively.

Lebachia and *Walchia* occur sparingly in the middle and late Pennsylvanian of North America but become more abundant in the early Permian where they are characteristic fossils of the "Red Beds" of Kansas, Oklahoma, Texas, and New Mexico, as well as of beds of similar age at other places. In contrast to the widely spread swamp conditions that prevailed throughout the greater part of the Pennsylvanian, the climate indicated by *Lebachia* is believed to be more arid. Its first appearance in the Pennsylvanian is in strata containing mud cracks and casts of salt crystals. The genus did not become widely distributed during the Pennsylvanian, but with the advent of the Permian and the gradual disappearance of the swamps, coupled with the decline of the swamp flora, *Lebachia* and some of the other members of the Voltziaceae expanded widely beyond their original restricted ranges to occupy a more prominent place.

Lebachia was a tree of unknown size with a straight slender trunk. Its pinnately arranged branchlets bore small needlelike leaves only a few millimeters long (Fig. 152). These leaves are spirally arranged, imbricate, decussate, spreading or appressed, and tetragonal in cross section. On the larger branches the leaves are longer and sometimes bifurcate. There are four bands of stomata, two long ones on the ventral surface and two shorter ones on the dorsal side. Occasionally the dorsal surface is without stomata. The stomata are irregularly distributed within these bands, and in this respect *Lebachia* differs from *Ernestiodendron*, a related genus, in which the bands consist of a single row of openings.

The cones of *Lebachia* are cylindrical or ovoid inflorescences borne singly at the tips of the ultimate branches (Fig. 152B). Seeds and pollen were borne within separate cones. In *Lebachia piniformis,* the most thoroughly investigated of the Paleozoic conifers, the cone axis bears spirally arranged bifurcated bracts which in the detached condition are known as *Gomphostrobus,* and in the axils of which there are short fertile shoots. These fertile shoots in turn bear several spirally arranged scalelike appendages of which one is fertile with a single terminal ovule (Fig. 153). This fertile appendage is in the same spiral sequence with

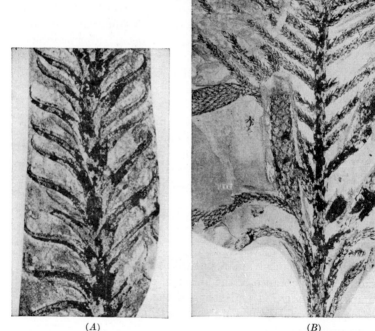

(A) (B)

Fig. 152.—(*A*) *Lebachia Goeppertiana.* Lower Rothliegende. Saar, Germany. (*After Florin.*) × 5. (*B*) *Lebachia piniformis.* Penultimate branch bearing pinnately arranged ultimate branches and cones. Lower Rothliegende. Lodève, France. (*After Florin.*) About ⅓ natural size.

the sterile ones. The seed cone of *L. piniformis* is therefore similar in construction to that of *Cordaianthus,* the essential difference being that the number of ovules on each axillary fertile shoot is reduced to one. The seeds are flattened and are of the *Cordaicarpus* type.

Rather little is known of the pollen-bearing organs of *Lebachia,* but they seem to be of similar construction to those which bear seeds. The pollen grains are similar to those of *Cordaianthus.*

The vegetative parts of *Ernestiodendron* are identical with those of *Lebachia* except for the single row of stomata making up the stomatal band, and before the discovery of this feature this genus was included with *Lebachia* in the form genus *Walchia*. *Ernestiodendron* contains one species, *E. filiciforme*, which is rather widely distributed in the Permian. Its inflorescences resemble those of *Lebachia* when viewed externally, but a careful study of the seed-bearing apparatus reveals an important step in the evolution of the coniferous inflorescence. In this genus the axillary dwarf shoot is a fan-shaped structure trilobate at the apex, and each lobe bears a seed on its adaxial side. This fan-shaped organ without separate seed-bearing stalks is looked upon as homologous with the appendage-bearing dwarf shoot of *Lebachia* and *Cordaianthus*, having reached this stage of relative simplification through the process of fusion and reduction of parts.

Pseudovoltzia is a genus from the upper Permian of Europe. The leaves on the ultimate branches are spirally arranged and dimorphic. Those at the ends of the branches are long, linear, flattened, and obtuse, whereas those on the older portions are shorter, slightly incurved, and subtetragonal in cross section. The stomata are situated on both surfaces in numerous parallel rows with each row only one stoma wide. The ovuliferous cones bear spirally disposed bracts as do those of *Lebachia* and *Ernestiodendron*, but in the axil of each there is a flattened five-lobed appendage that bears two or three inverted ovules on the adaxial side (Fig. 154).

Lecrosia is a Paleozoic coniferous genus in which the branching is irregular in con-

Fig. 153.—*Lebachia piniformis.* Adaxial side of fertile dwarf shoot removed in its entirety from the axil of its subtending bract. The large oblate ovule has a deep apical notch and the two lateral lobes form the expanded wings of the mature seed. (*After Florin.*) × about 12.

trast to the pinnate arrangement of the genera just described. The leaves on the ultimate branches are needlelike and undivided. Two species are known, *L. Grand'Euryi*, from the late Carboniferous of France, and *L. Gouldii* from the Maroon formation of Colorado. In *L. Gouldii* the penultimate branches bear divided (*Gomphostrobus*) leaves. The fructifications of this species are unknown, but in *L. Grand-'Euryi* there are terminal ovoid cones made up of numerous overlapping scales, which appear to have borne two seeds each.

Ullmannia is a late Permian and early Mesozoic genus, which is frequently confused with other members of the Voltziaceae. The spirally arranged needlelike leaves are crowded and often keeled. The stomata are in numerous rows only one cell wide. In *U. Bronnii*, from the Upper Permian of Central Europe, the axillary fertile appendage complexes are completely fused into broadened or fan-shaped cone scales, which bear a single inverted ovule on the adaxial side.

F I G . 1 5 4 .— *Pseudovoltzia Lie-beana*. F e r t i l e dwarf shoot show-ing five terminal sterile appendages and below them two opposite seed pedi-cels. (*After Florin.*) Enlarged.

A few other coniferous types have been described from the late Paleozoic but those mentioned in the foregoing paragraphs are best known. Brief mention should, however, be made of *Araucarites Delafondi*, from the Lower Permian of France, which is often cited as the oldest known member of the Araucari-aceae. The material consists of broadly triangular cone scales, which are cuneate or truncate at the base and rounded with a small median indentation at the apex. In the middle is a shallow oblong depression that probably marks the attachment of the seed.

Evidence derived from studies of the Cordaitales and the Paleozoic Coniferales indicates that there were three overlapping stages in the evolutionary history of the higher gymnosperms of the Paleozoic. The oldest of the early members is the Upper Devonian pityean genus *Calli-xylon*, which was followed in the Upper Devonian and Lower Carboniferous by the related genera *Pitys* and *Archaeopitys*. The second stage began with the appearance of the Cordaiteae in the Lower Carboniferous. The Cordaiteae held forth during the remainder of the Paleozoic, but during or possibly before the middle of the Upper Carboniferous there appeared in some remote localities away from the swamps the first conifers (Voltz-iaceae). This marks the third stage. The course of the evolution of the conifers during the Paleozoic is not well understood, but it is believed that the group developed rapidly during the Triassic and Jurassic when some of the modern families became segregated, although it is not possible to trace many of the Recent genera much farther into the past than the later stages of the Mesozoic. In fact, only a small number of modern coniferous types are known with certainty to have existed before the Tertiary, and it is difficult to identify any modern genera in rocks older than the Lower Cretaceous. Looking backward beyond the early Cretaceous the various generic types tend to merge into a maize of inter-mediate forms among which distinct lines cannot be explicitly traced. These intermediate forms are aptly called the "transition conifers" because they embody characteristics that at present are expressed by distinct genera. The interpretation of these transition conifers remains one of the major problems of paleobotany.

MESOZOIC AND CENOZOIC CONIFERS

The coniferous remains of the Mesozoic and Cenozoic consist of a varied assortment of compressed twigs, foliage, seeds, cones, cone scales, and petrified and lignitized wood. Coal and carbonaceous shales often yield pollen in quantities.

TAXACEAE

The name *Taxites* has been applied to twigs bearing spirally arranged, distichously placed linear leaves from rocks ranging from the Jurassic to the Recent. Many so-called *Taxites* could, however, with equal propriety be referred to *Sequoia* or the artificial genus *Elatocladus*. No undoubted Taxaceae are older than the Tertiary. The use of the name *Taxites* is therefore not an accurate indication of the occurrence of the family during the past. Examples of some plants that have been more or less arbitrarily assigned to the Taxaceae are *Taxites longifolius* from the Rhaetic of Sweden and *T. brevifolius* from the Jurassic of Great Britain. *Taxus grandis*, from the Tertiary, shows epidermal structure similar to *T. baccata*, and may therefore be looked upon as the oldest authentically identified fossil species.

Several twigs from the Upper Jurassic, Cretaceous, and Tertiary have been referred to *Torreya* (*Tumion*) on evidence that is not conclusive. A few of these include *Torreya caroliniana* from the mid-Cretaceous of North Carolina, *T. Dicksoniana* and *T. parvifolia* from the Lower Cretaceous of Greenland, and *T. borealis* from the Eocene of Greenland.

Petrified wood with spiral bands inside the pitted tracheids is usually assigned to *Taxoxylon*. Such structures are conspicuous features in *Taxus*, *Torreya*, and *Cephalotaxus*, but they occur outside the Taxceae in *Pseudotsuga*, for example. The presence of spiral bands therefore is not conclusive evidence of affinity with the Taxaceae.

PODOCARPACEAE

This family of conifers is and always has been essentially southern although in India, Southern Japan, and Central America it does extend north of the equator. Although several times reported from the Mesozoic and Tertiary of Europe and the United States, its past occurrence in these nothern latitudes has never been confirmed. Some of these supposed podocarps from the north have been found to belong to *Amentotaxus* and others are dicotyledons. The family is an ancient one, the oldest members having been found in rocks of Rhaetic age in New Zealand and India. *Archaeopodocarpus germanicus,* described a few years ago from the Upper Permian of Germany, is believed to be misinterpreted, and the remains on which the restoration were based are believed to represent *Ullmannia.* Beginning with the Lower Jurassic the podocarps compete with the Araucariaceae for a position of dominance in the coniferous floras of the Southern Hemisphere. Remains resembling *Podocarpus* and *Dacrydium* have been found in the Middle Jurassic of Antarctica and have been described under such names as *Elatocladus* and *Sphenolepidium.* Throughout the Cretaceous and Tertiary rocks

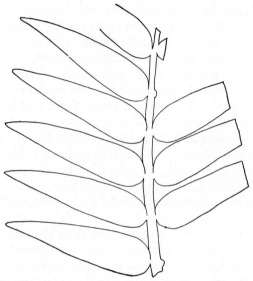

FIG. 155.—*Nageiopsis zamioides.* (*After Fontaine.*) ½ natural size.

podocarps have been found in numerous localities in Argentina, Chile, Australasia, South Africa, Tasmania, and India.

A Lower Cretaceous genus once thought to represent the Podocarpaceae, but which is of doubtful affinity, is *Nageiopsis* (Fig. 155), first described by Fontaine from the Potomac series of Virginia. The remains consist of branched shoots bearing two-ranked, linear to ovate-lanceolate leaves with parallel veins, which do not converge to any appreciable extent toward the apex. The leaf narrows at the base and is sometimes attached by a very short stalk. The parallel veins originate by a series of dichotomies. The leaves appear to have been persistent, and it is believed that they were spirally arranged on the shoots, although a distichous habit is strongly suggested by the compressions. *Nageiopsis* is so named because of its resemblance to the *Nageia* section of the genus *Podocarpus*, that section which includes those species with numerous parallel veins but no midrib. A midrib is present in the other sections of the genus. *Nageiopsis* also closely resembles *Podozamites* with which it is frequently confused. It differs, however, from *Podozamites* in which the leaves are deciduous and the veins converge to a more marked extent toward the apex. The fructifications of *Nageiopsis* are unknown.

The fossil records of *Phyllocladus* are inconclusive, although *P. asplenioides*, from the Eocene of New South Wales, resembles the living species. Other fossil leafy twigs that bear some resemblance to *Phyllocladus* are *Androvettia*, from the Raritan formation of Staten Island, and *Protophyllocladus* from the Dakota formation of Nebraska and the Atane beds of Greenland.

ARAUCARIACEAE

Since the majority of our readers may not be familiar with the araucarian conifers a brief account of the Recent members of this important and interesting family seems desirable. The family consists of the two genera *Agathis* and *Araucaria*. The former ranges from the Philippine Islands to New Zealand, and extends throughout the Dutch East Indies, Malaya, and Australia. The distribution of *Araucaria* is similar except that two species occur in South America. Although limited in its native haunts to the Southern Hemisphere, *Araucaria imbricata*, the "monkey puzzle" tree, thrives as an ornamental as far north as Scotland and along the western coast of North America some distance north of the Canadian boundary. *A. excelsa*, the Norfolk Island pine, is widely cultivated in greenhouses.

The genus *Araucaria* is divided into two sections. Section *Columbea*, which includes *A. imbricata*, has relatively large, flat, broad, coriaceous leaves traversed by several veins and arranged in a dense spiral. The ovulate cone scale is deep and relatively narrow and bears a single embedded seed. The seed contains two cotyledons, which remain beneath the ground when the seedling develops. The Section *Eutacta* has linear, subfalcate to falcate leaves that are more or less laterally compressed in cross section and traversed by one or occasionally more veins. The seed-bearing scale is laterally expanded to form a membranaceous wing, and the seed contains two or four cotyledons, which are lifted above the ground surface when the seed germinates.

In general form and habit the various members of the family are much alike. *Araucaria imbricata*, the species most commonly grown as an outdoor ornamental in the Northern Hemisphere, has a pyramidal form when young. The long, slender, lateral branches arise in false whorls of 4 to 16 and the lowermost ones often rest on the ground. The foliage is dark green. In their native haunts the older trees possess a tall columnar trunk which frequently reaches a height of 200 feet and which is devoid of branches except for the crown near the top. This crown forms a candelabralike head consisting of spreading horizontal shoots terminating in tufts of smaller branches. In some species the longest lateral shoots bend upward forming a saucer-shaped crown.

In the chapter on the Cordaitales mention was made of certain resemblances between the foliage and secondary wood of the Cordaitales and the Araucariaceae. Because of these resemblances the araucarians are often regarded as the oldest and the most primitive of the Coniferales, presumably having been derived from the Cordaitales. This assumed antiquity of the araucarians finds additional support in the persistence of alleged araucarian features in fossil woods that extend in almost unbroken sequence throughout the Mesozoic and Cenozoic to the present. Also throughout the late Paleozoic, the Mesozoic, and the Cenozoic there is an abundance of leafy twigs and stems that bear at least a superficial resemblance to the leafy branches of some of the species of *Araucaria*. Lack of organic connection, however, between parts usually renders determination of affinities difficult and uncertain.

A different view of the age and evolutionary status of the Araucariaceae has been advocated by a prominent school of American botanists headed by E.C. Jeffrey, which maintains that this family is not ancient but relatively modern, and that its apparent resemblances with the Cordaitales are characters which have been secondarily acquired. According to the protagonists of this theory the Pinaceae (Abietineae) are both more primitive and more ancient, and they, not the Araucariaceae, are the direct cordaitean relatives. This theory maintains that the evolutionary course of the conifers has been a complicated process wherein the multiseriate bordered pits in the tracheids of the cordaiteans became uniseriate, remote, and round in certain late Paleozoic and early Mesozoic members, and the thin-walled and unpitted ray cells became thick walled and pitted. Resin canals (lacking in the Cordaitales) also appeared. Then it is postulated that this rather strongly modified line split into two diverging branches, one continuing along the same direction and giving rise to the modern Pinaceae, and the other reversing its evolutionary course and regaining characters previously lost. This regressive branch would then constitute the Araucariaceae.

This rather elaborate and seemingly more difficult of the two theories of the origin of the Araucariaceae is founded for the most part on comparative studies of the transition conifers of the Mesozoic and on phenomena interpreted as ontogenetic recapitulations, traumatic reactions, and occurrences of vestigial structures in Recent members of both families. The essential weakness of this theory, however, is its disregard for the most direct resemblances as constituting the best evidence of affinity, and the substitution for them of implications and inferences of an indirect kind. Moreover, it rests to a considerable extent upon arbitrary choices, and it places too much reliance upon an imperfect and equivocal series of fossil forms. Its supporters have relied heavily upon two alleged occurrences of *Pityoxylon* (a pinelike wood) in the Carboniferous. These are cited in support of the argument that the geological record favors a close connection between the Cordaitales and the Pinaceae. However, when the circumstances surrounding these two supposed *Pityoxyla* were critically investigated, it was found that the structure of one of them had been misinterpreted (it is not *Pityoxylon*) and the origin of the other could be traced only to a rock pile of unknown source. Thus neither of them constitutes satisfactory evidence of pinelike conifers in the Carboniferous.

No effort is made in this book to give a complete analysis of the problem of the origin of the Araucariaceae, but it is believed that the final solution (if such is ever forthcoming) will show that both the Araucariaceae and the Pinaceae are descendants of ancient stock, and that both are derived from Paleozoic ancestors although the complete segregation did not come about until the late Mesozoic. Both lines were diversified within themselves during the Mesozoic and many of their members possessed characters found now within the separate families. The present segregation, then, is largely the result of the extinction of intermediate Mesozoic types.

The most abundant fossil material showing evidence of araucarian affinities is wood commonly referred to as *Araucarioxylon*. This genus is not a natural one, but the name is used for almost any fossil wood having uniseriate and flattened or multiseriate and polygonal bordered pits on the radial walls of the tracheids. Xylem parenchyma is rare or absent although resin may be present in some of the cells. The rays are mostly uniseriate, and the walls of the ray cells are thin and pitted only along the sides. Resin canals have never been observed.

Woods of the *Araucarioxylon* type merge with those commonly described as *Dadoxylon* of the Paleozoic which for the most part are cordaitean. Some authors object to the use of two generic names for woods which do not differ in essential morphologic characters, but in this instance they serve as convenient categories for the numerous miscellaneous woods of the Mesozoic and Paleozoic, respectively, of which the affinities can be expressed only in general terms.

The best display of trunks of the *Araucarioxylon* type to be found anywhere is in the Petrified Forest National Monument of Arizona. Scattered over an area of many square miles are thousands of silicified logs, which have been uncovered by weathering of the Upper Triassic Chinle formation. The so-called "forest" contains three known kinds of trees of which the most prevalent is *Araucarioxylon arizonicum.* The logs of this species are of considerable size, averaging 2 or 3 feet in diameter, but in some instances 5 feet. Some are 100 feet long. The trees are not in place, but represent drift logs that had floated for some unknown distance before burial. Since no foliage or fructifications have been found associated with these trunks, the anatomy is the sole clue to relationship. A small pith occupies the center of the trunk but because of the prevailingly poor preservation in this region very little is known of the primary wood. Nevertheless, it appears to be endarch as in modern conifers. The secondary wood is made up of thick-walled tracheids that appear roundish in cross section and rays ranging from 1 to 22 cells high. The pitting on the tracheid walls is uniseriate or rarely biseriate, and in case of the latter the arrangement is alternate and typically araucarian. Rather weakly defined growth rings are present in which the late wood consists of layers two to five cells in extent. *A. arizonicum* is not confined to this one locality, but is found in the Triassic of other parts of Arizona and at places in Utah, New Mexico, and Texas.

Woodworthia, with one species, *W. arizonica*, is associated with *Araucarioxylon arizonicum*, but differs in the presence of persistent short shoots in the secondary wood. As the trunk increased in size, these short shoots lengthened in a manner similar to those in the genus *Pinus.* Near the central portion of the trunk the short shoot is subtended by a leaf trace. This shows that the short shoots were axillary structures which persisted on the outside of the branches after the leaves had fallen. In this respect *Woodworthia* differs from the living araucarians because in them the leaf trace is persistent and increases in length as the trunk grows in diameter. The persistent shoot of *Woodworthia* is interpreted as an abietinean feature in combination with araucarian characteristics, and the genus is one of those Mesozoic types which was not completely segregated. The logs of *Woodworthia* are smaller than those of *Araucarioxylon arizonicum* and seldom exceed 2 feet in diameter.

Araucariopitys, with one species, *A. americana*, from the mid-Cretaceous of Staten Island, is another type that appears transitional between the araucarian and abietinean conifers. In this genus most of the pits on the tracheid walls are contiguous and flattened (an araucarian feature) but at places they are circular and separated as in Recent pines. Another abietinean feature is that all the walls of the ray cells are pitted, and traumatic resin canals were formed. The genus is further character-ized by the presence of deciduous shoots which apparently were shed annually and which bore fascicles of linear leaves. Although *Araucar-iopitys* is usually classified with the Araucariaceae, it is an excellent example of a transition conifer.

The name *Araucarites* has long been in common usage for leafy shoots, twigs, cones, and cone scales believed to possess araucarian affinities, and assignment to this genus is usually based upon resem-blances to certain of the living species of *Araucaria*. The oldest fossil to be referred to *Araucarites* with any degree of probability is *A. Delafondi* from the Permian of France, although associated with the remains are vegetative shoots of *Ullmannia frumentaria*, a member of the Voltz-iaceae. *A. Delafondi* consists of broadly triangular cone scales bear-ing on the upper surface the scar of a single seed. Throughout the Triassic and Jurassic, especially the latter, cones and cone scales resem-bling those of the *Eutacta* section of *Araucaria* are commonly encountered.

There is ample evidence that the Araucariaceae extended into the central part of North America during the Mesozoic era. The existence of *Araucarioxylon* in the Triassic has already been mentioned. In addition to this, sterile branches bearing elliptical-ovate leaves reminiscent of *Araucaria imbricata*, from the Cretaceous of New Jersey, were named *Araucarites ovatus* by Hollick, and Berry described similar material from the Upper Cretaceous of Alabama and North Carolina as *Araucaria bladensis*. *Araucarites Hatcheri* came from the Upper Cretaceous of Wyoming, and *A. hesperia* is the name of a portion of a seed-bearing cone from the Montana formation of South Dakota.

Occurring throughout the Mesozoic are many kinds of unattached shoots bearing short, spirally arranged, falcate leaves seemingly arau-carian but difficult to identify. *Pagiophyllum* is a type that is widely distributed throughout the Jurassic but extends from the late Permian to the Cretaceous. Its leaves are imbricate, crowded, broadly triangular, tetragonal in cross section, and possess a distinct dorsal keel. Many authors use this name for all Mesozoic coniferous shoots lacking fructi-fications. *Elatides* is the name often used for twigs bearing cones resembling *Picea or Abies* and leaves resembling *Araucaria excelsa*.

The most important fossil araucarian remains to be found within recent years came from Southern Argentina. In the Cerro Cuadrado region of Patagonia, well-preserved silicified cones and branches were found in volcanic ash beds probably of Eocene age. The largest of these cones is *Proaraucaria mirabilis* (Fig. 156), which is spherical or ellipsoidal, and varies up to 6 by 9 cm. in breadth and length. Except for being smaller, it resembles the cone of a living *Araucaria*, but an important structural difference lies in the fact that in the fossil cones there is a deep notch between the cone scale and its bract as in the Pinaceae, whereas in the modern araucarians the two structures are fused together into a single organ.   In *Proaraucaria* a single seed is embedded within the cone scale on its adaxial surface.   The seeds are oval, slightly flattened, and measure about 11 mm. in length.   Some of them have been observed to contain an endosperm and an embryo with two coty ledons and a well-developed hypocotyl.   In many respects *Proaraucaria* appears related to the *Eutacta* section of *Araucaria*.

Fig. 156.—*Proaraucaria mirabilis.*   Silicified cone.   Probable Eocene.   Cerro Cuadrado Petrified Forest, Patagonia.   Natural size.

The other cone genus found in the Cerro Cuadrado forest is *Pararaucaria*, which differs from *Proaraucaria* in its smaller size and more cylindrical form.   A specimen of average size will measure about 2 by 5 cm. in breadth and length, but only a few of them have been found complete.   It is essentially similar to *Proaraucaria* in showing the deep sulcus between the cone scale and the bract.

Associated with both of the cone types just described from Patagonia are impressions of twigs bearing leaf scars resembling the *Columbea* section of *Araucaria*.   If these stems belong to the associated cones, the whole complex shows not only a combination of araucarian and abietinean features but also those of the *Eutacta* and *Columbea* sections of the living genus, a further indication that the family has undergone extensive evolution within comparatively recent times.

Cone scales of *Agathis* are rather abundant in the Cretaceous and Tertiary rocks. *Agathis borealis*, first discovered in the Arctic, occurs also in the Cretaceous of the Atlantic coastal plain. Smaller forms which intergrade with *A. borealis*, and which are of similar age, have been described as *Protodammara speciosa*.

A Mesozoic family now believed to be close to the Podocarpaceae, but probably related to the Araucariaceae, is the Cheirolepidaceae, which consists of the four genera *Cheirolepis*, *Hirmeriella*, *Indostrobus*, and *Takliostrobus*. The only parts known are the megasporangiate cones, which bear spirally arranged seed scales subtended by partly fused bracts. Extending backward over the pair of ovules on the upper surface of the seed scale, and joined by its margins to the edge of the scale, is an epimatium that covers the ovule except at the micropylar end. The seeds are therefore produced within a deep pocket. In *Cheirolepis* (Fig. 157), the type genus of the family, the bract is digitately divided into five points, a feature that recalls certain other conifers such as *Voltzia* or *Swedenborgia*. *Cheirolepis* is the most widely distributed genus of the family, occurring in rocks belonging either to the uppermost Triassic or the lower-

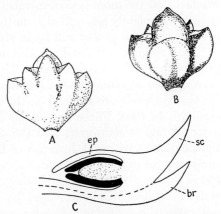

Fig. 157.—*Cheirolepis Muensteri.* (*A*) Outer surface of cone scale. (*B*) Inner surface showing two ovules. (*C*) longitudinal section to one side of median line showing bract (*br*), scale (*sc*), and ovule and epimatium (*ep*). (*After Hirmer.*)

most Jurassic of Western Europe, Eastern North America, southern China, and Southern Australia. The other genera are more limited in distribution. *Hirmeriella* is known only from the Rhaeto-Liassic of Germany, and *Indostrobus* and *Takliostrobus* are from the Upper Cretaceous of India.

CEPHALOTAXACEAE

The history of this small family consisting of the genera *Cephalotaxus* and *Amentotaxus* is not precisely known. Nevertheless, such appellations as *Cephalotaxopsis*, *Cephalotaxites*, and *Cephalotaxospermum* have been applied to foliage and seeds resembling those of the modern genus. *Cephalotaxopsis magnifolia* is abundant in the Patuxent (lowermost

Cretaceous) of Virginia, and it has been found in beds of similar age in South Dakota, California, and Alaska. *Cephalotaxopsis* occurs in the Upper Cretaceous of the Yukon Valley of Alaska.

Foliage identical in appearance and epidermal structure with that of *Amentotaxus* has been found in the Tertiary of Germany, and material showing only the form of the leaf occurs in the Eocene La Porte flora of California.

PINACEAE

This large family of gymnosperms has a long history that if perfectly understood would probably extend into the earlier stages of the Mesozoic, but it is not until the late Jurassic or early Cretaceous formations are reached that the more typical members show definite evidence of having become completely segregated from the transitional forms.   To simplify presentation, the genera will be treated individually.

The number of fossil species of pines that has been described probably exceeds two hundred.   The names *Pinus* and *Pinites* have been applied to remains as old as the Jurassic, but not until the Lower Cretaceous levels are reached do the determinations become reasonably certain.   In 1904, Fliche and Zeiller described two cones, one under the name *Pinites strobiliformis* and the other as *Pinus Sauvagei*, from the Portlandian (uppermost Jurassic) of France. *Pinites strobiliformis* is an elongate-oval object that somewhat resembles the cone of *Pinus strobus*. *Pinus Sauvagei* is a broader shorter cone.   The authors make positive statements about the generic identity of these cones but their figures are not entirely convincing.   Slender, needlelike leaves with a single vein extending the full length, and identified as *Pinus Nordenskjoldi*, have been found in the Jurassic of Oregon and California, but because none of them has been found in clusters their reference to *Pinus* is largely speculative.

Pines cannot be said to be abundant in the Lower Cretaceous rocks although there is conclusive evidence that they existed at that time. *Pinus vernonensis*, from the Patapsco formation (upper Potomac) of Virginia and Maryland, consists of seeds and cone fragments.   The cones are about 1.8 cm. in diameter and 7 cm. long.   The seeds are elliptical bodies which, with the wing, vary from 5 to 15 mm. or more in length.   The foliage of this species is unknown.   Remains of *Pinus* have also been found in the Kome (Lower Cretaceous) of Greenland, in the Lakota of the Black Hills, and in the Kootenay of Western Canada. Several cones have been described from the Lower Cretaceous of Europe.

Pines increase in frequency in the Upper Cretaceous and are widely distributed in the Tertiary.   It seems, however, that during the early

Tertiary they were distributed mainly throughout regions of cooler climate. None has been reported from the Wilcox flora (Eocene) that bordered the Gulf of Mexico, and they are rare in the Green River and other formations of similar age. The Green River does, however, contain pollen of *Pinus*, and a single leaf cluster with three needles has been described from this formation. In the Miocene and Pliocene, seeds, cones, and clusters of leaves are frequently encountered, although there is seldom much variety in any given locality. One of the most prolific localities for fossil pines is in the clays of the early Miocene Latah formation near Spokane, Washington, where four species have

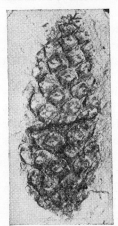

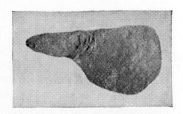

(A) (B) (C)

FIG. 158.—(A) *Pinus coloradensis.* Ovulate cone. Tertiary. Creede, Colorado. ½ natural size. (B) *Abies longirostris.* Cone scale bearing long terminal bract. Tertiary. Creede, Colorado. Natural size. (C) *Abies laticarpus.* Winged seed. Trout Creek diatomite. Harney County, Oregon. Natural size.

been named. The remains consist of needle clusters, seeds, and staminate cones. Three or four pines also occur in the Miocene Creede formation in central Colorado. One of these, *Pinus coloradensis* (Fig. 158*A*), consists of well-preserved cone imprints believed to represent the modern hard pine group. In the Great Basin area of the Western United States remains of pines are scattered throughout the middle and late Miocene and the Pliocene, although they are rare or nonexistent in the Eocene beds of the same regions. A Miocene pine of common occurrence is *P. Knowltoni*, a three-needle form represented by foliage, seeds, and cones in the Payette, Mascall, Bridge Creek, Eagle Creek, and other floras in eastern Oregon and adjacent parts of Idaho.

Prepinus is the name given by Jeffrey to a conifer with clusters of leaves borne on short shoots from the Raritan formation (lowermost Upper Cretaceous) at Kreischerville on Staten Island. Only the leaf-bearing short shoots are preserved, and the supporting branches are unknown. The short shoots are less than 1 cm. long, and they bear 20 or more spirally disposed leaves. In cross section the leaves are polygonal, contain two lateral resin ducts, and have a single median mesarch vascular bundle, which is surrounded by a complex system of transfusion tissue. The mesophyll surrounds the bundle as in Recent pines but the photosynthetic cells lack the characteristic infolded walls. *Prepinus* is regarded as an ancestral form of the Pinaceae, and it may constitute a link between the pines, the cedars, and the larches.

Little is known of the genus *Abies* previous to the Tertiary, although foliage from the Cretaceous is sometimes assigned to *Abies* and *Abietites*. Because of the ease with which the cone scales become detached from the cone axis, well-preserved complete cones are rare. In the Tertiary, the Miocene in particular, the seeds and characteristic bracts are often found (Figs. 158*B*, and *C*).

The genus *Keteleeria*, at present restricted to China, has been recognized in the Pliocene of Germany, and also in the Latah formation at Spokane, Washington. At the latter place the material was originally mistaken for *Potamogeton*. Wood identified as probably *Keteleeria* has been found in the Lower Cretaceous of Alberta.

Pseudotsuga has been reported from the Tertiary of Europe, but the best preserved remains are probably those from the Miocene of Western North America. Seeds, cones, and foliage have been found in the diatomaceous earth and volcanic ash beds of eastern Oregon.

The genus *Tsuga* appears to have been widely distributed throughout Europe during the Tertiary, but there are few references to it from North America. Wodehouse has identified pollen of this genus in the Green River shales of Colorado.

Remains resembling *Picea* occur in the Lower Cretaceous of Belgium. Needles and seeds are widely scattered throughout Tertiary rocks, and have been found in abundance at a number of places in the Miocene of Western North America. The names *Piceoxylon* and *Protopiceoxylon* have been applied to woods ranging from the Jurassic to Recent, but they are not necessarily all related to the modern genus.

Fossil records of *Larix* are meager. In North America the genus has been reported only from the Pleistocene, and in Europe our knowledge of it is equally scant. Remains superficially resembling the leafy shoots of *Larix* or *Cedrus* and described under the form genus *Pityocladus* have been found in the Jurassic.

Cedrus is difficult to distinguish from *Larix* in the fossil condition. Wood showing characteristic pitting of *Cedrus* has recently been recognized in the Auriferous gravels of Miocene age in California. Cones

(*Cedrostrobus*) resembling those of *Cedrus* are known from several places in Europe and range in age from Lower Cretaceous to Tertiary. Wood showing the characteristics of *Cedrus* but probably belonging to a variety of other conifers is often described as *Cedroxylon*. *Cedroxylon* ranges from the Triassic to the Recent.

Literature dealing with the fossil Pinaceae usually contains descriptions of form genera or organ genera which refer to detached foliage, leafy shoots, cones, cone scales, or seeds, which range from the late Triassic to the Tertiary. The use of these various generic designations is far from being consistent, with some authors applying them broadly and others restricting them to a few forms. *Pinites*, for example, was proposed long ago by Endlicher for leaves and cones allied to *Pinus*, *Abies*, *Larix*, and *Picea*. Seward, on the other hand, applies the name *Pinites* only to types which appear definitely related to *Pinus*, and suggests the name *Pityites* "for Abietineous fossils which cannot with confidence be referred to a more precise position." *Pinites* has in the past been widely used for fossil coniferous wood regardless of its exact affinities, but more recently the use of the name *Pityoxylon* has become customary for woods resembling the pines and other members of the Pinaceae. Other names that have been proposed for the various organs in the detached condition are *Pityanthus* for staminate cones, *Pityostrobus* for seed cones, *Pityolepis* for cone scales, *Pityospermum* for seeds, *Pityocladus* for leafy shoots, *Pityophyllum* for unattached leaves, and *Pityosporites* for pollen. These terms are frequently useful from a nomenclatorial standpoint for placing on record plant parts believed to be related to the Pinaceae, but they signify little or nothing concerning affinities with modern genera.

It is beyond the scope of this book to mention and describe the various types of wood that have been assigned to the Pinaceae. The interpretations are often uncertain because many of them belong to the transition conifers, some fourteen of which have been named and described during the past three decades. In describing these intermediate forms stress has been placed upon those characteristics which bear upon the controversial matter of the relative antiquity of the Pinaceae and the Araucariaceae, but the results have for the most part tended to increase the confusion rather than to clarify the situation. Identical sets of facts are sometimes used by different authors in support of opposite theories.

A simplified classification of fossil abietinean woods was recently offered by Bailey who proposes three form genera as follows:

1. *Pinuxylon*, for woods that fall within the range of structural variability of *Pinus*.

2. *Piceoxylon*, for woods that fall within the range of structural variability of *Picea, Larix*, and *Pseudotsuga*.

3. *Cedroxylon*, for those woods which should be included within the structural range of *Keteleeria, Pseudolarix, Cedrus, Tsuga*, and *Abies*. This group would contain a number of the so-called "transition" conifers such as *Planoxylon, Pinoxylon, Protopiceoxylon, Thalloxylon, Protocedroxylon, Araucariopitys*, and *Metacupressinoxylon*. This classification is admittedly artificial but has the advantage of simplicity. Regardless of the fact that the range of structural variability in some of the living genera may be wide, the proposed classification reduces to a reasonable number the list of generic categories with which one has to deal.

Taxodiaceae

The members of this large and diversified family have a varied history, and as with the other families of the Coniferales, the fossil representatives are often confused and misinterpreted.

The name *Elatides* has been used for Mesozoic coniferous remains bearing spirally arranged falcate leaves and small cones externally resembling those of some species of the Pinaceae. The resemblance of the vegetative twigs to *Araucaria* is, however, deceptive, and the affinities of the genus are now believed to be with the Taxodiaceae.

Probably the best known species of *Elatides* is *Elatides Williamsonis*, from the Middle Jurassic Gristhorpe plant bed in Yorkshire. It is the oldest known member of the Taxodiaceae. The stiff angular needle like leaves arise from decurrent bases beyond which they bend away from the stem, and then near the tips they bend forward again. The stomata form bands on the two abaxial faces. The abaxial surfaces are without stomata, an unusual feature in a thick coniferous leaf. Within the bands the stomata are irregularly arranged.

The staminate cones are sessile and borne in small clusters at the tips of leafy twigs. Mature specimens are about 23 mm. long and 5 mm. wide. The stamen consists of a stiff stalk bearing an expanded somewhat diamond-shaped head. Extending from the head toward the cone axis are three elongated pollen sacs, two of which lie parallel to the stalk and attached to it, the third lying below and between and attached to them but not to the stalk.

The seed-bearing cones are terminal and range up to 6 cm. in length. Each cone scale consists of a stout stalk and an expanded terminal portion with a tapering apex, and near the juncture of the stalk and terminal part are as many as five ovules in an arched row. The oval seeds are about 2 mm. long and 1.4 mm. wide.

Elatides Williamsonis is believed to be closest to *Cunninghamia*, among Recent conifers although the vegetative shoots are intermediate in appearance between those of *Sequoia gigantea* and *Cryptomeria japonica*. The fused pollen sacs also invite comparisons with *Sciadopitys*.

Sciadopitys, the umbrella pine, is represented at present by one species in Japan. Several fossil species have been described, and if the material assigned to it is correctly interpreted, the genus extends to the Lower Jurassic. *Sciadopitys Hallei* and *S. Nathorsti* are from the Jurassic and Cretaceous of Greenland, and *S. macrophylla* has been described from the Middle Jurassic of Norway. *S. scanica* occurs in the Rhaetic of Sweden. The genus may be old as well as once having been widely distributed throughout the Northern Hemisphere.

The living sequoias are remarkable for their present limited distribution and for the large size of some of the trees, the latter fact rating them among the natural wonders of the world. As with most conifers, the early history of the sequoias is beclouded in a maze of transitional forms. In rocks older than Miocene the remains tend to become confused with such genera as *Taxus*, *Taxodium*, and others in which the arrangement of the leaves on the twigs presents a distichous effect. Fossils of the *Sequoia sempervirens* type are often indistinguisable from *Taxodium distichum* if the cones are absent. The confusion between the fossil members of these genera is well illustrated by such names as *Taxodioxylon sequoianum* for Tertiary wood which, as far as the affinities are revealed by the structure, might belong to either *Sequoia* or *Taxodium*.

The sequoias were probably in existence by the middle of the Mesozoic era although the oldest fossil to be assigned to that genus with any degree of probability is a small cone named *Sequoia problematica* from the Portlandian (late Jurassic) of France. However, it is quite probable that some of the older shoots of the *Elatocladus* category from the Rhaetic and Liassic may ultimately prove to be early sequoias.

In the Potomac and Kootenai series of the Lower Cretaceous there are at least six foliage and cone types that have been assigned to *Sequoia* although the identification of many of the fragments is open to grave doubt. Probably a more satisfactory example of a Lower Cretaceous *Sequoia* is a petrified wood fragment from the Lower Greensand of the Isle of Wight which shows a close resemblance to *S. gigantea*. Numerous fragments believed to be *Sequoia* occur in the early Upper Cretaceous Raritan and Magothy formations of the Atlantic coastal plain and in formations of similar age along the Gulf of Mexico. In the Dakota sandstone of Kansas, Iowa, and Minnesota there are remains of undoubted sequoias, but probably the most significant discoveries within this period have been made in the Patoot series of Western Greenland. A number of conifers have been found in these beds of which the commonest is *Sequoia*

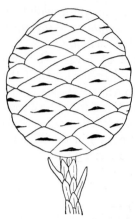

concinna, a form with slender branchlets bearing straight or slightly curved, decurrent,
acuminate leaves. The oval cones measure
about 20 by 23 mm. (Fig. 159). It is probably
related to *S. gigantea*.

The most widely distributed Tertiary
Sequoia is *S. Langsdorfii* (Fig. 160), which is
not essentially different from the living coast
redwood, *S. sempervirens*. The type of tree
designated by these two names has at one time
or other inhabited most of the Northern
Hemisphere, including the Artic regions, but
there is no evidence that it ever existed simul-
taneously throughout all this vast area. In

FIG. 159.—*Sequoia Reichenbachi*. The probable cone of *S. concinna*. Kome series. Lower Cretaceous. Greenland. (*After Heer.*) × about 2.

North America at least, where the Tertiary
history of this species has been studied in considerable detail, it has migrated according to
changing climatic conditions. *S. sempervirens*
occurs at present in regions of ample rainfall, frequent fogs, and relatively moderate winter tempera-
tures. *S. Langsdorfii* existed in
Alaska and Greenland during the
early Tertiary (Eocene) at which
time the area it now occupies was
covered with subtropical forests.
It probably extended farther
south in the central part of the
continent than it did along the
coast. By early Miocene time it
had migrated southward along the
Pacific Coast to its present latitude but it extended inland for
several hundred miles, at least as
far as the Rocky Mountains and
maybe farther. In the Bridge
Creek flora of central Oregon a
typical redwood association exists
in which *S. Langsdorfii* occurs
with such hardwood types as
Alnus, Lithocarpus, and *Umbellu-*

FIG. 160.—*Sequoia Langsdorfii*. Oligocene. Bridge Creek, Oregon. Natural size.

laria. By the end of the Miocene epoch, however, the rising Cascade

Range had so reduced the rainfall to the eastward that the redwoods were able to survive only along the coast where they have since persisted. In Europe, due to the coldness accompanying the Pleistocene glaciation, the redwoods were entirely eliminated from that continent.

At many places, notably at Florissant, Colorado, the Yellowstone National Park, and near Calistoga, California, there are numerous silicified trunks of sequoias that approach the living trees in size. At Florissant one standing trunk (*Sequoioxylon Pearsallii*) is 10 feet high and 17½ feet in diameter at about the original ground surface. *S. magnifica* of the Yellowstone National Park was about the same size, and some of the trunks still stand to a height of 30 feet. It is believed that *S. magnifica* is the trunk which bore the foliage called *S. Langsdorfii*. The trunks at Calistoga are similarly large but were buried in a prostrate position beneath a thick layer of Pliocene ash. The largest trunk is 126 feet long with a mean diameter of 8 feet. Another specimen at the same place is 12 feet in diameter and 80 feet long. Although not of the greatest diameter these are probably the most massive fossil trunks known, rivaling even the huge logs of *Araucarioxylon arizonicum* of the Petrified Forest National Monument.

Taxodium is abundant in the Tertiary of North America, Europe, and Asia, although it is frequently confused in the fossil state with *Sequoia*. Wood referred to as *Taxodioxylon* is similar to *Cupressinoxylon* and ranges from the Jurassic to Recent. *Taxodium* wood can be distinguished from *Sequoia* only by the presence in the latter of traumatic resin canals. *Taxodium dubium* (Fig. 161*A*) is probably the most widely distributed *Taxodium* species in North America, as it occurs in the early and middle Tertiary of Alaska, Washington, Oregon, British Columbia, Tennessee, Virginia, and other places. Exceptionally well-preserved twig compressions are found in the Latah formation at Spokane, Washington. *T. distichum*, the Recent species, has been reported several times from the Oligocene and the late Tertiary. At present it occurs mainly along Atlantic and Gulf Coast regions, but it once existed in British Columbia.

Glyptostrobus is represented in the Recent flora by a single species growing along the coast of Southeastern China. Remains doubtfully belonging to *Glyptostrobus* have been recorded from the Lower Cretaceous of Greenland. It has been reported from beds as late as Pliocene in Europe but the determinations are questionable. Since the Pliocene it seems to have confined itself to its restricted range in China. *Glyptostrobus* has been recognized several times in North America, the three best known species being *G. dakotensis* based upon cone scales from the Fort Union (Eocene) beds of North Dakota, *G. oregonensis* consisting of twigs with cones from the Miocene of Oregon, and *G. europaeus* (Fig. 161*B*), from several Tertiary localities.

Cryptomeria is sparsely represented in the fossil record, but has been reported from the Tertiary of Great Britain and the Upper Cretaceous of Japan. Because of its resemblance to *Araucaria excelsa* it has probably been overlooked at other places.

Arthrotaxis has been recorded from the Lower Cretaceous of Patagonia, but otherwise little is known of the history of this genus. Material

(A)   (B)

Fig. 161.—(*A*) *Taxodium dubium*. Latah formation. Spokane, Washington. Natural size. (*B*) *Glyptostrobus europaeus*. Tertiary. Cripple Creek, Colorado. Natural size.

described as *Arthrotaxis* from Bohemia and other places north of the equator is of questionable affinity.

Equally obscure is the history of the genus *Cunninghamia*, although a cone named *Cunninghamiostrobus yubarensis* from the Upper Cretaceous of Japan, resembles the living genus. Leafy branches assigned to *Cunninghamites* are found in rocks as old as the Lower Jurassic, but most of them should be referred to the form genus *Elatocladus* until their affinities are more accurately determined. They furnish no reliable data concerning the age of the genus. Pollen of *Cunninghamia* occurs in the Green River shales of Colorado and Utah.

CUPRESSACEAE

This family is for the most part a rather recent one although wood referable to *Cupressinoxylon* occurs in the Jurassic and formations of later age. *Cupressinoxylon*, however, embraces woods belonging to many genera, including some of the transition conifers. In *Cupressinoxylon* resin canals are absent except in wounded areas (as in *Sequoia*,) and xylem parenchyma occurs scattered throughout the annual growth layers. It is a form genus with structure similar to *Cedroxylon*, *Sequoioxylon*, and *Taxodioxylon*, and its generic limits are variously set by different authors. As with other woods bearing the name of a living genus as a prefix, mere citation is not dependable evidence of the existence of the genus of which the name serves as the root word. *Cupressinoxylon lamarense*, from the Yellowstone National Park, shows some points of resemblance to the Taxaceae.

Callitris has been found in the Upper Cretaceous and Tertiary of Europe. The characteristic winged seeds, described as *Callitris potlachensis*, have been found in the Latah formation of Idaho, but the determinations are unsupported by recognizable vegetative remains.

Thujiopsis is not known in North America but vegetative material has been found in the Tertiary of a few localities in other continents. *Thuja* occurs throughout the Upper Cretaceous of the Atlantic coastal plain and has been recognized at several places in the Tertiary of the Western United States and in British Columbia. Leafy twigs called *Thujites sp.* come from the Latah formation, and *Thuja interrupta* is abundant in the Fort Union and other beds of similar age.

The history of *Libocedrus* is not well known although it has been reported from the Miocene of France and the Oligocene of Germany and Italy. It is also found in the Pliocene of Western North America.

Satisfactory accounts of *Cupressus* are rare. A cone, *Cupressus preforbesii*, was found in the Pliocene of California. The genus has been reported several times from the Tertiary of Europe. Remains commonly designated as *Cupressites* and *Cupressinocladus* are difficult to distinguish in most instances from foliage compressions of *Libocedrus*, *Thuja*, and *Juniperus*.

Juniperus is thought to be an important constituent of the Miocene lignites of Germany. Wood fragments from these deposits are referred to as *Juniperoxylon*. Vegetative twigs of *Juniperus* are frequent in the Tertiary of both Eastern and Western North America. Until recently *Chamaecyparis* was unknown in beds older than Pliocene, but it has lately been reported from the Miocene of California.

Some Mesozoic Conifers of Uncertain Affinity

Many, in fact probably most, of the coniferous types found in the Mesozoic rocks are difficult to classify in a completely satisfactory manner. Intermediate forms are especially prevalent in the Triassic and Jurassic, and it is not until the later series are approached that material can be assigned to modern families and genera with reasonable facility. The difficulties encountered in the older forms arise mainly from the fact that they belong for the most part to transitional types in which the essential characteristics of modern genera are not distinctly

Fig. 162.—*Elatocladus elegans.* Upper Cretaceous. Judith River, Montana. Natural size.

segregated. It has been pointed out that in genera like *Woodworthia*, a Triassic genus, the wood is essentially araucarian but the included short shoots suggest affinity with the pine family. In *Araucariopitys*, abietinean and araucarian features are exhibited in the pitting, and the leafy shoots called *Elatides* bear araucarian foliage along with cones of a different type. These are but a few of the examples that could be cited. If these ancient conifers which at present evade attempts at permanent classification are ever found more completely preserved, many of them will find their places in or near modern families, but others will have to be referred to families having no Recent representatives. No attempt is made here to give a comprehensive account of all the fossil conifers of uncertain position, but reference is made to a few of those most frequently encountered in the literature.

Elatocladus (Fig. 162) was proposed by Halle as a comprehensive genus for sterile coniferous shoots that cannot be otherwise satisfactorily assigned, and the name is used when it is undesirable to give implications of affinity. Many fossil conifers originally assigned to *Taxites* because of the seemingly distichous arrangement of the leaves on the twigs can be placed in *Elatocladus* if it is desirable to disassociate them with the Taxaceae and at the same time avoid the suggestion of affinity with

Fig. 163.—*Podozamites lanceolatus*. Jurassic. New South Wales. Slightly reduced.

any other family. Usage of the name has varied considerably as with most form genera. Some authors apply it broadly whereas others use it only for forms that cannot be included in any other generic group, either natural or artificial.

Podozamites (Fig. 163) was originally instituted for leaves believed to belong to cycadophytes. The leafy twigs resemble pinnate fronds with long leaflets having parallel or slightly spreading veins. Their coniferous affinities were discovered by the Swedish paleobotanist Nathorst who observed that the supposed pinnae of some of the species are spirally arranged showing the supposed frond to be a leafy shoot. Moreover, the individual leaves are frequently deciduous and are often found detached from the stems on which they grew. *Podozamites* cannot be placed within any particular family, and like *Elatocladus*, is a form genus. Some forms resemble *Nageiopsis*. Certain authors have considered the genus intermediate between the cycadophytes and conifers.

Cycadocarpidium is a seed-bearing fructification often associated with *Podozamites* and believed to belong to certain species of it. The organ consists of an ovate sporophyll with converging longitudinal veins borne on a slender stalk, and at the base of the lamina portion are two lateral bractlike projections that bear a single seed each. The sporophylls are produced in cones. The best known species are *Cycadocarpidium Erdmannii* (Fig. 164), and *C. Swabi*, both from the Rhaetic of Eastern Greenland and Western Europe.

Swedenborgia is another coniferous fructification often associated with *Podozamites* and other conifers, but its foliage is unknown. The cone bore spirally arranged cone scales, each consisting of a slender stalk and five rigid distal lobes. A small bractlike scale projects outward from the stalk and ends in a concavity on the lower side of the scale. Each of the five lobes of the scale appears to have borne a single inverted ovule in which the micropyle faced the cone axis.

Fig. 164.—*Cycadocarpidium Erdtmanni*. Megasporophyll bearing two seeds. (*After Nathorst*.)

PHYLOGENY OF THE CONIFERS

The problem of deciphering the phylogeny and the geologic history of the conifers has been muddled by a lack of unanimity of opinion on the part of investigators as to which anatomical and morphological features are primitive and which are derived. This applies particularly to the Araucariaceae and the Pinaceae (Abietineae), and the rather curious situation has developed whereby identical sets of facts have been put forth in support of opposite interpretations. Because of certain obvious similarities in the structure of the secondary wood of *Araucaria* and *Cordaites*, a close connection between the two has long been assumed and *Araucaria* was looked upon as the most ancient of living conifers. The supposition that the araucarian line extends in almost unbroken sequence from the Paleozoic to the Recent has been kept alive partly by the somewhat arbitrary application of the name *Araucarioxylon* to Mesozoic gymnospermous woods that resemble both *Araucaria* and *Cordaites*. However, it must be borne in mind that none of the early

Mesozoic species of *Araucarioxylon* have been found with cones or foliage attached, and their affinities are subject to reconsideration at any time.

Opposed to the widely held theory of the antiquity of the araucarian conifers is the view briefly discussed earlier in this chapter, and sponsored mainly by Jeffrey, that the abietinean line is both more primitive and more ancient. Some of the arguments set forth in support of this premise seem rather dogmatic but the essential ones are as follows. First, the pitting in the xylem of an araucarian seedling is of the pine type. The pits are in a single row on the tracheid walls and well spaced, as opposed to the crowded pitting in the later formed xylem. The condition in the seedling is assumed by the advocates of this theory to reveal the ancestral form. Second, the pits in the cone axis of *Araucaria* are often separated by bars of Sanio, or crassulae, structures that occur regularly in the stem wood of pines and related genera. These two characteristics are regarded by proponents of abietinean affinity as ancestral, that is, their presence in the seedling and cone axis is interpreted as proof that they are vestigial structures held over from ancestral forms in which they were generally distributed throughout the plant body as in the Pinaceae. The weakness of this argument lies in the uncertainty whether the type of pitting in the seedling and the occurrence of crassulae in the cone axis are features held over from original forms or whether they might have arisen independently as a direct response to certain physiological requirements. There is no proof that the latter is not the case, and, in fact, the identical structure of the mature vegetative stem is hardly to be expected in the very limited amount of tissues in the seedling and the cone axis. A third argument advanced in favor of the antiquity of the abietinean line is that the transition conifers of the Mesozoic show the characters of the two families combined. *Araucarioxylon noveboracense*, for example, from the Cretaceous Raritan formation, has resin cells of the abietinean type in association with wood that is otherwise araucarian. Ray tracheids have been found in still other species of *Araucarioxylon*, and occasionally pinelike pitting is present in the first annual ring. Traumatic resin canals have been found in others, a feature recalling *Sequoia*. Two misleading reports of pinelike conifers in the Carboniferous are mentioned earlier in this chapter.

It is obviously impossible in view of the contradictory and equivocal nature of the evidence furnished by the transition conifers to settle the question of the relative antiquity of the Araucariaceae and the Pinaceae. Data contributing to an ultimate solution may be forthcoming when these intermediate Mesozoic forms are more thoroughly investigated with the view of determining the exact course followed by certain developmental lines that began in the Paleozoic. It seems almost certain, however, that the whole problem should be approached from a slightly different angle. One of the prime difficulties has always been that the investigators were preoccupied with the notion that one or other of the two groups under consideration is older than and ancestral to the other and that the remaining one is its derivative. It would be more reasonable to admit the possibility that both are ancient and both sprang from more primitive ancestral forms and that one is not necessarily the offshoot of the other. It seems that the transition conifers support this view rather than the other. The transition conifers are not a unified assemblage, but rather an artificial plexus of widely divergent forms that may have originated independently in the remote epochs of the Paleozoic. The later conifers are then the residue of this great Mesozoic complex from which many types have disappeared.

On the basis of the assumption that the araucarians and abietineans both belong to ancient lines, it has been suggested that the former may extend back to the Upper Devonian and Lower Carboniferous Pityeae and the latter to the Cordaiteae. The Araucariaceae resemble the Pityeae in the possession of a relatively broad pith and in certain details of the leaf trace structure. Moreover the fleshy leaf, or phyllode, of *Pitys Dayi* bears a closer resemblance to the araucarian leaf than to that of any other conifer, and the assumed habit of *Callixylon* finds a close parallel in almost any araucarian species. However, the fructifications of the Pityeae, which are at present entirely unknown, will have to be discovered before much more can be said in support of an alliance between the two groups.

In the chapter on the Cordaitales mention was made of a possible homology between the cordaitean inflorescence and the abietinean cone, and now it may be pointed out that certain of the Paleozoic conifers, namely, *Lebachia*, *Ernestiodendron*, and *Pseudovoltzia* reveal a series of intermediate stages through which the abietinean cone scale might have been derived. It is postulated that by a series of evolutionary changes involving elimination of some of the sterile appendages and in some cases a reduction in the number of ovules, some of which underwent a change in position, the dwarf shoot of the cordaitean inflorescence became transformed into the abietinean cone scale. The possible evolutionary sequence may be outlined as follows:

The cordaitean inflorescence as it is now understood consists of a slender axis bearing rather distantly placed slender-pointed bracts probably in a modified whorl. In the axil of each bract is a dwarf shoot that bears in spiral arrangement two or more fertile and several sterile bracts in close order. The short shoot with its bracts is therefore the strobilus. In *Lebachia piniformis* the inflorescence is similarly constructed but the number of fertile stalks (seed or pollen sac stalks) has been reduced to one per strobilus. Then in *Ernestiodendron filiciforme*, the axillary dwarf shoot is a fan-shaped object three-lobed at the apex, and each lobe bears a single seed on the adaxial side. In *Pseudovoltzia Liebeana* there is in the axil of each bract on the inflorescence axis a flattened five-lobed dwarf shoot that bears two or three inverted ovules on the adaxial side. Then in *Ullmannia Bronnii* the number of ovules is reduced to one. The homology suggested by this series of late Paleozoic conifers is that the individual cone scale and seed is the equivalent of the cordaitean dwarf shoot and its appendages, and that the bracts subtending the cone scales of the Pinaceae are identical with the bract of the cordaitean inflorescence. The ovules, whether there be one or more, became inverted in *Pseudovoltzia*, and they have remained in this position in all the later forms. In some modern genera, the bracts subtending the short shoots have disappeared or have become fused with the seed scales.

If the interpretation just given of the homology between the cordaitean inflorescence and the abietinean cone is supported by future discoveries, we may consider the phyletic relationship of the Cordaiteae and the Pinaceae well established. There are, however, certain matters that will require further investigation, one being the prevailingly araucarian type of wood in the Cordaiteae and the Pityeae. Then there is the matter of the foliage. How did the long strap-shaped leaf become transformed into the needle of the pine or any of the other members of the Pinaceae? The suggested phylogeny may fail to stand for lack of more adequate proof, and the series just described may be the result of parallel development. However, it may be confidently assumed that the Cordaiteae have played an important role in the evolution of the higher gymnosperms at some stage of the process.

References

BAILEY, I.W.: The cambium and its derivative tissues, VII, Problems in identifying the wood of Mesozoic Coniferae, *Annals of Botany*, **47**, 1933.

BARGHORN, E.S., and I.W. BAILEY: The occurrence of Cedrus in the Auriferous Gravels of California, *Am. Jour. Botany*, **25**, 1938.

BERRY, E.W.: A revision of the fossil plants of the genus Nageiopsis of Fontaine, *U. S. Nat. Mus. Proc.*, **38**, 1910.

————: A revision of several groups of gymnospermous plants from the Potomac group of Maryland and Virginia. *U. S. Nat. Mus. Proc.*, **40**, 1911.

BURLINGAME, L.L.: The origin and relationships of the araucarians, I-II., *Bot. Gazette*, **50**, 1915.

CHAMBERLAIN, C.J.: "Gymnosperms: Structure and Evolution," Chicago, 1935.

CHANEY, R.W.: A comparative study of the Bridge Creek flora and the modern redwood forest, *Carnegie Inst. Washington Pub.* 349, 1925.

DARROW, B.S.: A fossil araucarian embryo from the Cerro Cuadrado of Patagonia, *Bot. Gazette*, **98**, 1936.

FLORIN, R.: The Tertiary fossil conifers of south Chile and their phytogeographical significance, *K. Svenska vetensk. akad. Handl.*, III, **19**, 1940.

———: Die Coniferen des Obercarbons und des unteres Perms, *Palaeontographica.*, **85**, Abt. **B**, pt. 2, Lief. I–V, 1938–40.

GOTHAN, W.: Zur Anatomie lebenden und fossilen Gymnospermenhölzer, *Abh. K. preuss. Geol. Landesan. und Bergakad.*, **44**, 1905.

———: Die fossilen Hölzer von König Karls Land, *K. svenska vetensk. akad. Handl.*, **42**, 1908; **44**, 1910.

HARRIS, T.M.: The fossil flora of Scoresby Sound, East Greenland, pt. 4, Ginkgoales, Coniferales, Lycopodiales, and associated fructifications, *Meddelelser om Grönland*, **112**, 1935.

HIRMER, M., and L. HÖRHAMMER: Zur weiteren Kenntnis von *Cheirolepsis* Schimper und *Hirmerella* Hörhammer mit Bemerkung über deren systematische Stellung. *Palaeontographica.*, **79**, Abt. **B.**, 1934.

HOLDEN, R.: Jurassic wood from Yorkshire, *Annals of Botany*, **27**, 1913

JEFFREY, E.C.: Araucariopitys, a new genus of araucarians, *Bot. Gazette*, **44**, 1907.

———: "The Anatomy of Woody Plants," Chicago, 1917.

——— and M. A. CHRYSLER: Cretaceous Pityoxyla, *Bot. Gazette*, **42**, 1906.

KRÄUSEL, R.: Die fossilen Koniferenhölze, *Palaeontographica.*, **62**, 1919.

———: Gymnospermae, in R. Pilger, "Die natürlichen Pflanzenfamilien, Band 13, Leipzig, 1926.

KNOWLTON, F.H.: A catalog of the Mesozoic and Cenozoic plants of North America, *U. S. Geol. Survey Bull.* 696, 1919.

SEWARD, A.C.: "Fossil Plants," Vol. 4, Cambridge, 1918.

——— and S.O. FORD: The Araucarineae, recent and extinct. *Royal Soc. London Philos. Trans.*, **B**, **198**, 1906.

THOMSON, R.B., and A.E. ALLIN: Do the Abietineae extend to the Carboniferous? *Bot. Gazette*, **53**, 1912.

WIELAND, G.R.: The Cerro Cuadrado petrified forest, *Carnegie Inst. Washington. Pub.* 449, 1935.

WODEHOUSE, R.P.: Tertiary pollen-II. The oil shales of the Green River formation, *Torrey Bot. Club Bull.*, **60**, 1933.

CHAPTER XIII

ANCIENT FLOWERING PLANTS

The relatively small amount of space allotted to the fossil angiosperms in this book may seem disproportionate to the position occupied by the flowering plants in the world today. Therefore it seems advisable to give an explanation for such brevity. In the first place any account of the flowering plants in which they are treated as fully as other groups would greatly enlarge the volume. Secondly, the nature of the subject matter itself is not such that it integrates well with the manner of presentation of the other groups. Regardless of the voluminous literature on the fossil angiosperms, the history of the group is too short to enable us to approach it with the same objectivity with which we view the lower groups such as the lycopods, the ferns, or the cycadophytes. The flowering plants are probably still in the initial stages of their expansion, hence developmental trends are not clearly expressed in the fossil series. Coupled with this difficulty in interpreting their evolutionary status is the nature of the fossil record itself, which consists mainly of wood and impressions of leaves and leaf fragments with only a sprinkling of seeds and fruits to aid in their determination. Thousands of angiosperm species have been named on the basis of foliar remains, but errors in generic and specific assignments are so prevalent, especially in the older works, that the value of a large proportion of them is practically nil. Careful studies within the last two decades, in which more attention has been paid to the range of variation of the leaves within species of living plants and to the natural associates of Recent genera and species, have resulted in considerable clarification of certain phases of the geologic history of the angiosperms, but the subject is so vast that it will be many years before we are in possession of accurate data sufficient to give a comprehensive view of any major portion of the group.

One of the most remarkable phenomena in biological evolutionary history is the rapidity with which the angiosperms arose to a position of dominance in the plant world during the latter part of the Mesozoic era. Many explanations have been offered but none are wholly acceptable. The problem is not a recent one because as early as 1879 Darwin is said to have written to Hooker, "The rapid development, so far as we can judge, of all the higher plants within Recent geological time is an

abominable mystery. I would like to see the whole problem solved."
Since Darwin's time the number of fossil angiosperms that have been
brought to light has increased many times, but as to their origin, the
same deep mystery remains. It seems almost certain, however, that
the angiosperms have had a longer and more extensive pre-Cretaceous
history than so far has been revealed by the fossil record and that the
scarcity of fossils is due to their predominatingly upland habits where
the remains were not readily buried and preserved. It has been pointed
out in connection with other plant groups that they are seldom well
represented in the fossil series until they had become firmly established
as elements of the flora, and in most cases in habitats not too remote
from swamps or other bodies of water where preservation of large quantities of parts was possible. These ecological considerations, on the
other hand, do not fully explain the sudden (though probably more
apparent than real) rise to a place of dominance of the angiosperms
during the Cretaceous because the oldest angiosperms known to us with
certainty are water lilies, not remains brought in by accident from the
uplands. And then, in the late Cretaceous and Tertiary deposits, upland
species are often preserved in considerable numbers. Our paucity of
knowledge of pre-Cretaceous angiosperms indicates that they were subordinate to gymnosperms as concerns numbers of both species and
individuals.

Not only are plant evolutionists at a loss to explain the seemingly
abrupt rise of the flowering plants to a place of dominance, but their
origin is likewise a mystery. It is hypothetically assumed that the
angiosperm line took shape at some unknown time during the Mesozoic
era, and all the naked-seeded groups (the pteridosperms, the Cordaitales,
the conifers, the cycadophytes, the Gnetales) and even the ferns, have at
times been proposed by various authors as the possible precursors of
the flowering plants. The gross result of these postulations, however,
has been to stress our ignorance of the subject more than anything else.

The question may very reasonably and pertinently be asked, "Since
the flowering plants are so prominently displayed in the rocks of the
late Cretaceous and Tertiary, why are we so ignorant of their origin and
evolution?" Several reasons may be given, but the essential one is
that we know of no series of fossil forms connecting the flowering plants
with lower groups. Considerable attention has been centered on the
cycadeoid fructification and its resemblance to the flower of *Magnolia*.
Without going into detail, it suffices to say that the cycadeoids cannot
be looked upon as the progenitors of the angiosperm class. Granted
they could be, the question of recognizing the intermediate stages would
remain. It is a long stride from the gymnospermous ovule of *Cycadeoidea*

to the carpel-enclosed seed of any flowering plant, or from the cycadophytic frond to the magnoliaceous leaf. The cycadeoids are a highly specialized order of plants that in all probability lacked the plasticity required to undergo the further changes necessary to give rise to such a diversified plexus as the flowering plants. Arber and Parkin, writing in 1907, attempted to supply a missing link with the hypothetical "Hemiangiospermae" in which the fructifications were built according to the cycadeoid "flower." This fructification is alleged to have possessed a primitive perianth of spirally arranged parts and an androecium of numerous stamens in similar sequence. The carpels were open leaflike structures with marginal megasporangia and were seated upon a dome-shaped receptacle. Were there any proof that a fructification of this type ever existed, it would possess vast evolutionary significance. However, workable phylogenies cannot be built upon forms that exist only in the mind.

With the discovery a few years ago of the Middle Jurassic Caytoniales, hopes ran high of bringing some light to bear upon the question of angiosperm origin. However, continued investigations finally showed undoubted affinity with the pteridosperms, and they are now classified as Mesozoic remnants of that group. The gap between the Caytoniales and the flowering plants is probably as great as that between any of the other vascular plants and the angiosperms.

In looking for possible ancestors of the flowering plants, attention has long been directed toward the Magnoliaceae because of the considerable number of primitive characters that converge within that family. In the magnoliaceous flower the sepals and petals are numerous, large, and often colored alike. The numerous stamens are spirally arranged, and the separate carpels are in spirals on an elongated conelike receptacle. In the secondary xylem the vessels are predominantly of the scalariform type, and the pitting on the walls of the vessels and wood fibers is scalariform. Primitiveness of a slightly lesser degree is also expressed in the Ranunculaceae, Nymphaeaceae, and Calycanthaceae. The assumption that these families are of ancient lineage is supported to some extent by the fossil record, but comparative studies of the living members of these families furnish the best evidence of their primitive character.

Another family of importance in angiosperm evolution, but of which our knowledge of its geologic history is meager, is the Winteraceae, a family restricted at present to the Australasian area and Mexico and South America. The family embraces the six genera *Belliolum*, *Bubbia*, *Drimys*, *Exospermum*, *Pseudowintera*, and *Zygogynum*. These are placed by some authors in the Magnoliaceae, but their relation to this family is remote. They all agree in having vesselless xylem, a feature shared by the two magnoliaceous genera *Trochodendron* and *Tetracentron,* but as regards both vegetative and floral structures they are less specialized than the Magnoliaceae. Whether they are more ancient is quite unknown.

A family in which a very primitive condition is revealed by the pistil

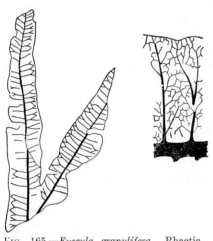

of the flower is the Degeneriaceae, a monotypic family inhabiting the Fiji Islands. In this flower the gynoecium consists of one pistil of one carpel, which is basally attached and with three vascular bundles. The carpel is folded in a conduplicate manner to enclose the ovules, which are not marginal but on the ventral (adaxial) surface well away from the margin. There is no style, but a stigmatic surface extends along each lateral margin of the carpel where the edges come in contact. These edges are not fused, and in the young condition they are even separated to a slight

Fig. 165.—*Furcula granulifera.* Rhaetic. Eastern Greenland. (*After Harris.*)

extent. This is probably the closest approach to gymnospermy known among the true flowering plants.

As with most of the other plant groups, the exact place in the geologic sequence at which the angiosperms can be first recognized is uncertain. It can be definitely stated, however, that all references in the literature to angiospermous plants in the Paleozoic can be set aside. Either the formations were wrongly determined or the identification of the plants was at fault. The oldest plant known to science that bears any undisputable resemblance to an angiosperm is *Furcula granulifera* (Fig. 165) discovered by Harris a few years ago in the Rhaetic (late Triassic) rocks of Eastern Greenland. The leaves of this plant are 7 to 15 cm. long and 6 to 8 mm. broad and are usually forked at about the middle. The petiole is short and the apex is usually acute. The leaf has a well-developed midrib and secondary veins which arise about 2 mm. apart and which pass directly toward the margin. These veins usually fork before the margin is reached. The veinlets form an irregular network between the larger veins, and small veinlets end free in the meshes. Numerous stomata of the dicotyledonous type are irregularly scattered and slightly sunken in the lower surface. This leaf is different from any other known in the older Mesozoic, and were it to be found in the

Cretaceous or Tertiary, it would be placed among the dicotyledons without the slightest hesitation.

An example of a possible Jurassic angiosperm is a piece of secondary wood, *Homoxylon rajmahalense*, from the Rajmahal Hills of India. Although its exact source is unknown, all evidence points to its Lower Jurassic age. The wood resembles that of a gymnosperm in that no vessels are present, but the tracheid walls bear a mixture of scalariform and circular pits as do the Winteraceae and the homoxylous Magnoliaceae such as *Trochodendron* and *Tetracentron*. If the age of *Homoxylon* were to be proved Jurassic beyond all doubt, it would lend strong support to the assumed antiquity of these families.

Probably the most positive evidence of the existence of pre-Cretaceous angiosperms is pollen recently found in coal of Jurassic age in Scotland. These pollen grains bear three longitudinal grooves like those of the nymphaeaceous genus *Nelumbo*. Accompanying them are also pollen grains resembling those of *Castalia*.

FIG. 166.—*Populus primaeva*. Probably the oldest leaf to be ascribed to a modern angiosperm genus. Kome series, Lower Cretaceous. Western Greenland.

The three examples just given show at least that there were plants in existence during the Jurassic with angiospermous characteristics. The decipherable history, however, of the flowering plants begins with the Lower Cretaceous, for it is not until this period that we find them extending in unbroken sequence to the present.

Some of the oldest undoubted angiosperms have come from the Lower Cretaceous Kome beds of Western Greenland where flowering plants are represented by a few leaf impressions. One of them, *Populus primaeva* (Fig. 166), is probably the oldest leaf to be referred to a modern angiosperm genus. Only slightly, if at all younger, are the Potomac beds of Maryland, which contain the most extensive of known Lower Cretaceous angiosperm floras (Fig. 184). Flowering plants in this flora are decidedly in the minority, with ferns, cycadophytes, and conifers in the lead. Within the Potomac formation an increasing number of angiosperm genera appear in passing from the lower to the higher beds. From the lowest, or Patuxent, about 100 species of plants have been identified, of which 6 appear to be angiosperms. These have been given such generic designations as *Ficophyllum*, *Quercophyllum*, and *Juglandiphyllum* in allusion to their resemblance to the living genera. In the succeeding bed, the Arundel, there is a small flora of about 30 species, of which 5 are referred to angiosperm genera. In the Patapsco, on the other hand, the flora of about 100 species contains about 25 undoubted angiosperms, which is a remarkable increase over the preceding stages. A few of these may be assigned to such modern genera as *Populus* and *Sassafras*, although most of them are given names that merely recall the modern genera they most closely resemble. Examples are *Nelumbites, Menispermites, Sapindopsis,* and *Celastrophyllum.* At no place in the Lower Cretaceous do angiospermous types exceed 25 per cent of the species.

Additional evidence of the existence of early Cretaceous flowering plants is the occurrence in England of petrified wood in the Lower Greensand. About five genera of dicotyledonous woods that possess vessels and rays typical of living members have been described. All available evidence, therefore, tends to show that although flowering plants were still in the minority during the Lower Cretaceous, they were, nevertheless, well established, and had been preceded by a long period of development.

The very meager fossil record of the Lower Cretaceous angiosperms reveals a few significant facts. It shows that during these times flowering plants were well established in northern latitudes. Furthermore, the flora of the Cretaceous formations in Western Greenland suggests the possibility that the Arctic region may have served as a general distribution area from which species spread to other parts of the earth. Secondly, the early Cretaceous angiosperms indicate the antiquity (although not necessarily primitiveness) of certain families such as the Nymphaeaceae, Salicaceae, Ebenaceae, Menispermaceae, and others. Thirdly, the occurrence of undoubted monocotyledons in the early Cretaceous in direct association with dicotyledons shows that as far as the fossil sequence is concerned the two groups are of about equal antiquity, and that neither is noticeably of more recent origin than the other.

It seems advisable to digress at this point and to comment upon some of the problems concerned in the interpretation of the remains of flowering plants. The fossil record of the angiosperms consists mostly of impressions of leaves, and the few fruits, seeds, and flowers present are seldom in organic connection with vegetative parts. Leaves, especially those which show a wide range of variation, are often difficult to identify correctly, and too often there has been a tendency to place normal variants in different species. For this reason the number of fossil species of such genera as *Sassafras, Populus, Salix, Ficus,* and *Liriodendron,* for example, may far exceed the number of actual species present in a given flora. Then, too, there has been a tendency to place too many forms within a comparatively few familiar genera. For example, the list of so-called "species" of *Ficus* from the Upper Cretaceous and

Cenozoic of North America includes more than 150 names. Those which may ultimately prove to belong to *Ficus* probably will not exceed 10 per cent of this number. Likewise, the one hundred odd fossil species of *Populus* not only include leaves belonging to other genera (such as *Cercidiphyllum*) but many variants of the same species. The tendency to multiply fossil species is especially prevalent when dealing with highly variable leaves such as *Sassafras* and *Quercus*.

The difficulty and frequently the impossibility of making correct determination on leaves alone has resulted in a rather general feeling of distrust of paleobotanical data among students of the Recent flora who rely mostly upon floral structures for criteria of affinity. Incorrect determinations of fossils may lead not only to erroneous ideas concerning the composition of a flora but will give inaccurate notions of the history of modern genera. Also, interpretations of past climates may be entirely wrong. Many of the earlier investigators of the Upper Cretaceous and Cenozoic floras, although placing the plants in modern genera, failed to take into consideration the important factor of the natural associations of these genera, with the result that the extinct floras were looked upon as composed of a heterogeneous conglomeration of ecologically unassociated types. Furthermore, insufficient consideration was often given to seeds and fruits associated with the leaves. Even when not in organic union, a fruit or seed may constitute the reliable clue to the identity of a dominating leaf type in a flora. For example, certain leaves in the Miocene of Oregon and Washington were variously assigned to *Sapindus*, *Rhus*, *Ficus*, and *Apocynum*, until the highly characteristic seeds of *Cedrela* were found in abundance among them. Now the occurrence of *Cedrela* at certain localities and in certain horizons is definitely established. In like manner some leaves that were once assigned to *Populus*, *Zizyphus*, *Grewia*, *Paliurus*, *Cissus*, and a number of other genera, have since been shown, because of associated seeds, to belong to *Cercidiphyllum*. These are but isolated examples chosen at random. It is hardly to be hoped that all or even most of the myriads of leaves occurring in the Upper Cretaceous and Tertiary rocks will ever be assigned to the correct genera. However, a few forms correctly placed are far more valuable as indicators of past climatic and other ecological factors than a multitude of wrongly identified specimens. When considered from all angles, the problem of elucidation of the angiosperm assemblages of the past is far from hopeless even though their identification must often contain a high percentage of errors.

A discussion of the several floral assemblages of the Upper Cretaceous and Cenozoic would center largely around enumerations of lists consisting almost entirely of modern genera. The Upper Cretaceous can positively be said to mark the beginning of modern plant life on the earth. The various climatic changes of the past, along with changes in altitude and surface characters, have at several times been strongly reflected in the floras, and recent studies of the floras have emphasized the assemblages in relation to possible environmental factors and to the distribution of genera now extant. One of the most remarkable of fossil floras is that of Western Greenland, where in rocks of Upper Cretaceous and early Tertiary age a large flora of temperate aspect exists. Accompanying the plants are beds of coal, which prove not only that there was an an abundant plant growth but that humid swamp conditions also existed. Thus we have direct floral evidence of marked climatic changes in the Arctic since early Tertiary times.

MONOCOTYLEDONS

More than 25 families of monocotyledons have been recognized in the Cretaceous and Cenozoic deposits of North America. The largest groups are the Gramineae, which includes the grasses, and the Palmaceae, or the palm family. Others that contain several representatives are the Cyperaceae and the Araceae. The remaining families for the most part are known by single genera.

Remains of monocotyledonous plants are always subordinate to those of the dicotyledons. This does not indicate that monocotyledons did not exist in considerable quantity during the late Cretaceous and the Cenozoic, but rather the paucity is explained on the basis of habitat. Many monocotyledons, especially those belonging to the grass family, inhabit the drier situations where there is less chance of their entering into the fossil series. Then, too, being for the most part low-growing forms, the detached foliage is not so readily carried by wind. A number of aquatic monocotyledons find their way into the accumulating sediments of swamps and bogs. Chief among these is *Typha*, the common cattail. *Typha* has been reported from the Magothy and Raritan (lower Upper Cretaceous), and from most succeeding formations up to the Recent. Then there is occasional reference to *Potamogeton*, *Sagittaria*, *Acorus*, and others. The occurrences of these types, however, is not especially

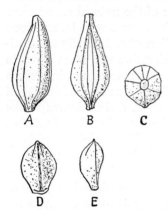

Fig. 167.—Caryopses of fossil grasses. (*A*) (*B*) (*C*) *Berriochloa glabra*. (*A*) Side view of lemma. (*B*) Ventral view of lemma with enclosed palea. (*C*) Basal view. All × 5. (*D*) (*E*) *Panicum elegans*. (*D*) Dorsal view of lemma. (*E*) Side view. × 12. Tertiary. Western Kansas. (*After Elias*.)

important as the identifications are often uncertain, and, moreover, their occurrence in water-laid deposits is something to be expected.

Leaves resembling *Smilax* have been found at several localities in the western part of North America. The oldest example of this genus is *S. grandifolia-cretacea* from the Dakota formation of Kansas. Several additional forms have been described from the Cenozoic.

Foliage, stems, and inflorescences of grasses are occasionally encountered in Upper Cretaceous and Tertiary rocks, although very few have been critically studied. Some of the genera that have been recognized are *Arunda*, *Poacites*, *Panicum*, and *Stipa*. Well-preserved fruitlets of *Panicum* (Fig. 167*D* and *E*), *Stipa*, and the extinct genus *Berriochloa* (Fig. 167*A*, *B*, and *C*), have been found in the Ogallala formation (late Tertiary) of the High Plains of western Kansas in association with remains of grazing mammals.

The most abundant and best preserved monocotyledons are palms. The remains consist of compressions of leaves, casts and compressions of fruits, petrified trunks, and roots (Fig. 168). Petrified palm trunks are usually assigned to *Palmoxylon*, a genus designed for stems with scattered bundles with large fibrous bundle caps. Foliage can sometimes be referred to Recent genera.

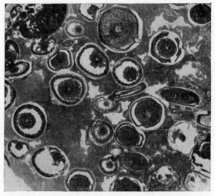

Fig. 168.—Silicified palm roots. Green River formation, Eocene. Wyoming. × about 5.

Palms first appear in the Upper Cretaceous rocks. *Palmoxylon cliffwoodensis* is from the Magothy formation of New Jersey, and a *Flabellaria* has been reported from the Dakota formation. During this time palms ranged as far north as Canada.

In North America the greatest display of fossil palms is found in rocks of Eocene age. Silicified trunks are common in many formations including the Yegua of Texas and the Green River of Wyoming. Foliage is also widely distributed. Large leaves of *Sabal* are abundant in the Raton flora of New Mexico, the Puget flora of Washington, and the Clarno of Oregon. In England numerous fruits of *Nipa* occur in the London Clay delta deposits.

Toward the close of Eocene time in North America shifting climatic conditions resulted in a migration of palms to the southward, which continued throughout the Pliocene. In the Mojave desert they occur in the middle Miocene Tehachapi flora, in the upper Miocene Barstow beds, and in the Ricardo beds of the lower Pliocene. Palms are not native in the Mojave desert except along the southern border at the present time because of deficient rainfall and low winter temperatures.

DICOTYLEDONS

The fossil record of the dicotyledons consists mainly of leaves and silicified wood, with a lesser assortment of seeds, fruits, and flowers. In North America dicotyledons occur principally in the Cretaceous and Tertiary deposits of the Atlantic coastal plain, along the coast of the Gulf of Mexico, and up the Mississippi River as far as Tennessee, in the Great Plains region east of the Rocky mountains, throughout the length of the Great Basin and into the Columbia Plateau, in the vicinity of the Puget Sound, and at numerous localities along the Pacific Coast from Mexico to the Yukon River. In other words, they are likely to be found wherever sedimentary rock formations of continental origin of Upper Cretaceous or Cenozoic age are exposed. Conditions during Tertiary time were exceptionally favorable in many localities for the preservation of large quantities of leaves. Along the eastern and southern coast of North America deposition took place principally in large embayments along the shallowly submerged continental shelf, where the extensive plant-bearing deposits of the Raritan, Magothy, Wilcox, Claiborne, and Citronelle formations were laid down. Inland, such extensive plant-bearing beds as the Dakota, Lance, Fort Union, Green River, Denver, and Raton were deposited in inland lakes and shallow seas. In such places as the Lamar Valley of Yellowstone National Park, at Florissant, Colorado, and the Great Basin and Columbia Plateau regions, the plants occur principally in volcanic ash deposits or in shale or diatomaceous earth beds formed in lakes produced by the damming of streams by lava flows. Of course the floras thus preserved are not made up exclusively of dicotyledons, but contain in addition ferns, gymnosperms, and monocotyledons.

The dicotyledons preserved in greatest number are the arborescent forms such as oaks, willows, poplars, sycamores, birches, alders, elms, hackberries, maples, lindens, dogwoods, and figs. Many of the larger shrubs are represented, but the low-growing herbs are a decided minority. Their occurrence is more or less accidental, and their rarity as fossils is not an indication of scarcity during the past. Leaves of forest trees, on the other hand, can be carried for great distances by high winds, and logs may float many miles from their source of origin to a place of deposition. As a rule, however, the majority of the specimens found in a fossiliferous deposit are from near-by sources. Often a single fossil flora will contain types from several local habitats. For example, the Trout Creek flora of southeastern Oregon, of Miocene age, which is preserved in a diatomaceous earth deposit laid down in a mountain lake about 7 by 10 miles in area, contains leaves, seeds, and fruits of a large number of plants which ranged from the shallow borders of the lake to wooded uplands some distance away. Thus the leaves and rhizomes of *Nymphaea* represent an aquatic habitat. Stream side or riparian habitats are indicated by leaves and fruits similar to a recent *Acer saccharinum* and *A. negundo* and by foliage resembling recent species of *Betula, Carpinus, Rosa, Salix,* and *Vitis.* Types more definitely upland are represented by leaves of which the modern equivalents are *Acer macrophyllum, Arbutus Menziesii, Quercus myrsinaefolia, Lithocarpus densiflora, Amelanchier alnifolia,* and *Sorbus sitchensis.* The association of species at present occupying these several habitats leads to the conclusion that the region during a part of the Miocene at least was of diversified relief. A level plain would be expected to yield a more uniform assemblage.

In the instance just mentioned the dicotyledons are accompanied by a few monocotyledons as *Typha* and *Lysichiton,* and a number of conifers including *Pinus, Thuja, Abies,* and *Picea.* Dicotyledons, however, constitute the majority of species.

SELECTED FAMILIES OF DICOTYLEDONS[1]

Magnoliaceae.—Of the seven living genera of the Magnoliaceae only two, *Magnolia* and *Liriodendron,* are well known in the fossil condition. The family is considered one of the most primitive of the flowering plants, and it is probably a derivative of the still more ancient ranalean stock from which other related families developed. It is also an interesting example in which primitiveness as indicated by the geological record is paralleled by the anatomy and morphology of the recent members. Mention has been made of certain resemblances between *Magnolia* and the cycadeoids.

Magnolia has been widely distributed throughout the Northern Hemisphere since early Upper Cretaceous time. It was especially abundant in the northerly latitudes during the warm Eocene epoch, but during the Pleistocene ice age it retreated toward the south and disappeared entirely from Europe. In common with other angiosperm genera the most abundant remains of *Magnolia* are leaf compressions. However, the conelike fruit clusters are sometimes found, and they provide the most positive means of identification of the genus.

Liriodendron was also abundant during the Upper Cretaceous although it has lately been found that a widely distributed leaf long known as *Liriodendron Meekii* is a leaflet of the legume genus *Dalbergia.* As

[1] These are arranged according to Bessey's system.

with *Magnolia,* the existence of *Liriodendron* in Europe was terminated by the Pleistocene glaciation. However, up to that time a species indistinguishable from the American *L. tulipifera* existed in the lowlands of Western Europe.

Cercidiphyllaceae.—This family is founded upon the genus *Cercidiphyllum,* the katsura tree, of Eastern China and Japan. Although now of limited distribution, investigations show that during the Cretaceous and early Tertiary its range extended throughout Europe and across the Arctic into North America. The oldest occurrence of *Cercidiphyllum* is in the Potomac group of Maryland. In west-central North America it is rather widely scattered in the early Tertiary Fort Union and other formations of similar age in Montana, Wyoming, Colorado, and the Dakotas. It is found in younger beds farther west and south, and its latest occurrence appears to be in the early Miocene along Bridge Creek in Oregon.

Leaves now believed to belong to *Cercidiphyllum* were originally referred to such genera as *Populus, Paliurus, Grewia, Zizyphus,* and several others. In 1922 Professor Berry proposed the name *Trochodendroides* for certain leaves resembling *Trochodendron,* which were found to belong to *Cercidiphyllum* when associated seeds and fruits were discovered.

Five fossil species of *Cercidiphyllum* have been named from North America. The leaves of the earliest species (*C. ellipticum*) are elliptic to broadly ovate-elliptic with rounded or cuneate bases. In the later forms they become more deltoid (*C. arcticum*) and finally cordate (*C. crenatum*). The fruits are small, blunt-pointed pods, and the seeds bear a superficial resemblance to the small winged seeds of some conifers. Considerable variation exists among the leaves, seeds, and fruits of the fossil forms just as in the one living species.

Berberidaceae.—There are no satisfactory fossil records of *Berberis* from North America but in Europe the genus has been recorded from the Oligocene and Miocene. The so-called *Berberis gigantea* from the Miocene of the John Day Valley in Oregon is believed to belong to the Araliaceae. Several species of *Mahonia* (*Odostemon* of authors) are known from the Eocene, Miocene, and Pliocene of Western North America.

Menispermaceae.—A dicotyledonous family frequently mentioned in paleobotanical literature but of which little precise information is available is the Menispermaceae. Many years ago the name *Menispermites* was proposed for a large, broadly deltoid, three-lobed leaf with entire or undulate margins and with three to five prominent primary veins, which diverge from a peltate or subcordate base. The finer veins resemble those of *Menispermum*. The genus was based upon a group of Upper Cretaceous leaves that had previously been assigned to *Acerites*, *Dombeyopsis*, and *Populites*. There is, however, no proof that any of the several original species of *Menispermites* are related to the Menispermaceae although the name has also been used for later material, which probably does belong to this family.

Lauraceae.—The most readily recognizable of the several lauraceous genera that have been found in the fossil condition is *Sassafras*. Although lauraceous foliage is easily confused with genera belonging to other families, the leaves are characteristically oval or broadly canoe-shaped, being broadest at the middle and tapering to an almost acute base and apex. In some genera two prominent lateral veins leave the midrib at an acute angle above the base, and either curve upward maintaining about the same distance from the margin, or in the lobed leaf of *Sassafras* pass nearly straight into the lobes. These three main veins are often interconnected by a system of prominent and nearly transverse cross veins. Most of the genera except *Sassafras* and *Benzoin* prefer mild climates and even these two range into the warmer latitudes. In the Eocene London Clay flora the Lauraceae appear to be the best represented of all families of dicotyledons where about 40 species of seeds and fruits have been recognized. Many of them belong to extinct genera.

Of the various fossil representatives of the Lauraceae *Sassafras* is best known. It is also one of the oldest of living forest tree genera. Leaves unquestionably belonging to this genus have been found in the late Lower Cretaceous of both Europe and North America. During the Upper Cretaceous it spread into the Arctic. In the Dakota sandstone nearly a dozen species have been named although it is certain that many of them represent normal variations of a lesser number (Fig. 169*A*). However, it is altogether probable that at the time of its widest distribution over the Northern Hemisphere, there were more species than at present.

Although now extinct in Europe, *Sassafras* existed there as late as the Pleistocene. It is now restricted to Eastern North America and Eastern Asia, but was present in the Great Basin area during the Miocene. *S. hesperia* (Fig. 169*B*), is a common species, having been found at several places in Washington and Oregon.

Other fossil representatives of the Lauraceae include *Cinnamomum*, *Persea*, *Laurus*, *Machilus*, and *Umbellularia*.

Dilleniaceae.—This family, which has living members in all continents, is mainly Oriental, being most copiously represented in Australia. Its rather meager fossil record indicates that during the early Tertiary it was expanded considerably beyond its present limits in response to milder climatic conditions. The fossil remains also show that certain genera which are confined at present to the Western Hemisphere existed there during the early Tertiary as, for example, *Empedoclea* and *Doliocarpus,* which are represented by two species each in the early Tertiary of Chile. Other genera of the family have been found in the early Tertiary of England, France, Holland, Belgium, and other places in Europe.

In North America the fossil occurrences of the Dilleniaceae are mostly in the Eocene. Berry has described five species of *Dillenites* from the

| (A) | (B) |

Fig. 169.—(A) *Sassafras acutilobum.* Dakota sandstone. Salina, Kansas. Slightly reduced, (B) *Sassafras hesperia.* Miocene. Malheur County, Oregon. Slightly reduced.

Wilcox and other Eocene beds of the Southeastern United States. *Dillenites, Dillenia,* and *Saurauja* have been reported from the Tertiary of Alaska, and *Tetracera* has been found in the Goshen beds of Oregon, the Weaverville beds of California, and the Latah beds of Washington, the last named being its most recent occurrence. *Actinidia* also occurs in the Weaverville beds. Members of the Dilleniaceae may be more prevalent in the fossil condition than is generally realized because of the similarity of the leaves to those of *Castanea* and certain species of *Quercus.*

Sterculiaceae.—This family, which at present is widely distributed throughout the tropics, is probably an old one, and its history is similar to that of many others which had spread far beyond their present limits during the late Cretaceous and early Cenozoic. A fragment of a deeply iobed leaf from the Patapsco formation of the Potomac group was named *Sterculia elegans* by Fontaine, but it hardly constitutes acceptable

(A)

(B)

Fig. 170.—(A) *Sterculia coloradensis.* Green River formation. Fossil, Wyoming. Slightly reduced. (B) *Ficus mississippiensis.* Green River formation. Fossil, Wyoming. Slightly reduced.

evidence of this genus in the Lower Cretaceous. In the Magothy and Dakota groups of the Upper Cretaceous, and in the Eocene, a number of so-called "species" of *Sterculia* have been named (Fig. 170*A*). The genus is rare in the Miocene of North America.

Tiliaceae.—The genus *Tilia* has been reported from the Upper Cretaceous, but the oldest remains that have much claim to authenticity are from the Paleocene Raton and Fort Union deposits of Colorado and Montana. The genus spread widely during the Eocene and extended into the Arctic. Leaves and the characteristic bracts have been found at a number of localities in the Miocene of Western North America. The leaves have often been confused with those of *Platanus* and *Vitis*.

There are many references to *Grewia* and *Grewiopsis* in the older literature on the fossil plants of Western North America, but the status of most of these is doubtful. Some of them belong to *Cercidiphyllum*.

Ulmaceae.—Although the remains of *Ulmus* are easily confused with those of such genera as *Ptelea*, *Zelkova*, and *Ostrya*, there is ample evidence that it was an abundant genus during the Eocene. Its Cretaceous history is uncertain although there is no reason to doubt that it was in existence then. During the Miocene *Ulmus* exceeded its present distribution although it was not as abundant at that time in North America as in Europe. A few species have been reported from the John Day Valley in Oregon.

The history of the genus *Celtis* is not well known although it is recorded from the Upper Cretaceous of Europe. A few specimens have been found in the Eocene of Wyoming and Georgia and in the Miocene of Oregon. In the late Tertiary of western Kansas and Oklahoma, leaves and fruits of *Celtis* occur in deposits with the remains of grasses and grazing mammals, which shows that the genus had long been able to thrive where the rainfall was below the optimum for most forest trees.

Moraceae.—The genera of this family having the most paleontologic importance are *Artocarpus* and *Ficus*. The large deeply cut leaves of the breadfruit have been found in the early Upper Cretaceous Cenomanian series of Greenland and in the Upper Cretaceous Vermejo beds of Colorado. Objects believed to be the fruits have also been found in Greenland. In the Eocene *Artocarpus* occurs in the Wilcox flora of the Gulf States and the Raton flora of Colorado.

Ficus is well represented in the Cretaceous and early Tertiary, although it is extremely doubtful whether more than a small proportion of the compressions that have been assigned to it are correctly placed. From North America alone about 150 species have been recorded. Assignment to *Ficus* is often made mainly upon apparent coriaceous texture of the leaves, the distinct reticulate pattern often produced by the finer veins, and the elliptical or broadly oval or ovate outline. Some leaves are entire, but others are deeply cut. In the Lower Cretaceous

Potomac group of Maryland, leaves somewhat resembling *Ficus* are called *Ficophyllum*, but these do not constitute positive evidence of the existence of figs at that time. Most of the leaves assigned to *Ficus* occur in the Upper Cretaceous and early Tertiary rocks. *F. planicostata* is an index species of late Cretaceous deposits in the Rocky Mountain region. Its leaves are less than 10 cm. long, elliptical or broadly oval, broadest near the middle, and the apex is abruptly narrowed to a blunt point. *F. mississippiensis*, a species characteristic of the Eocene Wilcox flora, is larger, ovate or ovate-lanceolate, broadest below the middle, and has an acute tip (Fig. 170*B*). Leaves have been referred to *Ficus* from most of the Eocene localities of North America and also those from the Arctic and Europe. It is occasionally mentioned from the Miocene, but seldom from the temperate latitudes.

Rutaceae.—Several genera of the Rutaceae have been reported from the Tertiary, but one most frequently encountered and most easily recognized is *Ptelea*. This genus at the present time is confined to North America, and although it has been reported from the Tertiary of Europe, Professor Berry doubts that it ever existed there. The large, circular-winged, compressed fruits of *Ptelea* are more easily identified than the foliage although they are sometimes confused with the samaras of *Ulmus*, which they closely resemble. Three or four species of *Ptelea* have been described from North America, but the most important ones are *P. eocenica* from the lower Eocene Wilcox flora, and *P. miocenica* which is widely distributed throughout the Miocene of the western states.

Simarubaceae.—The genus *Ailanthus*, limited at present to eastern Asia, is represented in the Eocene and Miocene of Western North America and in the Oligocene and Miocene of Europe. Its most diagnostic remains are the long-elliptical samaras, which have been found in Colorado, Oregon, Wyoming, and other places.

Meliaceae.—Leaflets of *Cedrela* occur in the Eocene Wilcox and Lagrange formations of the lower Mississippi Valley, and leaflets, seeds, and capsules are found in the Miocene of Colorado, Oregon, Idaho, Nevada, and Washington (Fig. 171). The fossil leaflets of this genus have been confused with those of *Sapindus*, *Rhus*, *Apocynum*, and *Umbellularia*, and the seeds with *Pinus*, *Pseudotsuga*, *Libocedrus*, *Acer*, and *Gordonia*. It is mainly when the seeds are present that the foliage can be identified with certainty.

Although the seeds of *Cedrela* bear a superficial resemblance to the winged seeds of some conifers, they are readily distinguished upon close inspection. The tip of the wing is rounded and usually shows a crescent-shaped bevel, and the lower margin often bears a short, backward-pointing, spinelike point (Fig. 171*C*).

Malpighiaceae.—Several genera of the Malpighiaceae have been identified from the characteristic winged fruits. Samaras resembling *Banisteria* and *Hiraea* occur in the Wilcox and Lagrange floras.

Euphorbiaceae.—This large family has comparatively few fossil representatives although such names as *Euphorbiophyllum* and *Crotonophyllum* are occasionally encountered in the literature. The fossil occurrences are mostly Upper Cretaceous and Eocene. Euphorbiaceous

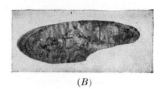

(B)

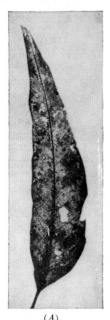

(A)   (C)

Fig. 171.—(A) *Cedrela oregonensis*. Miocene. Malheur County, Oregon. Slightly reduced. (B) *Cedrela pteraformis*. Seed. Miocene. Malheur County, Oregon. Slightly enlarged, (C) *Cedrela sp.* Capsule. Miocene. Malheur County, Oregon. Natural size.

fruits have been found in the London Clay. Leaves referred to *Crotonophyllum* occur in the Lagrange, Magothy, and Tuscaloosa beds, and *Drypetes* has been reported from the Lagrange and Wilcox of the Southeastern United States. Along the Pacific Coast *Aporosa* and *Mallotus* have been found in Oregon, and *Aclypha*, *Aleurites*, and *Microdesmis* occur in the La Porte flora of California.

Nymphaeaceae.—Pollen grains recently discovered in Jurassic coals in Scotland render the Nymphaeaceae the oldest recognizable family of the dicotyledons. The probable angiosperm *Furcula* from the late Triassic is older, but its family relationships are undetermined.

The Jurassic nymphaeaceous pollen is practically indistinguishable from that of the modern *Castalia* and *Nelumbium*, the resemblance being so close that it seeems impossible that they could belong to any other family.

Literature on Upper Cretaceous and Cenozoic floras frequently includes descriptions of leaves, rhizomes, and seed capsules referable to such genera as *Nymphaea*, *Castalia*, *Nelumbo*, and *Brasenia*. Being aquatic plants, they grow in situations favorable for preservation, although large quantities of the remains are seldom found at any one place.

Salicaceae.—Regardless of the uncertainty shrouding the identity of many fossil leaves assigned to *Populus*, there is ample evidence that this genus has a long history and that it was widely spread and abundant throughout most of the late Cretaceous and Tertiary. More than 125 species of fossil poplars have been described, and at least 100 of these occur in North America. However, some of the most common fossil leaves, such as the supposed *Populus Zaddachi*, are now believed to belong to *Cercidiphyllum*.

The oldest plants thought to be poplars are found in the late Lower Cretaceous. Poplarlike leaves are abundant in the Dakota group of the Upper Cretaceous, and throughout the Eocene and Miocene they are everywhere present in the continental deposits of Western and Central North America and the Arctic. The favorite habitat for poplars throughout the past has been along streams and on mountain slopes and other places where they were not shaded during their early stages of growth by taller trees, and for this reason they are absent from deposits formed under tropical conditions. They often characterize wind-blown deposits that were carried into open lakes.

The history of *Salix* is similar to that of *Populus*, but it seems to have reached its maximum development in postglacial times. Willow-like leaves occur in the Lower Cretaceous and many have been found in the Upper Cretaceous. By Eocene time the genus had migrated into the Arctic where it has persisted to the present day. About 60 fossil species of willows have been described from North America, but this number is excessive due to the frequent confusion between willows and other trees with narrow, taper-pointed leaves.

Unfortunately, the geologic record has yielded nothing of the structural details of the inflorescences of the ancient willows so that it throws no light upon the question of whether the flowers are primitive or reduced. However, the ecological implications are that the Tertiary willows and poplars possessed adaptations similar to those of the present-day species, which would lead one to suspect that the simple flower structure so strikingly expressed is at least an ancient feature.

Ebenaceae.—During the late Cretaceous and early Tertiary the genus *Diospyros* extended considerably to the north of its present southerly range in the Northern Hemisphere. Its leaves are readily confused with those of several other genera, but the characteristic leathery four-lobed calyces frequently supply a reliable clue to the occurrence of the genus. As would be expected, *Diospyros* is well represented in the Wilcox and other Eocene floras, and it is occasionally encountered in the Miocene.

Ericaceae.—Of the Ericaceae, several species of *Arbutus* have been reported from the Tertiary of the Western States, *A. Trainii*, from the Miocene of Oregon, resembles the living madrona of the Pacific Coast. Fossil species of *Arcotostaphylos, Vaccinium, Andromeda,* and other genera have also been reported.

Oleaceae.—The geological history of this family centers mainly around the genus *Fraxinus* whose history goes back at least to the dawn of the Cenozoic. Several species occur in the Eocene of North America. In Europe *Fraxinus* does not appear until the Oligocene, which suggests that the genus is of American origin. It became widely spread during the Miocene but apparently declined somewhat during the Pliocene. The leaflets of *Fraxinus* are sometimes confused with those of other dicotyledons, but fortunately they are often accompanied by the characteristic fruits.

Rosaceae.—This large and diversified family is represented in the Cenozoic by a wealth of leaf compressions belonging for the most part to such genera as *Amelanchier* (Fig. 172*A*), *Crataegus* (Fig. 172*C*), *Cercocarpus, Cydonia, Chrysobalanus, Rosa, Pyrus, Spiraea,* and *Prunus.* Less frequent genera are *Chamaebatia, Chamaebataria, Holodiscus* (Fig. 172*B*), *Lyonothamnus,* and *Photinia.* Some members of the Rosaceae doubtlessly were in existence during the late Cretaceous, but in general it may be assumed that the family did not become widely distributed or abundant until the Miocene. Eocene occurrences are relatively few. In the western part of North America, rosaceous types are almost invariably present in those deposits which show evidence of slackened precipitation, as in the Miocene Tehachapi flora of the western part of the Mojave Desert where *Cercocarpus, Chamaebataria, Holodiscus, Lyonothamnus,* and others occur in a typical chaparral association with *Ceanothus, Arctostaphylos,* and species of *Quercus.* At other Miocene localities, such as at Trout Creek in southeastern Oregon or at Creede, Colorado, where there is evidence of a cool and temperate but not an arid climate, we find an abundance of leaves of *Amelanchier, Crataegus, Prunus, Pyrus,* and other genera. Paleobotanical evidence seems to indicate that the Rosaceae have long been important members of the floras of drier situations within the temperate zone, and that they

(A)

(B)   (C)

FIG. 172.—(A) *Amelanchier dignatus.* Slightly reduced. (B) *Holodiscus harneyensis.*
Natural size. (C) *Crataegus microcarpifolia.* Slightly reduced. All from the Miocene.
Harney County, Oregon.

ranged from areas receiving moderate rainfall to those predominantly semiarid.

Leguminosae.—This large family, like the Rosaceae, has many fossil representatives. Being mainly a tropical family, and well represented in the temperate zone, its range during the warm Upper Cretaceous and Eocene epochs was considerably beyond its present limits. The genus *Dalbergia,* for example, existed in Greenland during the Upper Cretaceous. In the Eocene Wilcox flora of the Southeastern United States, the legumes constitute the largest assemblage with about 86 nominal species belonging to 17 genera. The best represented genera are *Cassia, Dalbergia,* and *Sophora.* Some investigators believe that the legumes were the largest angiosperm alliance during the Eocene.

The legumes occupy a less conspicuous place in the Miocene assemblage although both foliage and pods are frequently found. Some genera reported from the North American Miocene are *Cassia, Sophora, Cercis, Robinia,* and *Pithecolobium.*

Saxifragaceae.—The Saxifragaceae is represented in the fossil series by several genera including *Hydrangea* and *Philadelphus.* The four-parted sterile flowers of *Hydrangea* furnish undisputed evidence of the existence of this genus in the Miocene of Washington and Oregon, and leaves believed to belong to it occur at Florissant, Colorado. A few species of *Philadelphus* have been reported from the western Miocene.

Hamamelidaceae.—A few years ago a portion of a *Liquidambar* leaf was found in the Upper Cretaceous Aspen shale of Wyoming, but previous to this discovery, the oldest authentic remains were from the late Eocene of Oregon and the subarctic. This genus was widely distributed during the Miocene, the remains having been found in Japan, Asiatic Russia, Central Europe, Oregon, Idaho, and Colorado. It persisted in these latitudes until the Pleistocene glaciation when it became extinct in Europe and in Western North America.

The genus *Hamamelis* is probably an ancient one but its history is not well known probably because of the uncertainty in distinguishing its foliage. Leaves referred to the form genus *Hamamelites* are found in the Dakota and Fort Union formations.

The genus *Fothergilla,* consisting at present of three species in the Southeastern United States, has recently been found in the Miocene of Shantung, China. *F. viburnifolia,* the fossil species, closely resembles the living *F. Gardeni.* This is the only known occurrence of the genus in the Tertiary.

Platanaceae.—The genus *Platanus* is one of the oldest of the dicotyledons and has a nearly continuous fossil record from the middle of the Cretaceous to the present. Its first definite appearance is in the early

Upper Cretaceous Raritan formation of New Jersey and the Tuscaloosa formation of Alabama. Four so-called "species" have been named from these beds. The genus appears to have originated in North America. During the late Cretaceous *Platanus* spread into the Artic, and from there, either during the late Cretaceous or the early Tertiary, into Europe.

The wide range in form of the leaves of *Platanus* renders determination of the fossils rather difficult, and, as with other genera, there has been a pronounced tendency toward multiplication of species. There has

Fig. 173.—*Platanus dissecta.* Miocene. Malheur County, Oregon. ½ natural size.

been considerable confusion between *Platanus* and certain members of the Aceraceae, Araliaceae, and Vitaceae. Nevertheless, the fossil record seems to indicate quite clearly that the evolutionary trend has been from the shallowly lobed leaf with a cuneate base to those with larger lobes and a rounded or cordate base. Thus in the early Tertiary forms such as *P. coloradensis* the terminal lobe is strong but the lateral ones are weak or absent, and the base is almost as acute as the tip. In the middle Eocene *P. appendiculata* the leaf is more nearly round and may or may not possess short obtuse lateral lobes. The more modern type of leaf is expressed by the well-known *P. dissecta*, a Miocene form quite similar to the modern *P. occidentalis* and *P. orientalis*. *P. dissecta*

(Fig. 173), is widely distributed throughout the Miocene lake beds in Washington, Oregon, California, Nevada, Idaho, and neighboring areas,

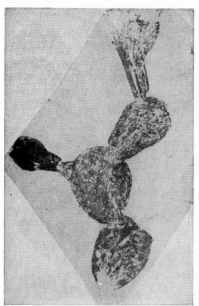

with the most prolific locality for the leaves being along Sucker Creek in eastern Oregon. At about the end of the Pliocene the sycamores of the *P. dissecta* type disappeared from Western North America to be superseded in some places by a deeply cut form, *P. paucidentata*, which is probably the forerunner of the western *P. racemosa*.

Cactaceae.—The fossil record of this family is limited to a single specimen from the Eocene of Utah named *Eopuntia* (Fig. 174), because of its resemblance to the modern prickly pear cactus. The specimen consists of three joined sections of which one bears a large open trumpet-like flower, and another a pear-shaped fruit. This one discovery shows that the Cactaceae are

Fig. 174.—*Eopuntia Douglassii.* Eocene. Utah. (*After Chaney.*) Reduced.

not a recent development but one that has probably been in existence for some time. Its paucity in the fossil series is probably to be explained by its occurrence largely on ridges or arid basins where its stems did not enter readily into the fossil record.

Rhamnaceae.—This family is rather well represented in the fossil series. *Rhamnus, Paliurus,* and *Zizyphus* are the most frequently encountered genera but a number of others are known. *Rhamnus* appears in the early Upper Cretaceous, and is widely distributed though not always abundant in the North American Eocene. Six species have been described from the Denver formation. A few species are known in the Miocene, one being *R. idahoensis* from the Latah formation, which is comparable to the living *R. Purshiana,* the cascara tree, of the Pacific Northwest.

Leaves believed to belong to *Paliurus* are found throughout the Upper Cretaceous of North America including Alaska. The genus is most certainly recognized when the peltate, circular-winged fruits are present. The oldest of these are from the lower Eocene. During the Eocene *Paliurus* existed throughout North America and extended into the Arctic. It was still present in North America during the Miocene when it was preserved in the Florissant lake beds and in the Latah clays at Spokane, Washington. *Zizyphus* also appeared in the early Upper Cretaceous, but has not been found in rocks of that epoch in Europe. However, in North America several species are known. In the Eocene *Z. cinnamomoides* is an abundant species in the Green River formation, and the genus is well represented elsewhere in North America and also in Europe. Six species have been described from the Denver formation alone.

Rhamnidium, one of the rarer genera, existed in California during the Eocene, and then migrated to Central and South America where three Miocene species are known. *Berchemia* is reported from the Upper Cretaceous and Eocene of North America but is best represented in the Miocene and Pliocene.

Of other fossil representatives of the Rhamnaceae mention should be made of *Ceanothus*, *Colubrina*, *Condalia*, and *Karwinskia*, which occur in the late Miocene and early Pliocene in association with species indicative of semiarid conditions such as are portrayed by the Tehachapi flora of the Mojave Desert.

Vitaceae.—The history of this family rather closely parallels that of the Rhamnaceae to which it is closely related, but it is often difficult to distinguish the foliage of the Vitaceae from that of such genera as *Acer*, *Aralia*, *Platanus*, and some members of the Menispermaceae.

Fossil leaves resembling *Cissus* are often referred to the artificial genus *Cissites*. *Cissus* and *Cissites* are common in the Upper Cretaceous and the Eocene although many of the leaves assigned to these genera undoubtedly belong to something else. Half a dozen or more species of *Ampelopsis* have been described from the early Eocene of North America and the late Oligocene and early Miocene of Europe. *Vitis* is believed to be present in the Upper Cretaceous and early Tertiary of Western North America but in Eastern North America it is not known before the Pliocene. In the Miocene, it has been found in Colorado, Oregon, and Washington.

Celastraceae.—A large number of simple serrate leaves from the Paleocene, Eocene, and Miocene have been assigned to *Celastrus* but the proportion of inaccurate determinations is probably large.

Sapindaceae.—The principal fossil genus of this family is *Sapindus* of which more than 130 species have been named. However, not more than two-thirds of this number are considered authentic. The genus dates from the Upper Cretaceous, but it is best represented in the Eocene and Miocene. Our knowledge of the history of the genus, however, is not entirely satisfactory because of the ease with which the leaves are confused with those of such genera as *Carya*, *Juglans*, and *Cedrela*.

Winged fruits resembling those of *Dodonaea* have been found in the

Eocene of the Southeastern United States and other places. *Cupanites* and *Cupanoides* are names for leaves and fruits, respectively, that resemble *Cupania*. Both occur in the Eocene, the latter in the London Clay. Two species of *Cupania* occur in the Goshen flora of Oregon.

Aceraceae.—As compared with most dicotyledonous families, the history of the genus *Acer* is well known. Although there is some danger of confusing the leaves with other lobed, palmately veined types, the leaves are often accompanied by fruits that render the determinations certain beyond all question.

The maples first appear in the Upper Cretaceous, and in the Eocene they occur in the Arctic. Their greatest development, however, was

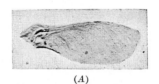

(A)

(B)

Fig. 175.—(A) *Acer glabroides.* Samara resembling that of living *A. glabrum.* Miocene. Harney County, Oregon. Natural size. (B) *Acer Osmonti.* Samara believed to be equivalent to that of living *A. saccharum.* Miocene. Harney County, Oregon. Natural size.

during the Miocene. Maples are likely to occur in any Miocene plant-bearing deposit, and they rank high in species as well as in number of individuals. At least half a dozen positively determined species occur in the Miocene of the Great Basin and the Columbia Plateau regions, especially in the Bridge Creek, Mascall, and Latah formations. They are all similar to living species although the living equivalents of some are confined to areas remote from Western North America. *Acer Bendirei* is the name given large leaves and fruits that closely resemble the living *A. macrophyllum* of the Pacific Coast. *A. gladroides* (Fig. 175A), resembles *A. gladrum*, and *A. Osmonti* (Fig. 175B) appears indistinguishable from the recent *A. saccharinum*. *A. Scottiae* includes leaves and fruits resembling those of *A. pictum* of Eastern Asia, and the modern box elder, *A. Negundo*, finds its Miocene equivalent in *A. negundoides*.

Unlike many other trees, the maples were not completely eradicated from Europe by the Pleistocene glaciation. They retained a foothold along the Mediterranean Coast.

The paired but more or less circular winged fruits of *Dipteronia* have been found in the Miocene of Colorado and Washington. This genus is at present confined to Central and Western China.

Anacardiaceae.—The most important fossil members of this family belong to the genera *Rhus*, *Pistacia*, and *Anacardites*, the latter being a form genus for anacardiaceous foliage of uncertain affinity. Many fossil species of *Rhus* have been named but not all of them can be regarded as authentic. *Rhus* dates from the Upper Cretaceous, but it seems to be most prevalent in the Oligocene and Miocene.

Juglandaceae.—Several members of the Juglandaceae are known in the fossil condition. The leaf record is fraught with numerous mis-identifications, but the characteristic fruits and seeds often provide clues to identification not available for some other families. In the Eocene London Clay flora, inflorescences and seeds resembling those of *Juglans* and *Platycarya* are described as *Pterophiloides* and *Juglandicarya*. Nuts undoubtedly of *Juglans* or *Carya* occur in the lake deposits at Florissant, Colorado, in the Miocene Braunkohle of Germany, and in the White River Oligocene of Nebraska.

The former distribution of *Juglans* was formerly much wider than at present. It inhabited Europe from the Upper Cretaceous to the Pliocene. *Pterocarya*, which is limited to a few species in Transcaucasia, China, and Japan, was present in North America and Europe during the early Tertiary, and *Engelhardtia*, characterized by its conspicuous fruit bract, was even more widely dispersed. Well-preserved fruits occur in the Wilcox flora near the Gulf of Mexico. *Carya* is unknown in the Cretaceous but is present in the early Eocene in Western North America. The genus appeared in Europe during the late Eocene and attained its widest distribution during the Miocene. The Pleistocene glaciation drove it from Europe and Western North America, but it has persisted east of the Great Plains.

Betulaceae.—The genus *Betula* was abundant and widely spread throughout the Northern Hemisphere during the Eocene. Unlike most of the angiosperms that extended into the Arctic at that time, some of its members became adapted to the boreal climate and have survived there. *Betula nana*, a dwarf birch of the far north, is a minute woody plant which in some habitats does not exceed in height the mosses among which it grows. *Betula* is also abundant in Miocene deposits, especially those of Southern Europe. The oldest known birches are found in the Dakota sandstone and the Upper Cretaceous rocks of Greenland. The bilaterally winged fruits and pollen grains are commonly met with in Pleistocene peat.

The foliage of *Betula* is not easy to distinguish from that of *Alnus*, and it is only when the characteristic conelike catkins of the latter are present that positive determinations can be made. The history of *Alnus* closely parallels that of *Betula* except that it has extended into the Southern Hemisphere. It has been found in the Pliocene of Bolivia.

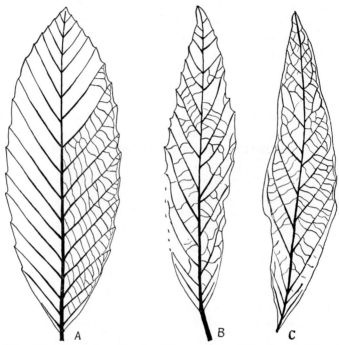

Fig. 176.—(*A*) *Dryophyllum Moorii.* Wilcox group. (*After Berry.*) ½ natural size. (*B*) and (*C*) *Dryophyllum subfalcatum.* Medicine Bow formation, Upper Cretaceous. Wyoming. *Drawn from retouched collotype by Dorf.* ¾ natural size.

The genus *Corylus* is not known from the Cretaceous. It extended into the Arctic during the Eocene but had retreated to approximately its present range by Miocene time.

The foliage of *Ostrya* and *Carpinus* are often confused and identification of these genera is difficult unless the fruits are present. In *Ostrya* the small nutlike fruit is enclosed within a saccate involucre, whereas the involucre of *Carpinus* is a deeply lobed, leaflike structure. The history of the two genera is quite similar although that of *Carpinus* is less known. The present discontinuous distribution of *Ostrya* in

North America can be explained from the fossil record. As late as the Pleistocene it extended over at least the southern half of the continent with a continuous range from the Atlantic Coast to Central America, but the advent of arid conditions in Texas and Northern Mexico separated the southern extension of the genus from the northern.

Fagaceae.—It is believed that the modern members of this family were derived during the Upper Cretaceous time from the extinct genus *Dryophyllum* (Fig. 176). *Dryophyllum* is not known in rocks older than Upper Cretaceous, but since it appears to have been widely spread and highly diversified at the beginning of this epoch, there is reason to believe that it existed earlier. Leaves of *Dryophyllum* are lanceolate to oblong, and they usually have dentate margins. The secondary veins are pinnately arranged and more or less parallel, and the veinlets are transversely decurrent. The leaves resemble the chestnut or the holly oak type, which suggests that the deeply lobed leaves are derived. *Dryophyllum* is not restricted to the Cretaceous but is abundant in the Eocene of the Gulf Coast. It probably became extinct in the Miocene.

Fagus was probably derived from the *Dryophyllum* complex during the late Cretaceous, and it became widely spread during the Tertiary. During the early Tertiary it existed along with the closely related *Nothofagus* in Southern South America and Australia.

Quercus is one of the most commonly encountered genera of the Upper Cretaceous and Cenozoic. The fossil record consists of numerous leaf compressions occasionally accompanied by acorns and quantities of petrified wood. Because of the wide variation in leaf form, not only between species but within the same species, the geological history of *Quercus* is somewhat confused, and the accuracy of the records has suffered not only from erroneous determinations but also from an over-multiplication of fossil species. As long ago as 1919 Knowlton listed 184 species of oaks (including the form genera *Quercinium* and *Quercophyllum*) from North America alone. As several more have been subsequently named, the total probably exceeds 200. Professor Berry, however, has shown that some of the supposed lower Eocene oaks of the Southern States belong to the Dilleniaceae, and numerous others, upon critical examination, have been found to belong to genera unrelated to the Fagaceae.

The oaks are believed to have been derived from the *Dryophyllum* plexus early during the Upper Cretaceous, and a number may be recognized in rocks of this age from Alaska and Greenland. Many others, including fruits, are known from the Eocene. Practically all the early oaks are of the unlobed or chestnut type, and the prominently lobed leaf did not become common until the Miocene.

The volcanic ash and diatomaceous earth deposits of the northern Great Basin and Columbia Plateau regions and adjacent areas of the Western United States have yielded enormous numbers of unlobed leaves. These include among others *Quercus simulata* (Fig. 177) and the very similar (some authors believed identical) *Q. consimilis*, which are probably the fossil equivalents of the living *Lithocarpus densiflora* or *Q.*

FIG. 177.—*Quercus simulata.* Equivalent to the living *Q. myrsinaefolia.* Miocene. Harney County, Oregon. ⅔ natural size.

myrsinaefolia. There are others that resemble *Q. chrysolepis* and *Q. hypoleuca.* The deeply lobed leaf which appeared at this time is expressed in *Q. pseudolyrata* which is the equivalent of the living *Q. Kelloggii.* Probably the ultimate in lobation is revealed by *Q. cruciata* from the Miocene of Switzerland.

Castanea and *Castanopsis* also emerged from the *Dryophyllum* stock during the late Cretaceous. Their leaves often occur along with those of *Quercus* and in some cases they are difficult to distinguish. Such names as *Quercus castanopsis* are frank admissions of this problem.

Myricaceae.—The Myricaceae is represented by a few genera including *Myrica* and *Comptonia.* The latter genus is best represented in the Miocene although it is reported from formations as old as the Raritan. *Comptonia* is sometimes confused with the rosaceous genus *Lyonothamnus.*

Araliaceae.—The existence of foliage resembling that of the Araliaceae at a number of Lower Cretaceous localities has led to the belief that this family is an ancient one. Its exact status, however, is greatly in doubt because of the difficulty in distinguishing the foliage from some other dicotyledons, such as *Acer, Platanus*, and *Vitis.* Several species of araliaceous leaves have been described from the early Upper Cretaceous Raritan, Magothy, and Dakota groups of the United States and from the Chignik and other Upper Cretaceous beds of Alaska. In the lower Eocene Wilcox flora, the Araliaceae is represented by several species of *Aralia, Oreopanax*, and *Schefflera.*

Considerable doubt has been cast upon several supposed species of *Aralia* from Eocene, Oligocene, and Miocene of the western half of North America. Some specimens of *A. Whitneyi* from the Eocene and

Miocene are now believed to belong to the Platanaceae, and other species have been confused with *Acer*. *Aralia dissecta* is a large deeply cut leaf from the Florissant beds, which may be correctly assigned. The existence of *Oreopanax* in the Miocene of eastern Oregon and northern Nevada is unquestioned, and two and probably three species are known. One of these, *O. precoccinea*, was so named because a fragment of a leaflet was originally mistaken for a leaf of *Quercus coccinea*, the scarlet oak. The leaf is large and consists of seven to nine spiny, deeply cut leaflets attached to the summit of the petiole. *O. Conditi* is a similar species from northern Nevada, and *O. gigantea* is based upon a fragment of a very large leaf from the Mascall beds of central Oregon. All these leaves are leathery and are believed to be holdovers of the earlier Tertiary floras, which have since receded into the mountainous regions of Central America and Northern South America.

Cornaceae.—The leaves of *Cornus* first appear in the Upper Cretaceous rocks. During this time, however, the genus appears to have been restricted largely to North America, and not until the Eocene did it exist to any extent in Europe. In the late Tertiary it spread completely across Europe and Asia. The leaves and the characteristic bracteolate inflorescences are widely scattered though seldom abundant in the Miocene of the Western States. During the Pliocene it seems to have waned somewhat in North America.

The history of the genus *Nyssa* is similar to that of *Cornus* except that during the Miocene it became extinct in Europe. The small hard seeds of *Cornus*, *Nyssa*, and *Mastixia* are often found in Tertiary lignites, and in the Eocene London Clay the seeds of several apparently extinct genera of the Cornaceae abound.

Caprifoliaceae.—Numerous species of *Viburnum* have been recorded from the early Tertiary, but the accuracy of the determinations in the majority of instances is questionable. The genus is easily confused with *Acer*, *Ribes*, and others.

Other dicotyledons.—A number of form genera have been proposed to embrace fossil dicotyledonous leaves of which the family affinities are in doubt. A few of these are briefly described.

Under the name *Protophyllum* Leo Lesquereux grouped a number of leaves from the Dakota sandstone which, although they bear some resemblance to the Ulmaceae, do not agree exactly with those of any known living genus. *Protophyllum* leaves are large, coriaceous, oval or round-pointed, subpeltate with entire or undulate borders and pinnately arranged, more or less parallel secondary veins. The apex of the petiole is covered by two basal lobes of the blade into which pass somewhat downwardly directed branches of the lowermost secondary veins.

Closely resembling *Protophyllum* and sometimes confused with it is *Credneria,* an Upper Cretaceous leaf believed by some to be allied to the Platanaceae. The margin rather recalls the leaf of *Platanus* but the lateral lobes are lacking or only very weakly developed and the apex is blunt. The first pair of major secondary veins departs at some distance above the base, and below this pair are several pairs of weaker secondaries that pass out nearly at right angles to the midrib.

Dombeyopsis is the name that has been given large leaves that have been variously referred to the Buettneriaceae, Bignoniaceae, and Sterculiaceae. The leaves are entire, usually cordate-ovate, and bear considerable resemblance to the modern *Catalpa.*

The term *Phyllites,* which has no precise meaning, is applied to the foliage of any leaf of unknown family relationship. It is generally applied to dicotyledonous leaves but can be used for leaves of any group. It is purely a form genus, which is used only when it is desirable to give a certain leaf type taxonomic status without committing oneself as to its affinities. Similarly, *Carpolithus* and *Carpites* are used for seeds and fruits and *Antholithus* for unassigned fructifications or inflorescences.

Of the some 250 recognized families of the dicotyledons, probably less than half are known to have fossil representatives. In 1919 Knowlton listed 93 families from North America, but subsequent discoveries would probably raise the total to more than 100. A considerable number of these, however, are represented by only a few dubiously identified forms, and a complete revision of the late Cretaceous and Tertiary floras would result in the removal of several families and the inclusion of others not at present known to be represented.

There are a number of families which, although very important in the study of modern floristics, are either without known fossil representatives or are so sparsely represented that we can claim no precise knowledge of their history. Some of these are Campanulaceae, Compositae, Cruciferae, Labiatae, Ranunculaceae, Scrophulariaceae, Solanaceae, and Umbelliferae. The scarcity of fossil members of these and other families is partly because some of them are the recent products of evolution and were either rare or nonexistent previous to the Pleistocene ice age, and partly because they consist for the most part of low-growing annual or perennial herbs, which do not periodically shed their foliage. As such foliage does not accumulate in quantities where sediments are deposited, it does not enter strongly into the fossil record. The preponderance of deciduous species in the rocks therefore does not necessarily mean that the majority of the species of the surrounding region was of that type. This introduced an error in our interpretations of past floras, and the discrepancies resulting from the accident of preservation extend throughout the Mesozoic and Paleozoic as well as the Cenozoic.

References

ARBER, E.A.N., and J. PARKIN: The origin of the angiosperms, *Linnean Soc. London Jour.*, Botany, **38**, 1907.

BERRY, E.W.: "Tree Ancestors: A Glimpse into the Past," Baltimore, 1923.

——: Revision of the Lower Eocene Wilcox flora of the southeastern states, *U. S. Geol. Survey Prof. Paper* 156, 1930.

BROWN, R.W.: Additions to some fossil floras of the western United States, *U. S. Geol. Survey Prof. Paper* 186J, 1937.

——: Fossil leaves, fruits, and seeds of *Cercidiphyllum, Jour. Paleontology*, **13**, 1939.

CHANEY, R.W.: Paleoecological interpretations of Cenozoic plants in Western North America, *Bot. Rev.*, **4**, 1938.

——: Tertiary forests and continental history, *Geol. Soc. America Bull.*, **51**, 1940.

EDWARDS, W. N., Dicotyledones (Ligna), *Foss, Cat.* II: Plantae, Pars 17, The Hague, 1931.

——, and F. M. WONNACOTT: Sapindaceae, *Foss. Cat.* II: Plantae, Pars 14, Berlin, 1928.

——: Anacardiaceae, *Foss. Cat.* II: Plantae, Pars 20, The Hague, 1935.

HARRIS, T.M.: The fossil flora of Scoresby Sound, East Greenland, pt. 2, Description of seed plants *Incertae Sedis* together with a discussion of certain cycadophytic cuticles, *Meddelelser om* Grönland, **85** (3), 1932.

KIRCHHEIMER, F.: Umbelliflorae: Cornaceae, *Foss. Cat.* II: Plantae, Pars 23, The Hague, 1938.

——: Rhamnales I: Vitaceae, *Foss. Cat.* II: Plantae, Pars 24, The Hague, 1939.

KNOWLTON, F.H.: A catalog of the Mesozoic and Cenozoic plants of North America, *U. S. Geol. Survey Bull.*, **696**, 1919.

——: "Plants of the Past," Princeton, 1927.

LAMOTTE, R.S.: Supplement to catalog of Mesozoic and Cenozoic plants of North America, 1919–1937, *U. S. Geol. Survey Bull.* **924**, 1944.

MACGINITIE, H.G.: The Trout Creek flora of southeastern Oregon, *Carnegie Inst. Washington Pub.* 416, Washington, 1933.

NAGALHARD, K.: Ulmaceae, *Foss. Cat.* II: Plantae, Pars 10, Berlin, 1922.

NAGEL, K.: Juglandaceae, *Foss. Cat.* II: Plantae, Pars 6, Berlin, 1915.

——: Betulaceae, *Foss. Cat.* I: Plantae, Pars 8, Berlin, 1916.

POTONIÉ, R., and W. GOTHAN: "Lehrbuch der Palaobotanik," Berlin, 1921.

REID, E.M., and E.J. CHANDLER: "The London Clay Flora," London, 1933.

SAHNI, B.: Homoxylon rajmahalense, gen. et sp. nov. a fossil angiospermous wood devoid of vessels, from the Rajmahal Hills, Behar, *Palaeontologia Indica*, New Series, **22**, 1932.

SEWARD, A.C.: The Cretaceous plant-bearing rocks of Western Greenland, *Royal Soc. London Philos. Trans.*, **B, 215**, 1926.

——: "Plant Life through the Ages," Cambridge, 1931.

SIMPSON, I.B.: Fossil pollen in Scottish Jurassic coal, *Nature*, **139**, 1937.

STOPES, M.C.: Petrifactions of the earliest European angiosperms, *Royal Soc. London, Philos. Trans.*, **B, 203**, 1912.

THOMAS, H.H.: The early evolution of the angiosperms, *Annals of Botany*, **45**, 1931.

ZIMMERMANN, W.: "Die Phylogenie der Pflanzen," Jena, 1930.

CHAPTER XIV

THE SEQUENCE OF THE PLANT WORLD IN GEOLOGIC TIME

In the preceding chapters the subject of paleobotany is organized around the separate plant groups, and although the geologic sequence is followed to some extent in the treatment of most of the groups, no effort is made to describe the floras of the separate eras and periods or to trace the course of development of the plant world as a whole. The purpose of the present chapter, therefore, is to present a brief chronological account of the subject. It is impossible within the confines of one chapter to delve into the intricacies of stratigraphic problems as they pertain to paleobotany other than to describe very briefly the floras of those rock series and groups in which fossil plants have played an important role in correlation. But because of the complexity of even this limited phase of the subject, the treatment will be restricted mainly to North America.

PRE-CAMBRIAN AND EARLY PALEOZOIC

In the Pre-Cambrian and those early stages of the Paleozoic preceding the Devonian, which together represent more than three-fourths of known geologic time, we know little of the plant life of the globe except for a few probable seaweeds, lime-depositing algae, and bacteria which might have been responsible for mineral deposition. A few years ago cutinized spores resembling those of modern land plants were reported from Cambrian coal, but since the age of this coal is a matter of question, this one occurrence of spores is not accepted as satisfactory evidence of land plants during that period. Information has been accumulating during the last two decades on a group called the Nematophytales which shows many attributes of the Thallophyta but which apparently was adapted to partial life on land. The best known members of the Nematophytales are *Prototaxites* and *Nematothallus*. Land plants probably evolved from aquatic ancestors during the early Paleozoic, but as yet we are quite ignorant as to how or exactly when this transformation occurred.

The Devonian has long been looked upon as having produced the first land plants, but lately the Silurian has yielded remains, which there is every reason to believe grew upon land. The oldest of these are from

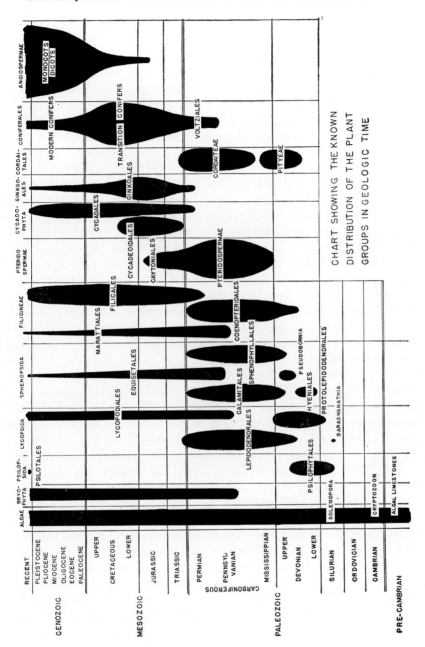

CHART SHOWING THE KNOWN
DISTRIBUTION OF THE PLANT
GROUPS IN GEOLOGIC TIME

beds containing *Monograptus*, presumably belonging to the Middle Silurian, of Victoria, Australia. *Baragwanathia*, evidently a lycopod, is the largest and the most completely preserved of these very ancient plants. In addition, there are two species of the psilophyte *Yarravia*. The Upper Silurian Walhalla series, also from the state of Victoria, has yielded a small flora which definitely connects the Silurian vegetation with that of the Devonian. The Walhalla plants consist of two species of *Sporogonites*, and one each of *Zosterophyllum* and *Hostimella*.

THE LATER PALEOZOIC

Devonian.—It is customary to divide the Devonian into three parts, the Lower, Middle, and Upper. The boundaries between these stages, however, are not always clearly defined and correlations over large areas are subject to frequent controversy. As a matter of convenience the Devonian rocks are frequently designated as "early" and "late," and from the standpoint of the floras, and to some extent from that of the faunas as well, a twofold division is probably as logical as the other.

The sedimentary deposits of the Devonian are of two kinds, continental and marine. The continental beds consist principally of sandstones, gray or red shales, and conglomerates, whereas the marine and brackish water counterparts are built up mainly of limestones and dark gray shales. Most of the plant remains of the Devonian are from the former. In Eastern North America the Lower Devonian rocks are almost exclusively marine, but during Middle Devonian time the great Catskill delta, which originated from sediments derived from the eroded Appalachian land mass to the east, began to extend westward. The growth of this delta continued throughout most of the Upper Devonian and by the end of the period extended as far as the western portions of New York and Pennsylvania. This delta supported a land flora. During the youthful stages *Aneurophyton*, primitive lycopods of the *Giiboaphyton* type and doubtlessly many psilophytes thrived, but near the end of Middle Devonian time these early types were largely replaced by the lycopod *Archaeosigillaria*, the fern *Archaeopteris*, and the supposed gymnosperm *Callixylon*. These persisted until the close of the period.

In the Perry Basin of eastern Maine, and at several places in Eastern Canada, there are similar beds of continental origin containing land plants. The Upper Devonian beds at Perry, Maine, have yielded *Aneurophyton*, *Archaeopteris*, *Barinophyton*, lycopods, and other plant remains. Material from Perry identified as *Psilophyton* belongs in all probability to *Aneurophyton*. The Gaspé sandstone at Cape Gaspé (probably Lower Devonian) contains many fragments of *Psilophyton*, and it is from there that Sir William Dawson secured the material from which he described the genus. The Upper Devonian "Fish Cliffs" at Scaumenac Bay contain *Archaeopteris* and other plants but the locality is best known for fossil fish.

In Northern Europe that part of the Devonian system deposited within the mountain troughs following the Caledonian revolution is called the Old Red Sandstone. It is of continental origin, being best developed in Scotland, although it extends into England, Wales, Ireland, and Scandanavia. For the most part it is unconformable with the marine Silurian upon which it rests. Correlations between comparable horizons of the Old Red Sandstone are difficult, but the whole series is believed to embrace the Lower and Middle Devonian and a part of the Upper.

The Downtonian of England and Wales was formerly considered a part of the Old Red Sandstone, but is now classified as a separate series. It apparently lies at the base of the Devonian, and by some is regarded as late Silurian. Vascular plants from the Downtonian include species of *Cooksonia* and *Zosterophyllum*. Other plants with cutinized structures but no vascular system are *Pachytheca*, *Parka*, and *Nematothallus*.

The sediments making up the Catskill delta and the "Fish Cliffs" at Scaumenac Bay in Quebec are often regarded as the North American equivalents of the Old Red Sandstone. Although not formed under exactly comparable conditions, they all, however, contain similar organic remains.

Several years ago Newell Arber applied the designations *Psilophyton* flora and *Archaeopteris* flora to the early and late Devonian floras on grounds that they were separate and distinguishable entities. We know now, however, that there is some overlapping of these floras. *Cladoxylon*, for example, a genus formerly supposed to occur only in the late Devonian and early Carboniferous, was found at Elberfeld, Germany, on the same rock slab with *Asteroxylon*, a Middle Devonian genus. In eastern New York *Aneurophyton* extends from the Middle Devonian into the lower part of the Upper Devonian, and *Archaeopteris*, long used as an index fossil of the Upper Devonian, occurs sparingly in the late Middle Devonian.

Although the general aspect of the early Devonian flora is predominantly psilophytalean, there are in it evidences of important evolutionary trends. It was during early Devonian times that the first sphenopsids and ferns are believed to have diverged from the psilophytalean complex, and it is likely that the precursors of the gymnosperms were in existence then. The Psilophytales, however, predominated during these times although the genus *Psilophyton* was not as prevalent in North America as has been generally supposed. There is no satisfactory evidence that *Psilophyton* survived into the Upper Devonian regardless of numerous references to it in the literature.   On the other hand, it was an important plant during the Lower and Middle Devonian at other places, as is shown by the numerous records of it from Western Europe.   *Zosterophyllum,* another member of the Psilophytales, is probably older than *Psilophyton* and has been found in the Upper Silurian. Other early Devonian plants belonging to the Psilophytales are *Cooksonia, Bucheria, Pseudosporochnis, Rhynia, Horneophyton, Asteroxylon,* and *Taeniocradia.*   *Arthrostigma,* once regarded as psilophytic, is now believed to be the same as *Drepanophycus,* which shows evidence of lycopodiaceous affinities.

Numerous primitive lycopods existed contemporaneously with the Psilophytales.   Although less simple in their bodily organization, the lycopods are probably as old.   Mention has been made of the lycopodiaceous characteristics of the Silurian *Baragwanathia.*

Hyenia, Calamophyton, and *Spondylophyton* are early Devonian plants usually classified under the Sphenopsida.   *Hyenia* has been found on Bear Island, in Germany, and in New York.   *Calamophyton* is known only from Germany and *Spondylophyton* is from Wyoming.

The Middle Devonian flora contains a few fernlike types.   *Protopteridium* is probably the most widely distributed.   Others are *Arachnoxylon, Reimannia,* and *Iridopteris* from the Hamilton group of western New York, all showing evidence of connection with the Psilophytales.

The early Devonian has yielded no seed plants.   *Eospermatopteris,* originally regarded as a seed plant, in all probability reproduced by spores, and its remains are indistinguishable from those of *Aneurophyton.*

The Upper Devonian has yielded several lycopods, such as *Protolepidodendron, Archaeosigillaria, Colpodexylon,* and *Cyclostigma.*   The Sphenopsida are represented by *Pseudobornia* and *Sphenophyllum,* and the ferns by *Archaeopteris, Asteropteris,* and *Cladoxylon.*   No objects that can be positively identified as seeds have been found anywhere in the Devonian, but *Callixylon* was undoubtedly a seed plant and a gymnosperm.   *Callixylon* and *Archaeopteris* existed contemporaneously throughout the Upper Devonian although their remains are seldom found together.

Carboniferous.—The prevailing practice among North American geologists is to divide the Carboniferous period into three epochs, the Mississippian, the Pennsylvania, and the Permian.   In actual practice, however, the name "Carboniferous" is often dispensed with and the epoch names are used in place of the period names.   According to European usage the Carboniferous is a period divided into the Lower and the Upper Carboniferous, and the Permian is a period equal in rank with the Carboniferous.   "Permo-Carboniferous" is often used for late

Carboniferous and early Permian rocks of which the exact age may be in doubt. The name is often applied to the late Paleozoic glacial tillites of the Southern Hemisphere.

The Mississippian of North America corresponds in general to the Lower Carboniferous of the British Isles and the Dinantian and Culm of continental Europe. In Great Britain the Lower Carboniferous consists of the Calciferous Sandstone of Scotland, the Carboniferous Lime-

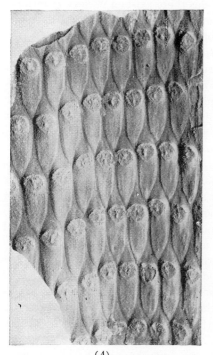

(A)   (B)

Fig. 178.—(A) *Lepidodendropsis cyclostigmatoides.* Price formation. Virginia. × about 2. (B) *Triphyllopteris lescuriana.* Pocono formation. High Bridge, Pennsylvania. Natural size.

stone, and the lower part of the Millstone Grit. The Calciferous Sandstone contains a large flora of lycopods, sphenopsids, and ferns, many of which are structurally preserved in volcanic ash. Two important localities are those on the Isle of Arran, where the large upright trunks of *Lepidophloios Wünschianus* were found, and at Pettycur, on the Firth of Forth, which has yielded numerous petrified ferns and other plants. The Mississippian series of North America, being predominantly marine, contain relatively few plant-bearing beds. The most prolific are the Pocono and Price formations of Pennsylvania and Virginia, and the

Horton shales of Nova Scotia. The Lower Carboniferous is usually separated from adjacent formations by unconformities.

The lycopods found in the Lower Carboniferous include such genera as *Lepidodendron*, *Lepidophloios*, and *Lepidodendropsis* (Fig. 178*A*). *Sigillaria*, if it was in existence then, was rare. *Lepidodendra* with long tapering leaf cushions are frequent. The Sphenopsida contain *Calamites*, *Asterocalamites*, *Sphenophyllum*, and *Cheirostrobus*.

Throughout the Lower Carboniferous there exists a variety of fernlike leaf types with rounded or assymetrical wedge-shaped or dissected pinnules in which the veins dichotomize uniformly throughout the lamina. Some examples are *Adiantites*, *Aneimites*, *Cardiopteris*, *Cardiopteridium*, *Sphenopteridium*, *Rhodea*, *Rhacopteris*, and *Triphyllopteris* (Fig. 178*B*). *Cardiopteridium* and *Triphyllopteris* are abundantly represented in the Pocono and Price formations. Some of these genera are pteridosperms and others are probably ferns.

The Lower Carboniferous ferns include well-known species of *Botryopteris*, *Clepsydropsis*, *Metaclepsydropsis*, and *Stauropteris*. The pteridosperms are represented by the petrified stems of *Heterangium*, *Rhetinangium*, *Calamopitys*, and *Protopitys*. Associated with *Heterangium* is the seed *Sphaerostoma*. Cupulate seeds referable to *Calymmatotheca* and *Lagenospermum* occur in the Pocono and Price formations.

The Lower Carboniferous gymnosperms include *Pitys* and related genera. The existence of *Cordaites* at this time is doubtful.

The Upper Carboniferous is marked by the prevalence of numerous arborescent genera, such as *Lepidodendron*, *Sigillaria*, *Calamites*, *Psaronius*, and *Cordaites*. Plants of smaller size were *Lycopodites*, *Sphenophyllum*, many ferns, and most of the pteridosperms. The seed-bearing lycopod, *Lepidocarpon*, flourished then as did also members of the Lyginopteridaceae and the Medullosaceae. The sphenopterids are particularly abundant in the lower stages but the pecopterids attained their greatest prominence during the middle and latter part of the epoch. The leading fern types were the Coenopteridales and forms probably related to the Marattiales. *Oligocarpia* and *Senftenbergia* suggest the existence then of primitive members of the Gleicheniaceae and the Schizaeaceae.

The Upper Carboniferous or Pennsylvanian is the great Paleozoic coal age, and fossil plants have played a more important role in the correlation of the strata of this epoch than of any other within the Paleozoic. In Eastern North America the Pennsylvanian is subdivided in ascending order into the Lee, Kanawha, Allegheny, Conemaugh, and Monongahela series. The name "Pottsville" was once employed for those groups now embraced within the Lee and Kanawha, but it is now used in a more restricted sense. In Illinois the division is into Pottsville, Carbondale, and McLeansboro, and in the Central States west of the Mississippi River into Morrow, Lampasas, Des Moines, Missouri, and Virgil series. The approximate correlations are shown in a table (see page 376). In Europe the three major divisions of the Upper Carboniferous are the Namurian, Westphalian, and Stephanian. The Namurian corresponds in general with the lower part of the Lee and Morrow of North America and to the upper part of the Millstone Grit in Great Britain. The Westphalian of Europe closely approximates the upper Lee, Morrow, and Allegheny and embraces the Lanarckian, Yorkian, Staffordian, and Radstockian of the British Coal Measures. In continental Europe the Westphalian has been divided in ascending order into A, B, C, and D, which corresponds roughly but not precisely to the British subdivisions named above. Other names have been applied to these various units in other parts of Europe. The European Stephanian corresponds roughly to the Conemaugh and Monongahela. Sharp differences of opinion exist, however, concerning the exact places where the boundaries of some of the units of the Carboniferous should be drawn and upon the extent of correlation between certain of the units in North America and Europe. On the whole it is generally agreed that the series of the American Pennsylvanian and those of the European Upper Carboniferous are essentially equivalent.

Floras of Lee and Kanawha age (still collectively referred to as "Pottsville") have been extensively studied in the anthracite coal regions of Pennsylvania and in the bituminous coal fields of West Virginia. In Canada floras of similar age exist in the Maritime Provinces. The flora of the Lee series is characterized by three zones, which are designated in ascending order as (1) the zone of *Neuropteris pocahontas* and *Mariopteris eremopteroides*, (2) the zone of *Mariopteris pottsvillea* and *Aneimites*, and (3) the zone of *Mariopteris pygmaea*. These floral zones are definitely related to the Namurian and Westphalian A floras of Europe. The flora of the Kanawha (formerly called upper Pottsville) has at its base the zone of *Cannophyllites* (*Megalopteris*), and above is the zone of *Neuropteris tenuifolia*. These two zones agree essentially with the Westphalian B of Europe and most of the Yorkian of Great Britain. Plants frequently encountered in the Kanawha are *Neuropteris tenuifolia*, *N. gigantea*, *Alethopteris grandifolia*, and species of *Eremopteris*, *Zeilleria*, and *Megalopteris*. *Megalopteris* is not known outside of North America. In rather constant association with this genus are the broadly winged seeds of the *Samaropsis Newberryi* type and the large cuneate type of cordaitalean foliage assigned to the form genus *Psygmophyllum*. In Nova Scotia plant-bearing groups equivalent to the Lee series are the

Canso and Riversdale, and above are the Cumberland and Lancaster groups, the equivalents of the Kanawha.

The flora of the Allegheny, and series of similar age, are revealed in well-known localities in Henry County, Missouri, Mazon Creek, Illinois, eastern Kansas, and in Oklahoma, Pennsylvania, and Nova Scotia. In the lowermost levels of the Allegheny there are a few holdover species

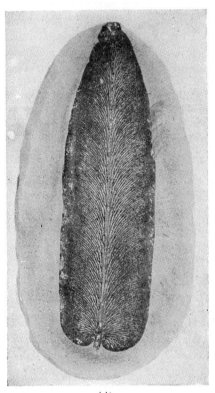 

(A)   (B)

FIG. 179.—(A) *Neuropteris Scheuchzeri*. Mazon Creek, Illinois. Natural size. (B) *Neuropteris rarinervis*. Mazon Creek, Illinois. Natural size.

of the late Kanawha and Pottsville, but for the most part the two floras are distinct. A widely distributed Allegheny species is *Neuropteris Scheuchzeri*, which has very large tapering or tongue-shaped hirsute pinnules (Fig. 179A). This species is somewhat variable and the different forms are often designated by separate names. Other Allegheny neuropterids that should be mentioned are *N. ovata* (a common species in the succeeding Conemaugh) and *N. rarinervis* (Fig. 179B). *Alethopteris Serlii* may be found at almost any *Allegheny* plant locality, and

Callipteridium Sullivantii, which is placed in *Alethopteris* by some authors, is frequent. *Callipteridium* is essentially like *Alethopteris* except that the pinnules are shorter in proportion to their length and more blunt. Pecopterids are numerous in the Allegheny, and at Mazon Creek the most abundant plant species is one usually identified as

Fig. 180.—*Walchia frondosa.* Staunton formation, Middle Pennsylvanian. Garnett, Kansas. Natural size.

Pecopteris Miltoni (Fig. 14). *P. unitus* and *P. pseudovestita* are also frequently found. The Allegheny formation contains numerous species of *Calamites, Sphenophyllum, Lepidodendron, Sigillaria,* and other genera, but these as a rule are less distinctive of this epoch than the fernlike foliage types.

Although the Conemaugh flora has not been intensively studied, it is notable for the appearance of certain late Pennsylvanian species such as

Pecopteris feminaeformis, P. polymorpha, and *Alethopteris grandini.* An important fact in connection with the Conemaugh is the occurrence of coal-balls and other types of petrifactions. In Great Britain the principal coal-ball horizon is in the Yorkian, and numerous other petrifactions come from the Calciferous Sandstone.

It was at about the middle of the Pennsylvanian epoch, or maybe even earlier in some places, that the first conifers as represented by *Lebachia* (*Walchia*) appeared (Fig. 180). Although usually regarded as typically a Permian genus, *Lebachia* occurs in the late Pennsylvanian at a number of places, and in the upper part of the Missouri series (Conemaugh) near Garnett, Kansas, it exists in association with *Alethopteris, Neuropteris,* and other typical Pennsylvanian plants. In central Colorado it occurs in rocks that may possibly be older. These early occurrences of *Lebachia* and other Permian types probably indicate local developments under conditions which were drier than those generally

APPROXIMATE CORRELATIONS OF LATE PALEOZOIC SERIES IN GREAT BRITAIN, WESTERN CONTINENTAL EUROPE, AND EASTERN NORTH AMERICA

EUROPE				NORTH AMERICA			
Great Britain	Western Europe			Maritime Provinces	Eastern United States	Illinois	Mid-Continent
	Rothegende	Permian	Permian	Dunkard			Red Beds Big Blue
	Stephanian	Upper Carboniferous (Coal Measures)	Carboniferous — Pennsylvanian	Monongahela			Virgil
					Conemaugh	McLeansboro	Missouri
Radstockian	D			Pictou Stellarton Morien	Allegheny	Carbondale	Des Moines
Staffordian	C Westphalian						
Yorkian	B			Cumberland Lancaster	Kanawha		Lampasas
Lamarckian	A			Riversdale	Lee	Pottsville	Morrow
Millstone Grit	Namurian			Canso			
Carboniferous Limestone	Dinantian Culm	Lower Carboniferous	Carboniferous — Mississippian	Horton Bluff	Mauch Chunk Greenbrier Pocono and Price	Chester	Mississippian
Calciferous Sandstone							

prevailing throughout the Pennsylvanian and which foreshadowed the more widely spread aridity of the Permian.

Little can be said of the Monongahela flora other than to note its general agreement with the late Upper Carboniferous flora of Europe. It possesses, in addition to species common also in the Conemaugh, *Taeniopteris* and other late types such as *Odontopteris Reichi, Lescuropteris Moorii, Mixoneura neuropteroides,* and *Callipteris.*

Permian.—The early Permian flora is essentially an extension of that of the late Pennsylvanian, and in many places no sharp division between

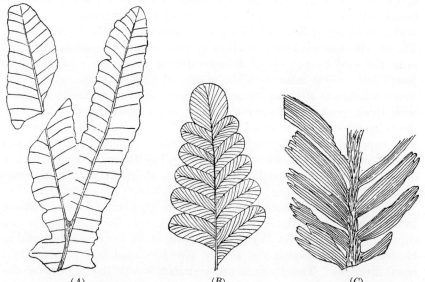

(A) (B) (C)

Fig. 181.—(A) *Gigantopteris americana.* Portion of forked leaf. Wichita formation. Lower Permian. Texas. Slightly reduced.

(B) *Callipteris conferta.* A characteristic Lower Permian species. Natural size

(C) *Tingia carbonica.* Lower Permian. Central Shansi, China. (*After Halle.*) Slightly reduced.

the two can be drawn. However, with the advance of Permian time, arid conditions that began to develop during the Pennsylvanian became more widely spread, and the effect on the flora was pronounced. The lush coal-swamp vegetation began to disappear and to be replaced by plants with smaller, more leathery, and more heavily cutinized leaves. The Permian in some places is marked by the presence of *Gigantopteris,* a plant with ribbonlike fronds (Fig. 181*A*). In North America the genus occurs in the "Red Beds" of Texas. It also occurs in Korea, and is especially abundant in the Shihhotse series of the province of Shansi in China. *Callipteris conferta* (Fig. 181*B*) is another species that is used as an "index fossil" of rocks of early Permian age.   Other characteristic Permian types are *Tingia* (Fig. 181*C*) and certain species of *Odontopteris*. *Calamites gigas* is a Permian species that occurs in the Pennsylvanian. The conifers are represented in the rocks of this epoch by *Lebachia* and related genera, *Pseudovoltzia*, and *Ullmannia*.   In the late Permian *Sphenobaiera*, the oldest member of the *Ginkgo* line, made its appearance.

The most recent of American Permian floras is that of the Hermit shale exposed in the Grand Canyon of Arizona.   The Hermit shale has a thickness of nearly 300 feet and consists of thin-bedded sandy shales and silts wth numerous current ripples.   Its color at many places is blood red.   That this formation was deposited under conditions of semi-aridity is amply indicated by deep mud cracks and casts of salt crystals. The climate was evidently one in which there were occasional torrential rains followed by periods of drought during which the pools dried up completely.   This interpretation of the climate is supported by the characteristics of the plant remains.   The plants are typical xerophytes with thick, probably heavily cutinized, often inrolled leaves in which the veins are deeply immersed in mesophyll tissue.   Many of the species were supplied with well-developed epidermal hairs, and the petioles and stems frequently bear scales and spines.   The foliage tends to be less deeply dissected and smaller than in typical members of the Pennsylvanian swamps.

The Hermit shale flora is notable for the number of forms unknown to occur at other places.   Genera peculiar to this flora include the two supposed pteridosperms, *Supaia* and *Yakia*, and a few others believed to be conifers.   In *Supaia* the simply pinnate frond is bifurcated into equal divisions.   The divisions are asymmetric with the pinnules on the outer side longer than those on the inner side (Fig. 182).   The pinnules are linear, decurrent, show alethopteroid venation, and are longest at the middle of the bifurcation.   They continue with diminishing length onto the short petiole below.   *Supaia* is believed to be a relative of *Dicroidium*, a typical late Permian genus of the Southern Hemisphere.   *Yakia* is a plant with large, much divided, frondlike leaves.   The ultimate branchlets bear rather distant, narrow, somewhat leathery lobes or pinnules. The fronds are often shriveled and present a skeletonlike aspect, which is probably the result of preservation.   On the surface of some of the impressions believed to belong to *Yakia* are clusters of elongated bodies resembling sporangia or small seeds.   They appear to be dorsally attached in small pits at the bases of the pinnules.   Their exact morphology, however, is uncertain.

No typical Pennsylvanian species are present in the Hermit shale flora and genera notable for their absence are *Cordaites* and *Calamites*.

The Sphenophyllales are represented by one species, but no undoubted ferns have been identified. There are a number of ancient coniferous genera such as *Walchia, Ullmannia, Voltzia, Paleotaxites, Brachyphyllum,* and *Pagiophyllum.* Aside from its new forms, the flora contains certain elements found in the lower Permian of the basins to the east of the Rocky Mountains, and in addition species that suggest a connection with the lower *Glossopteris* flora that succeeded the Permian ice sheets in India and in parts of the Southern Hemisphere.

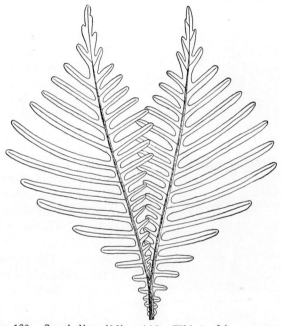

Fig. 182.—*Supaia linaerifolia.* (*After White.*) ¼ natural size.

The *Glossopteris* flora is very different from contemporaneous floras of northern latitudes. Its remains are found principally in India, Africa, Australia, and South America, covering the large area supposedly occupied by ancient Gondwanaland, and it developed in response to the glacial climate that prevailed over a large part of the Southern Hemisphere during Permian time. Whether the *Glossopteris* flora was in existence at the same time the glaciers covered the land had long been a question, but the recent discovery of winged spores in Australia in tillites evidently deposited by floating ice offers conclusive evidence that the flora developed during the glaciation rather than after it.

The leaves of *Glossopteris* are simple, entire, sessile or shortly petio-late, and spathulate to ovate or linear in outline (Fig. 183). They are often borne on characteristic rhizomelike structures known as *Verte-braria*. The midrib is distinct and the arched veins form a network. The fructifications have not been identified with certainty.

A constituent of the *Glossopteris* flora is *Gangamopteris*, a plant resembling *Glossopteris* except for the absence of a midrib. There are a few lycopods and scouring rushes in the flora, and several gymnosperms

and ferns. The silicified tree trunks in the Gondwanaland deposits show rather strongly developed growth rings, structures seldom present in the woods of earlier epochs.

Partly contemporaneous with the *Glossopteris* flora of the Southern Hemisphere is the Cathaysia flora of Eastern China, Indo-China, Malaya, and Sumatra, and the Angara flora of central and Eastern Siberia and Mongolia. The Cathaysia flora is for the most part the older of the two as its lower divisions embrace the equivalents of parts of the upper Carboniferous of Europe and America. The youngest division of the Cathaysia is represented in the upper Shihhotse series (lower Permian) in the province of Shansi in China and in the Kobosan series of Korea. Characteristic genera are *Gigantopteris*, *Lobatannularia*, and *Tingia*.

Fig. 183.—*Glossopteris Browniana.* Permo-Carboniferous. New South Wales. Natural size.

The later Angara flora contains many late Paleozoic types, such as *Callipteris*, *Ullmannia*, *Walchia*, *Odontopteris*, *Thamnopteris*, *Zalesskya*, and ancient members of the Ginkgoales. *Gangamopteriopsis*, a plant closely resembling *Gangamopteris*, and a member of the Angara flora, is suggestive of a connection with the *Glossopteris* flora to the south. The formations containing the Angara flora are not limited to the Permian, but include beds as late as Jurassic.

THE MESOZOIC

Triassic.—Because of the prevailing aridity at the beginning of the Mesozoic era, early Triassic floras are for the most part poorly preserved.

The only places in North America where plants have been found in quantity anywhere in the Triassic are near Richmond, Virginia, in New Mexico, and in Arizona. Ferns and primitive cycadophytes predominate but there are a number of conifers and a few lycopods and scouring rushes. *Pleuromeia*, a spectacular lycopod from the Triassic, is found in the Bunter (Lower Triassic) of Germany and Eastern Asia.

The Petrified Forest of Arizona offers ample evidence that huge forests grew during Triassic time in spite of the unfavorable climate that is supposed to have prevailed generally over the earth. The largest tree is *Araucarioxylon arizonicum*, which is believed to be an ancient member of the Araucariaceae.

In many parts of the earth there are thick successions of late Triassic beds known as the Rhaetic. The Rhaetic is not recognized as such in North America, and in Great Britain it is a comparatively thin series of beds separating the Triassic proper from the Jurassic. But in other places, as in Sweden, Germany, and Eastern Greenland it comprises an extensive formation, which at places contains large quantities of plant remains. The largest Rhaetic floras are in Sweden and Greenland. The two are similar in character and are probably contemporaneous or nearly so. Both contain numerous ferns, cycadophytes, ginkgos, and conifers along with lesser numbers of scouring rushes and lycopods. The Eastern Greenland flora is the larger with more than 50 species among which are members of the Caytoniales and one (*Furcula granulifera*) having apparent angiospermous affinities. During Rhaetic time the *Ginkgo* and cycadophyte lines became well established.

Jurassic.—The floras of this period are widely distributed and in general are more adequately preserved than those of the Triassic. In some places the lowermost Jurassic is called the Lias, a term adopted long ago by William Smith for the Lower Jurassic rocks in Great Britain. The plants of the Lias are similar to those of the Rhaetic. In Great Britain the Upper Jurassic is known as the Oölite, and it is in the Lower Oölite along the Yorkshire coast that the most important of all Jurassic plant discoveries have been made. It was in these beds that *Williamsonia gigas* and *Williamsoniella* were first found. These beds have also yielded foliage and fructifications of the Caytoniales, a group at first thought to be angiospermous but now believed to terminate the pteridosperm line.

The Jurassic rocks have yielded many forms of *Ginkgo* and related genera. Many of the ancient ginkgos are characterized by deeply cut leaves (Fig. 138). *Baiera*, an extinct relative of *Ginkgo*, has leaves similar to those of *Ginkgo* but the lamina is very deeply cut into slender fingerlike segments.

Angiosperms are first recognized with certainty in the Jurassic but they do not become important elements of the flora until the Cretaceous.

The Morrison formation, placed by some in the uppermost Jurassic but by others in the basal Cretaceous, extends from central Montana through Wyoming and Colorado into northern New Mexico and eastern Utah. This formation is famous for the dinosaur bones it contains. It also yields numerous cycadeoids, the best known locality being at Freezeout Mountain in Wyoming.

Cretaceous.—The Cretaceous is one of the critical periods in the history of the vegetable kingdom because it witnessed the most recent of the major transformations of the plant world. The best known Lower Cretaceous plant-bearing rocks in North America are those of the Potomac series of Maryland, the Lakota and Fuson formations of South Dakota and Wyoming, and the Kootenai formations of Montana and adjacent Canada. Although all these rock groups have yielded plant remains in considerable quantity, flowering plants constitute an important element only in the Potomac series (Fig. 184). In a previous chapter where the proportions of the major plant groups of the Potomac series are outlined, it was shown that the flowering plants increase from a mere 6 per cent of the total number of species in the Patuxent to about 25 per cent in the Patapsco. Some of the families believed to be revealed for the first time in geologic history are the Salicaceae, Moraceae, Araliaceae, Lauraceae, and Sapindaceae.

The Cretaceous rocks of Western Greenland have yielded a flora of exceptional interest. This flora was made known to us largely through the investigations of the great Swiss paleobotanist Oswald Heer who published between the years 1868 and 1882 a series of volumes on the fossil floras of the Arctic lands. The Arctic Cretaceous floras are remarkable for the early appearance of a large number of temperate zone genera, which has led to the supposition that the Arctic served as a region of distribution for many of them. Modern broad-leaved genera include among others *Quercus, Platanus, Magnolia,* and *Laurus. Sequoia* and needlelike leaves resembling *Sciadopitys* also occur there.

Heer recognized four plant-bearing series in Western Greenland. The Kome series of the Lower Cretaceous contains a flora mostly of ferns and gymnosperms but there are a few angiosperms. The Upper Cretaceous Atane and Patoot series are in turn overlain by plant-bearing Tertiary beds believed to be Eocene.

Upper Cretaceous floras that have been extensively studied in North America are the Raritan and Magothy of the Atlantic Coast and the Dakota of the Great Plains and the Fox Hills and Medicine Bow of the Rocky Mountains. The Raritan formation is best exposed in New

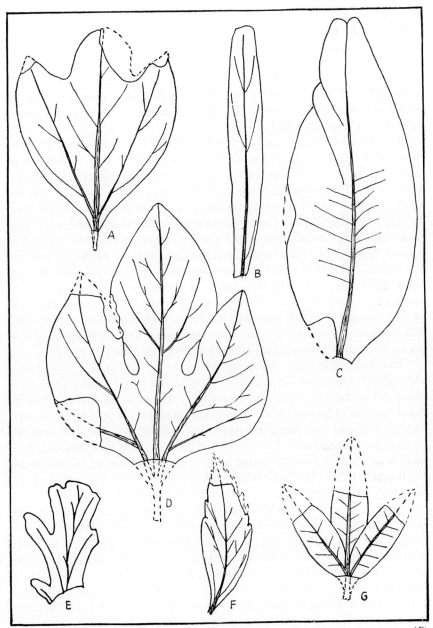

Fig. 184.—Dicotyledons from the Potomac series. (*A*) *Sassafras cretaceum.* (*B*) *Rogersia angustifolia.* (*C*) *Ficophyllum eucalyptoides.* (*D*) *Aceriphyllum aralioides.* (*E*) *Vitiphyllum crassifolium.* (*F*) *Quercophyllum tenuinerve.* (*G*) *Araliaephyllum obtusilobum.* (*After Fontaine.*) Slightly reduced.

Jersey. It contains about 300 species of which about 200 are angiosperms. The conifers number more than 50 species but there are less than 20 ferns and cycadophytes. Some of the genera appearing for the first time (outside of Greenland, just mentioned) are *Magnolia, Diospyros,* and *Eucalyptus.* The succeeding Magothy flora also has about 300 species of which about two-thirds are flowering plants. Here we find remains of *Liriodendron, Sassafras, Quercus,* and a few monocotyledons such as grasses and sedges.

The Dakota flora is one of the largest fossil floras known and the number of names embraces more than 500 species. It is said to range from Alaska to Argentina but is best revealed in Kansas, Nebraska, Iowa, and Minnesota, in deposits formed in a large sea that bisected most of North America during the early part of Upper Cretaceous time. In sharp contrast to the Lower Cretaceous floras, flowering plants comprise more than 90 per cent of the species. Cycads are here reduced to a mere 2 per cent, and the conifers are only slightly more. There are few ferns. Among the flowering plants there are numerous figs, hollies, tulip trees, poplars, willows, oaks, and sassafras. The number of species recorded for many of the genera, however, is probably in excess of the number actually present and future investigations of this flora will doubtlessly result in numerous changes. The presence of a few palms indicate a rather warm climate during this time.

The Fox Hills and Medicine Bow floras, which are distributed throughout southern Wyoming and southwestern Colorado, indicate a warm temperate or subtropical climate. These floras contain two or more species of palms, a screw pine (*Pandanus*), a few ferns, several figs, and in addition many dicotyledons now living in warmer regions. *Sequoia* is also represented. The age of the Fox Hills and Medicine Bow beds was long in doubt. Some geologists placed them in the Upper Cretaceous, others in the Tertiary. A recent critical study of the flora, however, has shown that their proper place is in the former.

Slightly younger than the Fox Hills flora is the flora of the Lance formation, which is best exposed in eastern Wyoming. This flora is a fairly large one of 74 species, mostly dicotyledons, of which the majority are only distantly related to Recent genera. Some of the most distinctive species are referred to *Dombeyopsis, Dryophyllum, Grewiopsis, Laurophyllum, Magnoliophyllum, Menispermites, Platanophyllum,* and *Vitis.* The structural characteristics of the leaves indicate a warm humid lowland with a climate approaching subtropical.

The Cenozoic

Paleocene.—In the Rocky Mountain region the Lance formation is succeeded by the Fort Union which, although not separated from it by any pronounced unconformity, possesses a fauna and flora which is distinct from that of the underlying Lance. Out of a flora of about 110 species only about 10 per cent occur in the Lance. The remainder are typically Tertiary and the species appear more closely related to modern forms than do the majority of those from the late Cretaceous.

Eocene.—Floras of Eocene age are widely distributed throughout the southern and western parts of the United States and in numerous other places. Only a few of the North American occurrences will be mentioned.

One of the largest and best known of Eocene floras is the Wilcox flora, which occurs along the coastal plain of the Gulf of Mexico from

THE MOST IMPORTANT CRETACEOUS AND TERTIARY PLANT-BEARING FORMATIONS AND FLORAS OF NORTH AMERICA

Pliocene	(16) Citronelle, Idaho, Mount Eden
	(15) Ricardo, Ogallala
Miocene	(14) Esmeralda, Barstow, San Pablo, Creede
	(13) Mascall, Payette, Latah
Oligocene	(12) John Day, Bridge Creek, Florissant (probable)
	(11) White River, Weaverville, Lamar River (?)
	(10) Jackson
	(9) Claiborne, Puget Sound, Green River, Goshen, Comstock, Clarno, Chalk Bluffs
Eocene	(8) Lagrange
	(7) Wilcox
Paleocene	(6) Fort Union, Upper Denver, Raton
	(5) Lance, Medicine Bow, Laramie, Denver, Fox Hills, Ripley, Vermejo
Upper Cretaceous	(4) Magothy
	(3) Tuscaloosa
	(2) Dakota, Woodbine, Raritan
Lower Cretaceous	(1) Potomac, Fuson, Kootenai

Alabama to Texas. It contains 543 described species belonging to about 180 genera and 82 families. The flora is typically a warm temperate one indicative of abundant rainfall. Conditions apparently were more tropical than those of the preceding late Cretaceous.

Succeeding the Wilcox flora is the Claiborne, which is more limited in its extent. It suggests a climate still warmer than the Wilcox. In the upper Eocene Jackson flora, which succeeds the Claiborne, the climate reached its maximum in warmth. Palms are abundant, as well as nutmegs and *Engelhardtia*, a tropical member of the Juglandaceae.

Other important and well-known Eocene floras of Western North America are the Green River of Wyoming, Colorado, and Utah; the Puget of western Washington and British Columbia; the Clarno of central Oregon; and the Comstock and Goshen of western Oregon. All indicate a relatively mild, equable if not tropical climate. Palms extended into

Canada during this time, and at places calcareous algae flourished in abundance, especially during the Green River stage.

Oligocene.—Several Tertiary floras of Western North America that were formerly included in the Eocene or Miocene are now assigned to the Oligocene. Outstanding among lower Oligocene floras are those of the White River and Weaverville beds. The White River beds are admirably exposed in the Big Badlands of South Dakota where numerous mammalian skeletons have been unearthed. The White River flora has not been fully investigated but it is known to contain many hardwoods, and in central Wyoming silicified wood showing unmistakable characters of *Casuarina* has been found.

The Weaverville flora, in northern California, contains abundant remains of *Taxodium* and *Nyssa*, which indicate a humid, warm-temperate climate. The plants are less tropical than those of the preceding Eocene but lived under warmer surroundings than those of the epoch that followed.

Notable upper Oligocene floras are the Bridge Creek and the John Day in central Oregon, and the Florissant of Colorado, although there is some question concerning the age of the latter. The Bridge Creek flora contains abundant remains of *Sequoia* of the *S. sempervirens* type and several other genera such as *Alnus*, *Cornus*, *Corylus*, *Acer*, *Mahonia*, *Umbellularia*, and *Lithocarpus*, which at present are associated with the redwoods along the narrow coastal belt of northern California. In addition to these redwood associates there are other genera no longer in existence in the wild state in the west, examples being *Tilia*, *Fagus*, *Castanea*, *Ulmus*, *Carya*, and *Carpinus*. Another is the katsura tree, *Cercidiphyllum*, now confined to China and Japan.

The plant-bearing beds at Florissant consist of volcanic ash, which filled an ancient fresh-water lake. Although nearly 200 species of plants have been identified from there, the exact age is questionable because the deposit is completely isolated from other sedimentary formations and rests entirely within a basin of volcanic rocks of unknown age.

Miocene.—The Miocene epoch in North America was characterized by widespread volcanic activity, which accompanied the uplifting of the Cascade Range. This uplift gradually deprived the region to the eastward of a large share of its rainfall. The oldest Miocene floras of this area indicate a moist and fairly mild climate, but increasing aridity accompanied by a marked elevation of the land surface is progressively revealed by the later floras.

Plant-bearing formations of Miocene age are widely scattered throughout the Columbia Plateau and Great Basin regions of Oregon, Washington, western Idaho, northern California, and Nevada. Other Miocene floras of lesser importance are the Alum Bluff of the Gulf Coast and the Brandon lignites of Vermont. The flora of the Lamar Valley in Yellowstone National Park is generally regarded as Miocene although the correctness of the age determination is often doubted. Recent authors believe it to be no younger than lower Oligocene or possibly as old as late Eocene.

A flora believed to be Miocene but retaining many earlier elements is the Latah, which is revealed at a number of places in the vicinity of Spokane, Washington. Among its members are *Cebatha* (originally identified as *Populus*), *Cedrela*, *Diospyros*, *Ginkgo*, *Pinus*, *Liriodendron*, *Keteleeria*, *Nyssa*, *Laurus*, *Salix*, and *Taxodium*.

The Mascall flora, which succeeds the Bridge Creek in central Oregon, contains fewer redwood associates. Instead there are species of *Acer*, *Quercus*, *Populus*, *Arbutus*, *Platanus*, *Oreopanax*, and others of which the living equivalents occur either at the borders of the redwood zone or in places where the moisture supply is more limited. As the Cascades continued to rise, the redwoods and their associates disappeared completely from the inland areas.

Pliocene.—During the Pliocene epoch conditions east of the Cascade Range were on the whole unfavorable for the growth of dense forests and for the preservation of their remains. The areas that supported the luxuriant forests of the early and middle Miocene now bore trees mainly along the streams. A few volcanic ash deposits contain leaves of such genera as *Alnus*, *Populus*, *Platanus*, and *Salix*. During this time many broad-leaved genera such as *Carpinus*, *Fagus*, *Ginkgo*, and *Tilia* disappeared from Western North America.

An important Pliocene flora has recently been described from the Mount Eden beds in southern California. It indicates a limited supply of rainfall and shows that the segregation now expressed by the floras of northern and southern California was in existence during Pliocene time. The Weiser flora of southwestern Idaho indicates a mild climate with dry summers and an annual rainfall of 20 to 30 inches. Conditions were probably less humid than in the preceding Miocene epoch but warmer and more moist than at present.

Of the few Pliocene floras of North America outside the Western States, the Citronelle of Alabama is the best known. It contains about 18 species, a few of which live in that region at present, and others which live near by. Fruits of *Trapa* (the water chestnut) are present. This represents the latest known occurrence of this Old World genus in North America except in areas where it has been recently introduced.

Pleistocene.—Remains of the vegetation of this epoch are preserved in unconsolidated stream and lake deposits and in peat bogs. Near

Toronto, Canada, a large flora of modern species is present in the thick mud bed that was laid down in the bottom of Lake Ontario when its level was higher than at present.

Plants have been found in the frozen Pleistocene muck at Fairbanks, Alaska, and at other places. Only modern species are represented. A Pleistocene flora on Santa Cruz Island off the coast of southern California contains *Cupressus goveniana, Pinus radiata, Pseudotsuga taxifolia,* and *Sequoia sempervirens,* thereby showing that these species once extended several hundred miles south of their present range.

The last remaining link between the floras of the past as described in the foregoing chapters and that of the present is revealed by plant remains in peat bogs of post-Pleistocene origin. This subject, however, belongs more fittingly to Recent botany, inasmuch as it is directly concerned with the distribution and climatic relations of modern species.

References

ARBER, E.A.N.: "Catalog of the Fossil Plants of the Glossopteris Flora in the Department of Geology of the British Museum," London, 1905.

AXELROD, D.I.: A Pliocene flora from the Mount Eden beds, southern California, *Carnegie Inst. Washington Pub.* 476, 1938.

BELL, W.A.: Fossil flora of the Sydney coal field, Nova Scotia, *Canada Dept. Mines, Geol. Survey Mem.,* **215,** 1938.

BERRY, E.W.: A revision of the flora of the Latah formation, *U. S. Geol. Survey Prof. Paper* 154 H, 1929.

———: A revision of the Lower Eocene Wilcox flora of the Southeastern United States, *U. S. Geol. Survey Prof. Paper* 156, 1930.

BROWN, R.W.: Additions to some fossil floras of the Western United States, *U. S. Geol. Surv. Prof. Paper* 186 J, 1937.

CHANEY, R.W.: A comparative study of the Bridge Creek flora and the modern redwood forest, *Carnegie Inst. Washington Pub.* 349, 1925.

———: The Mascall flora—its distribution and climatic relation, *Carnegie Inst. Washington Pub.* 349, 1925.

———: Paleoecological interpretations of Cenozoic plants in western North America, *Bot. Rev.,* **4,** 1938.

CROOKALL, R.: "Coal Measure Plants," London, 1929.

DARRAH, W.C.: American Carboniferous floras, *Compte rendu 2eme congr. stratr. Carb., Heerlen, 1935,* **1,** 1937.

DORF, E.: A late Tertiary flora from southwestern Idaho, *Carnegie Inst. Washington Pub.,* **476,** 1938.

———: Upper Cretaceous floras of the Rocky Mountain region. I. Stratigraphy and palaeontology of the Fox Hills and lower Medicine Bow formations of southern Wyoming and northwestern Colorado, *Carnegie Inst. Washington Pub.* 508, 1938; II. Flora of the Lance formation at its type locality, Niobrara County, Wyoming, *Carnegie Inst. Washington Pub.* 508, 1942.

GOTHAN, W.: In G. Gürich, "Leitfossilen," Berlin, 1923.

HALLE, T.G.: Paleozoic plants from central Shansi, *Palaeontologia Sinica,* ser. A, fasc. 1, 1927.

HOLLICK, A.: The Upper Cretaceous floras of Alaska, *U. S. Geol. Survey Prof. Paper* 159, 1930.

KNOWLTON, F.H.: "Plants of the Past," Princeton, 1927.

LESQUEREUX, L.: Contributions to the fossil floras of the Western Terretories, pt. I. The Cretaceous flora, *U. S. Geol. Survey Terr. Rept.*, **6**, 1874; pt. II. The Tertiary flora; *ibid.*, **7**, 1878; pt. III. The Cretaceous and Tertiary flora, *ibid.*, **8**, 1883.

MacGINITIE, H.D.: The Trout Creek flora of southeastern Oregon, *Carnegie Inst. Washington Pub.* 416, 1933.

MOORE, R.C. *et al.*: Correlation of Pennsylvanian formations of North America, *Geol. Soc. America Bull.* **55**, 1944.

SEWARD, A.C.: "Plant Life through the Ages," Cambridge, 1931.

WHITE, D.: The stratigraphic succession of the fossil floras of the Pottsville formation in the southern anthracite coal field, Pennsylvania, *U. S. Geol. Survey Ann. Rept.*, **20**, pt. 2, 1900.

————: The characteristics of the fossil plant *Gigantopteris* Schenk, and its occurrence in North America, *U. S. Nat. Mus. Proc.*, **41**, 1912.

————: The flora of the Hermit shale, Grand Canyon, Arizona, *Carnegie Inst. Washington Pub.* 405, 1929.

CHAPTER XV

FOSSIL PLANTS AND ENVIRONMENT

In order to achieve the fullest possible understanding of ancient plant life consistent with the fragmentary fossil record, it is necessary that some attention be given the environment under which the plants grew. Although the problems of ancient environments are difficult and in most instances impossible of more than partial solution, many inferences bearing on probable environments can be drawn from the distribution and structural modifications of ancient plants. Plants are sometimes called the thermometers of the past, which they are when their temperature requirements are known, but with most extinct species these requirements can only be inferred from what is known about the climatic relations of those living species which they most closely resemble. When living species are found in the fossil condition, we assume that they grew under conditions similar to those required by the species at present. Living species, however, decrease in numbers as we delve into the past and very few can be recognized in rocks older than Tertiary. Consequently the same criteria of climatic adaptations cannot always be used for both extinct and living species.

Geological history is punctuated with climatic fluctuations of major proportions. There were times when much of the surface of the earth was covered with ice, and at other times luxuriant vegetation thrived almost at the poles. The most intensive glaciation occurred during the pre-Cambrian, the Permian, and the Pleistocene. We known little of the effect of the pre-Cambrian glaciation upon vegetation because no land plants have been found in the rocks of this era. The Permian ice sheet, however, changed the aspect of the flora over the entire earth through the eradication of the majority of the Paleozoic coal-swamp plants and their replacement by more hardy types. The Pleistocene glaciation caused widespread changes in distribution and composition of modern floras as may be readily seen in the late Tertiary and Recent floras of places where they can be compared.

PALEOZOIC ENVIRONMENTS

Varieties of climatic conditions existed in the long intervals between the periods of major glaciation. By late Devonian time plants apparently adapted to temperate climates had spread to the Arctic. On Bear Island and in Ellesmere Land there are plant beds containing *Archaeopteris*, large lycopods, strange sphenopsids, and others that would appear ill-adapted to prolonged freezing temperatures and a land surface covered with ice throughout most of the year. Coal seams also occur in the fossiliferous strata on Bear Island.

There is evidence that there were seasonal changes during Upper Devonian time comparable to those of temperate North America today. Rings believed to be true annual rings occur in some Upper Devonian woods, the best examples being in *Callixylon erianum* (Fig. 185) from the

Fig. 185.—*Callixylon erianum*. Stem showing well developed growth rings. Upper Devonian. Erie County, New York. × 1¾.

Genesee shale of western New York. In a stem 4.6 cm. in diameter there are five complete rings. Similar layers are found in *C. Newberryi* from the New Albany shale. Additional evidence of mild winter weather is supplied by marks in certain Upper Devonian sandstones and shales believed to be the imprints of ice crystals.

The presence of annual rings in Devonian woods is especially significant in view of the fact that they are often absent in woods of the "Coal Measures." *Dadoxylon romingerianum*, a cordaitean wood from the middle Pennsylvanian of Ohio, shows near uniformity of growth in a radial extent of 5 cm. (Fig. 186). Likewise, *D. douglasense*, from the Virgil series of Kansas, is without true rings although to the unaided eye there are indefinite layers of slightly crushed cells that bear a superficial resemblance to rings. On the other hand, some woods of the Middle Carboniferous show ring development. Specimens of *Cordaites mater-iarum* and *C. recentium* from Kansas have rings varying from 3 to 8 mm. wide. In neither of these, however, are the rings sufficiently strong to suggest anything more than minor climatic variations, probably intervals of moisture deficiency rather than changes in temperature.

Fig. 186.—*Dadoxylon romingerianum.* Secondary wood showing practically uniform growth over a radial distance of about 3.5 centimeters. Middle Pennsylvanian. Coshocton, Ohio. × 2½.

In marked contrast to the fairly uniform growth of trees of the coal period, those which grew during and immediately following the Permian glaciation show strongly developed rings. In a stem of *Dadoxylon* from the Falkland Islands they are as pronounced as in modern trees (Fig. 187).

Investigators have long hoped that an examination of structurally preserved leaves would throw more light on questions of Paleozoic ecology, but the results have not been conclusive. Such features as epidermal outgrowths, size and number of intercellular spaces, arrangement and position of the stomata, cuticularization, and inrolling of the margins of the leaves are known to have definite environmental correlations in living plants but the degree of correlation is often peculiar to a species.   Only under extreme aridity, salinity, exposure to high winds, or other strenuous conditions do any of these features show specialization to the extent of being conspicuous.   The tolerance of structural variations under normal conditions of temperature and moisture is large, and different species in a similar habitat may show considerable variety of structure with respect to the features mentioned.   Most Paleozoic leaves of which the structure is known would apparently survive very well under modern north temperate zone conditions such as prevail in the eastern or southern portions of the United States or along the Pacific

Fig. 187.—Gymnospermous wood showing numerous sharply defined growth rings, Permo-Carboniferous.   Falkland Islands.   × 2½.

Coast as far north as Alaska.   They show no special adaptations to subdued light, high temperatures, or excessive humidity.   Seward makes an appropriate statement: "Broadly speaking we see no indication that these leaves were exposed to any condition of climate other than such as now obtain."

Whether temperate vegetation existed continuously in the Arctic from the Devonian into the Coal Measures is a moot question, but in the Lower Carboniferous (Culm) of Spitzbergen and Northeastern Greenland *Stigmariae* and *Lepidodendron* trunks 40 cm. in diameter are on record.   From Coal Measures time until the middle of the Permian moderate climates with only minor interruptions prevailed.   We know little of Arctic conditions during the late Permian, but judging from what is known of the Southern Hemisphere during these times there is no reason to doubt that temperatures were much lower than during the preceding epochs.

Climates throughout most of the Carboniferous period must have been very favorable for plant growth. That they were moist and rainy can hardly be questioned, but data bearing on possible temperatures are more debatable.

The theory that the Carboniferous coal beds were deposited under tropical conditions was advocated by the German paleobotanist H. Potonié who based his contention on three main arguments. In the first place he believed that the abundance of fernlike plants indicated tropical surroundings because of the prevalence of ferns in Recent tropical forests. His second argument was the apparent absence of annual rings in Carboniferous trees, and his third was the proximity in many Carboniferous plants of the fructifications to the stems.

Concerning the abundance of ferns as indicators of tropical conditions, the mere presence of fernlike plants in the Carboniferous is not in itself a reliable climatic indicator because of the remote relationship between any of the modern ferns and the Carboniferous species. Furthermore, present-day ferns grow in profusion in some places that are not tropical. In New Zealand, for example, fern forests extend to within a mile of the foot of the Franz Josef Glacier and approximately at the same elevation. We have already seen that growth rings are not entirely absent from Carboniferous trees. Their sparse development may indicate a generally equable climate but they are not proof of tropical conditions. With respect to the position of the fructifications of Carboniferous plants which Potonié believed was contributing evidence of tropical climates, we can only say that here again the relationships of the plants taken as examples are too remote from living ones to prove anything. Moreover, we know that in many Carboniferous plants the fructifications were not borne close to the stems, as is shown by the seeds of the pteridosperms or the sporangia of *Asterotheca* and *Ptychocarpus*.

Probably the strongest argument against tropical climates during the formation of the Carboniferous coal beds is the fact that temperate or cool climates at the present time furnish the most suitable environment for the accumulation of peat. Peat does form in tropical surroundings, but not on the large scale found in regions of lower temperatures. Except under extremely favorable circumstances for preservation, plant debris decays rapidly in the tropics, expecially if there is an annual dry season when the surface water evaporates. Extensive peat deposition is more of an indication of high humidity and heavy rainfall than of high temperatures.

Some of the most extensive peat deposits at present are found in the subarctic, examples being the tundras of Siberia and the muskegs of Alaska. In these prevailingly flat but slightly undulating areas the ground is covered to considerable depths with soggy, waterlogged vegetable matter. Drainage is poor, even on sloping ground, and there are numerous small lakes without inlets or outlets. Probably 80 per cent of the plants are mosses belonging to the genus *Sphagnum* which form the spongy substrate, and in these are rooted dwarf flowering plants of the heath family (*Chamaedaphnie*). Underneath the surface layer is a layer of well-formed peat, which may extend in some instances to a depth of 30 feet. Were the land surface to subside and were these tundras and muskegs to become inundated with sediment-bearing waters, they would furnish an ideal situation for the formation of coal beds.

Conditions in the Alaskan muskegs do not approximate those of the Carboniferous coal swamps as concerns vegetation types because all evidence points to heavy forestation of the latter. But they do show that there can be extensive accumulations of plant material in cold climates where the temperatures may be below freezing for much of the year.

There is very little doubt that all the major Carboniferous coal beds were laid down at elevations not far above sea level. This is indicated by the fact that the strata making up the formations lie nearly parallel to each other and that there are frequent layers containing marine faunas. Studies in the large coal-bearing areas of east central North America have revealed rather well defined sedimentation cycles beginning with deposition of sandstone and sandy shales following an uplift succeeded by erosion. Following this initial deposition of coarser sediments there is a long period of stability during which the surface became weathered and drainage became poor, with the consequent formation of underclay (*Stigmaria* clay). Heavy vegetation covered the ground and drainage was further impeded. This resulted in the formation of innumerable swamps varying greatly in extent in which peat accumulated. Further lowering of the land surface allowed flooding of the whole or a large part of the area, and the peat became covered with fine mud bearing much organic matter and with plant fragments from the vegetation that still persisted along the borders of the flooded areas and the higher places. When the last step in the sedimentation cycle began, salt water flowed in and killed the remaining land vegetation. Marine organisms replace land plants in the deposits formed at this stage. Calcium carbonate was deposited from the sea water, first forming calcareous shale and later limestone. At the same time the sea water seeped downward into the underlying peat where the minerals in solution encountered sulphurous gasses issuing from the buried plant material. As a result of this inter-action, we find, both in the coal and above it, coal-balls, bands of iron-stone nodules, phosphatic concretions, and other manifestations of chemical activity. These cycles were not always carried out to comple-tion. In many places the marine stage is lacking with fresh-water deposits succeeding the coal beds. Even the coal itself may be absent if the underclay stage happened to be followed either by elevation and erosion or by immediate inundation.

Although the Carboniferous coal swamps were situated near sea level, the inference is not made that there were no permanently elevated land masses during that time. In fact, the sediments that filled the swamps (aside from the vegetable matter) were mostly derived from higher land. The outlines of these ancient land masses are largely hypothetical, but of major importance in the formation of the coal-bearing strata of Eastern North America was Appalachia, a land mass to the east of the present Appalachian uplift. Likewise, Llanoria supplied a goodly share of the sediments of the central part of the continent, and in central Colorado extensive Pennsylvanian deposits were formed at the foot of the Ancestral Rockies, which occupied the approximate position of the present Front Range. In Europe the great coal fields of Upper Silesia, Westphalia, Belgium, Northern France, and England were laid down to the north of the high Variscan-Armorican Mountains, which during the Carboniferous period extended from Germany through central France and into Ireland.

The effect of these low-lying swamp-filled basins upon the climate of the times was profound. With so much of the land surface near sea level, an "oceanic" climate prevailed over the greater part of those portions of the earth now having "continental" climates. The moisture-laden winds encountering the mountains caused heavy precipitation and rapid erosion, all of which provided suitable environments for copious vegetable growth and ideal situations for preservation. Seasonal changes were moderate over the entire earth and climatic zones were not clearly marked.

Mesozoic and Cenozoic Environments

Climatic conditions during the Triassic, Jurassic, and Cretaceous were on the whole quite favorable for plant growth although there is evidence of moisture deficiency during the early part of the era.

The late Triassic (Rhaetic) of Eastern Greenland has yielded a large flora of lycopods, ferns, pteridosperms (Caytoniales), cycadophytes, ginkgos and conifers, and in the Jurassic rocks of Franz Joseph Land, one of the most northerly outposts for pre-Cenozoic plants, there is a diverse assemblage of ginkgoalean types. A flora containing several cycads, ginkgos, and conifers is also found north of the Arctic Circle at Cape Lisburne in Alaska. Similar plants occur in deposits believed to be of Jurassic age in the Antarctic Continent within 4 deg. of the South Pole.

A phenomenal development of later Arctic plant life is revealed in the Cretaceous rocks on Disko Island off the western coast of Greenland 300 miles north of the Arctic Circle. Of all Arctic floras of all ages this one is probably the most remarkable, as it shows an extensive northerly range of many modern warm-temperate types. The fossiliferous beds consist of shales, coal seams up to 4 feet in thickness, and sandstones all interleaved between massive sheets of basalt. The exact age of the plant beds has not been worked out with the same precision as in some other places, but from all available evidence they are nearly contemporaneous with the Wealden series of Europe. The over-all range is probably from early to late Cretaceous. Among the plants in these rocks are the leaves and fruits of *Artocarpus*, the breadfruit tree, and representatives of the Magnoliaceae, Lauraceae, Leguminosae, Platanaceae, and Fagaceae. In addition, there are several ferns and conifers including *Sequoia*. Such a flora at present could not thrive where the temperatures stayed below freezing for more than a few hours at a time, and although the plants might have been adapted to lower temperatures than their Recent relatives, they could not have existed in a climate such as Greenland has today. Being contemporaneous with or maybe only slightly younger than the Lower Cretaceous Potomac flora of the Eastern United States, it far exceeds the latter in number of types represented and presents strong evidence that the Arctic was probably the original center of dispersal of modern north temperate zone floras.

This luxuriant flora of dicotyledons and other plants at present restricted to lower latitudes persisted in the Arctic into the Eocene, but then we notice a gradual shift southward and their replacement by more hardy species. This recession toward the lower latitudes is well shown by the sequence of fossil floras along the western border of North America. If we select a given region, such as Washington or Oregon, and note the succession, we find in the Eocene rocks an abundance of palms and dicotyledons with large leaves having entire margins. By comparison with recent floras these plants must have endured an annual rainfall of not less than 80 inches and uniform temperatures free from frost. Such a climate now prevails in parts of Southern Mexico and Central America at elevations below 5,500 feet. But the contemporary Eocene floras in Alaska contain maples, birches, poplars, willows, sequoias, and a number of other more northern genera.

In the Miocene of Washington and Oregon the Eocene species are replaced by others suited to less rainfall and lower temperatures. The sequoias have arrived from the north and they are accompanied by alders, tanbark oaks, and other plants, which are at present a part of the redwood association. This migration of floras marks the establishment of *Sequoia* in its present locations.

The Problem of Distribution

Two far-reaching problems facing the paleoecologist are (1) an explanation for the existence of similar floras on opposite sides of wide ocean basins, and (2) a reasonable explanation for the existence at several times during the past of lush vegetable growth in the north- and south-polar regions now covered with ice and snow for the greater part or all of the year. Both of these are closely related and will be discussed together. Two theories have been proposed as possible answers to the first. Land bridges have been postulated as having connected at various times areas now remote from each other. Some of these bridges, if they ever existed, spanned wide ocean basins now thousands of feet deep. There is circumstantial evidence that Europe and North America were connected by a land bridge linking Great Britain with Iceland, Greenland, and Eastern Canada. During the late Paleozoic era the hypothetical land mass called Gondwana land supposedly provided direct connections between South America, Africa, India, and Australia over which the *Glossopteris* flora was able to spread. Land bridges of lesser extent existed during the later periods between Madagascar and Africa and between Siberia and Alaska. It was across the latter that many plants and animals, possibly including man, reached the North American continent during the Cenozoic.

As a counter proposal to the land bridge theory is the theory of continental drift proposed by the late Alfred Wegener. According to Wegener's hypothesis, disjunct land areas containing similar geologic features were once continuous, and have drifted apart as a result of great rifts in the crust of the earth. Thus Gondwanaland consisted of what is now South America, Africa, peninsular India, and Australia lying adjacent to each other as one continent. Similarly North America and Europe were once adjoining land masses, and the separation of the two continents formed the Atlantic Ocean.

Wegener's hypothesis, if it could be proven, would go far toward explaining some of the major problems of plant and animal distribution during the past, and the occurrence of temperate zone fossils in the Arctic would rest on the assumption that the lands bearing these extinct floras have merely drifted to their present northerly locations. On the basis of present knowledge, Wegener's hypothesis or a modification of it offers the only explanation for certain distributional phenomena. On the other hand, many or perhaps the majority of geologists reject the theory as improbable or even fantastic. It is claimed on good authority that the known forces causing deformation of the crust of the earth are wholly inadequate to move continents the distances necessary to satisfy the requirements of the theory, and although it may appear to offer the answer to certain questions, there are others for which it is entirely inappropriate.

In order to explain certain facts of plant distribution Wegener, in collaboration with Köppen, supplemented the theory of continental shifting with the additional assumption that the poles are not permanently fixed, and that they have moved from time to time. This idea, which may seem plausible in light of certain facts, collapses completely when applied to others. The theory of shifting poles prescribes that during Eocene time the North Pole was located in the North Pacific Ocean some 1,500 miles south of the Kenai Peninsula and 2,000 miles west of the present Oregon coast. With the pole situated thus, Spitzbergen is brought well within the north temperate zone, and no climatic hurdles need be overcome in explaining the presence of temperate zone plants in the Eocene rocks of that island. But the theory fails to take into account the actual distribution of Eocene plants, which is circumpolar and not congregated on the Siberian side of the Arctic Ocean. It is true that Spitzbergen represents a slightly more northerly extension of these floras than is to be found in the American Arctic, but to move the pole to the proposed location would subject the subtropical floras that thrived in Oregon during the Eocene to a climate still more unfavorable than they would find in Spitzbergen at the present time. As Chaney says: "In taking care of their own Tertiary forests, certain Europeans have condemned ours to freezing."

If the Wegenerian hypothesis of continental drift fails for lack of sufficient causal forces, and the shifting of the poles is opposed by well-established phytogeographical and astronomical evidence, what are the explanations that can be put forth to account for the repeated advance and recession of floras toward and away from the poles at several places in the geological series? In approaching this problem one is compelled, it seems, to seek first of all a thorough understanding of the causes and controlling factors underlying present climatic conditions over the earth. Foremost is the amount of heat received from the sun. An increase in the amount of radiant energy from the sun would result in a warmer earth, but an increase sufficient to bring about temperate conditions in the Arctic would render life impossible in the tropics. But we know that the effect of solar radiation is modified on a large scale by convection currents, which in turn are influenced by size and elevation of land masses and the direction of ocean currents. Alterations of any of these might result in far-reaching climatic changes. It has been calculated that the lowering of the mean annual temperature by only a few degrees Fahrenheit over the Northern Hemisphere would be sufficient to bring another ice age over most of North America, and that a corresponding rise would cause most of the Arctic ice to melt. Melting of great continental ice masses causes ocean levels to rise with a corresponding reduction in continental size and elevation. Both of these in turn tend to bring about further climatic amelioriation. The noticeable recession of the glaciers on the mountain slopes of Western North America indicates that our present climate is changing and that the next geologic epoch may herald another invasion of temperate zone plants into the Arctic. Our present climate, which because of a certain amount of inevitable bias we are inclined to regard as a "normal" climate, might not be the prevailing one were the heat received from the sun the undisturbed governing factor. With more uniform elevation of land surfaces, and a greater equality in distribution of land and water, climatic differences between high and low altitudes would unquestionably be less pronounced than they are at present on this earth of hugh oceans and continents and lofty mountain ranges. The Carboniferous coal-swamp climate with its equable temperatures and heavy precipitation was probably the nearest approach to a normal climate the world has ever seen. Where global climates are concerned, a very delicate balance exists between causal factors and effects, and slight changes in topography or shape or elevation of the dominant land masses may be expected to have far-reaching consequences.

References

ANTEVS, E.: The climatological significance of annual rings in fossil wood, *Am. Jour. Sci.*, **209**, 1925.

BERRY, E.W.: The past climate of the north polar regions, *Smithsonian Misc. Coll.*, **82**, 1930.

CHANEY, R.W.: "The Succession and Distribution of Cenozoic Floras around the Northern Pacific Basin, University California Press, Berkeley, 1936.

————: Tertiary forests and continental history, *Geol. Soc. America Bull.*, **51**, 1940.

COLEMAN, A.P.: Late Paleozoic climates, *Am. Jour. Sci.*, **209**, 1925.

GOLDRING, W.: Annual rings of growth in Carboniferous wood, *Bot. Gazette*, **72**, 1921.

HUNTINGTON, E.: Tree growth and climatic interpretations, *Carnegie Inst. Washington, Pub.* **352**, 1925.

NATHORST, A.G.: On the value of fossil floras of the Arctic regions as evidence of geological climates, *Geol. Mag.*, series 5, **8**, 1911.

SCHUCHERT, C.: Winters in the Upper Devonian, *Am. Jour. Sci.*, **213**, 1927.

SEWARD, A.C.: "Fossil Plants as Tests of Climate," London, 1882.

————: The Cretaceous plant-bearing rocks of Western Greenland, *Royal Soc. London Philos. Trans.*, **215**, 1926.

————: "Plant Life through the Ages," Cambridge, 1931.

STUTZER, O., and A.C. NOÉ: "Geology of Coal," Chicago, 1940.

WELLER, J.M.: Cyclical sedimentation of the Pennsylvanian period and its significance, *Jour. Geology*, **38**, 1930.

CHAPTER XVI

PALEOBOTANICAL SYSTEMATICS

The names of plants and animals have made up a large part of man's vocabulary as long as he has used language. He, however, uses many languages, but because science transcends racial lines and national boundaries it is necessary for users of scientific terminology to have some medium of common understanding. In chemistry this result is achieved by the use of standardized symbols for the elements, symbols that have the same meaning in any tongue. Biologists have attained this end by giving Latin and Greek names to all species of plants and animals, and in this way the confusion and lack of understanding that would result from the exclusive use of local or vernacular names is avoided. Language standardization, however, is only a partial solution to the intricate problems that arise, so in order that the greatest possible uniformity be brought about in the naming of plants and animals, they are named according to standardized rules which have been generally agreed upon. The rules used by botanists are different from those of the zoologists, and although the general plans of procedure outlined by both sets of rules are similar they differ in numerous details so that those pertaining to one group of organisms cannot be applied to the other.

Rules of Nomenclature.—Plants, both living and fossil, are named according to the *International Rules of Botanical Nomenclature*, which were drawn up at the International Botanical Congress held in Vienna in 1905. These rules constitute the *Vienna Code*, which was founded upon an older code formulated in Paris in 1867. The Vienna Code was modified at Brussels in 1910, at Cambridge in 1930, and again at Amsterdam in 1935. These modifications do not alter the basic scheme of the code, but consist of amendments, additions, and special provisions found necessary due to the expansion of botanical science over this period. The Vienna rules furnish the essential groundwork for the naming of fossil plants, but the peculiar circumstances brought about through fossilization have made certain modifications necessary. Difficulties in strict application of the rules still exist, however, and it is expected that further changes will be made in the future. Violation of the International Rules by paleobotanical authors almost always causes confusion. Some authors, particularly those trained primarily in animal paleontology, are prone to follow the procedure outlined for zoologists. In some cases editors of geological or paleontological journals who are unfamiliar with botanical rules insist on adherence to editorial policies that are not in accord with botanical practice. To such editorial insistence the author would strenuously object.

It is not the purpose of this chapter to describe the International Rules, but rather to point out some of the special provisions that have been made for fossil plants, and to discuss some of the problems that frequently arise in connection with the application of the rules to them.

The Species.—The unit in classification is the *species*, which is expressed in taxonomic language as the second term of a binomial. The implications of the term "species" are relative and are not absolute entities. For example, the French botanist, de Jussieu, defined a species as "the perennial succession of similar individuals perpetuated by generation." Pool has recently stated the same idea in the following words: "A group containing all the individuals of a particular kind of plant that exist now or that existed in the past, no matter where they were or where they may now be found." In the first definition the "similar individuals," and in the second "the individuals of a particular kind" are the variables. Under the ancient dogma of special creation whereby every "species" was supposed to have been designed independently of all others, these definitions would have presented no special difficulties, but because of the multitudinous variations that are constantly arising in the biological world due to hybridization, the effect of environment, and genetic factors of various kinds, and also to the fact that some of these variations are inherited and others are not, the rigid boundaries that species were once supposed to possess do not exist in many cases. For this reason it has become customary to speak of the "species concept," which of course is a simple admission of the fact that a species cannot be arbitrarily defined, and gives it whatever latitude is necessary.

In paleobotany as in modern botany the species name has the primary purpose of serving as a title for reference. The title may be used either to identify an individual with the larger group of which it is a part, or in a more inclusive way to designate the whole group of similar individuals as a unit without reference to any particular individual. For example, one may say, "This is a specimen of *Neuropteris Schlehani*," for the purpose of identifying the individual, or "*Neuropteris Schlehani* ranges from the Millstone Grit to the Yorkian," thereby referring not to one individual but to the whole group. A plant binomial, therefore, is not a name like *John Smith* to which the individual has exclusive claim. John Smith's taxonomic name is *Homo sapiens*.

According to the species concept those individuals which possess certain demonstrable characters in common are grouped together. Just which of the characters possessed by the individual are to be considered as pertinent to the species is a matter of which the investigator alone is the judge, and this is where the human factor enters strongly into the matter of specific delimitations. There are those, the so-called "splitters," who would segregate into separate species individuals showing the most minute differences. Such procedure, if carried to its logical extreme, would ultimately result in a separate species name for each individual. The "lumpers," on the other hand, are so conscious of the resemblances between organisms that they are reluctant to distinguish separate species except on the basis of very pronounced differences. The "lumpers'" species, therefore, tend to be large and unwieldy, and the descriptions are often in general terms lacking precision. Usually the most enduring results are produced by an intermediate course that avoids extremes. The "splitters" accuse the "lumpers" of being oblivious to the differences between organisms, and the latter charge the former with being hypermeticulous. The point of view of the "lumper" is often most favored by cataloguers, textbook writers, and others who have occasion to make practical use of the results, because its effect is to reduce the number of names to be used and it tends toward a simpler classification. The evils attending excessive "lumping" and "splitting" may be appreciated in a situation where some groups of organisms of two successive geological horizons are being monographed, one by a "lumper" and the other by a "splitter." The combined results might give an entirely erroneous picture of the evolution of that group throughout those two periods. The effects of this are probably more noticeable in animal paleontology, where large numbers of invertebrate forms are described, which reveal obvious evolutionary trends from one age to the next.

Some Guiding Principles.—Although it is neither possible nor desirable to eliminate the element of personal judgment from paleobotanical nomenclature, the difficulties that often arise from it would be materially lessened were authors always conscious of the benefits to be gained by adherence to certain principles, some of which are as follows:

1. Taxonomic writers should cultivate a sympathetic attitude toward the expressed and implied provisions of the rules as they have been set forth. The sole purpose of the rules is to produce uniformity and to prevent confusion. Failure to comply with them is usually the outcome of ignorance or indifference but sometimes of prejudice. Because the rules are drawn up at international gatherings of botanists they tend to be formal and arbitrary, and special cases often develop from one's individual research wherein the exact procedure to be followed may not

be explicitly prescribed. Under other circumstances the investigator may feel that strict adherence to the rules is pedantic and that he is justified in setting some particular provision aside. Intentional violation can hardly be condoned under any circumstances, but there may be times when the individual's best judgment may dictate a course seemingly at variance with them. Under such conditions the author should explain the situation fully and give the reasons for the procedure he chooses to follow. In this way readers are less likely to be misled as to his ultimate intentions than if no explanation is offered for what may seem to be a taxonomic anomaly. On the other hand one should always bear in mind that names and diagnoses not drawn up as specified are subject to abrogation by subsequent writers.

2. Characters used to distinguish genera and species should preferably be those which are revealed through the simplest techniques. Within recent years studies of the cutinized epidermis have revealed important facts concerning the affinities of certain fossil leaf types, especially those of the cycads and conifers. But in making routine field and laboratory determinations it is seldom possible to examine the cuticle, and in fact it is rare that this structure is sufficiently preserved for examination at all. While epidermal fragments may sometimes appear to furnish a sound basis for taxonomic categories, in many plants preservation imposes a limiting factor. However, spores, epidermal fragments, and other microscopic objects obtained by maceration may be classified and named according to whatever scheme may seem most desirable.

3. Specimens so fragmentary or so poorly preserved that their essential morphological features are undeterminable should not be given new names. The older literature is replete with figures of casts lacking diagnostic surface markings, bark imprints, and compressions of stem and leaf fragments, which can be neither adequately described nor satisfactorily compared with organs with known affinities. To describe an object as new merely because it is too poorly preserved to permit identification serves no useful purpose.

4. New proposals (genera and species) should be made to fit, if at all possible, into existing groups. New genera or species should not be created unless there are good reasons for doing so. Taxonomic literature contains hundreds of generic and specific names which are unnecessary and which are never used. However, if inclusion of new material in existing genera and species would expand them beyond reasonable limits, one should not hesitate to institute new groups. On the other hand, a genus or species should never be broken simply because it is large.

5. Seldom should one attempt the revision of an entire group of plants (for example a genus or family) unless he has access to or is familiar

with the greater part of the material which has been previously assigned to that group. A single museum collection, no matter how well the specimens are preserved, is seldom adequate to make one an authority on the whole group, especially if it is a large one. Special reticence should be exercised when it comes to passing judgment on species not well described and figured when the original specimens have not been examined. Newell Arber's attempt to revise the genus *Psygmophyllum* is a concrete example. After examining a few specimens he concluded that the leaves are spirally arranged in all members of the genus, and he further proposed to include within it *Archaeopteris obtusa*, known to him only from the rather crude figures in Lesquereux's "Coal Flora." But had he been able to examine the original specimen of *A. obtusa*, he would have seen that the leaflets are in two ranks. Consequently his revision fails completely with respect to this essential point. A much abused privilege is that of reducing names to synonymy on cursory evidence, especially when proposed in foreign countries or applied to specimens not well figured.

Naming Fossil Plants.—The basic purpose in naming a fossil plant is the same as when naming a living plant, and that is to supply a universal and generally applicable means of reference. Other than being merely a name, the binomial should preferably possess a definite meaning and should be so constructed that it can be translated directly into a phrase in some modern tongue. For this reason it is important when constructing binomials to heed such grammatical attributes as number, gender, and case of the names to be used. Technical names are written in either Greek or Latin, with care being taken that the root word and prefixes and suffixes are from the same language. An example of a "word hybrid" is *Protoarticulatae*, which was intended to designate certain ancient representatives of the scouring rush class, and which is a combination of the Greek *protos* and the Latin *articulatus*. It is of course permissible to construct generic and specific names from words unrelated to Latin or Greek if Latin endings can be attached. Thus we have generic names such as *Dawsonites* and *Zalesskya*, and specific epithets as *Smithii* and *Knowltonii*. The generic name is always a noun in the singular written with a capital letter, and usually denotes some prominent feature of the plant or some person associated with it. The specific name may be an adjective agreeing in gender with the generic name (that is, *Sequoia magnifica*), a noun in apposition (that is, *Cornus Mas*), or a noun in the genitive case (that is, *Sequoia Langsdorfii*). The name is usually descriptive, or refers to some person, place, or (in case of a fossil plant) geologic formation associated in some way with the species. However, names which are anagrams or merely meaningless combinations of letters are permissible under the rules. Largely as a matter of euphony place names that do not harmonize well with Latin suffixes should not be utilized. Examples are *Aristolochia whitebirdensis, Larix churchbridgensis,* and *Salix wildcatensis.* There is no way of unburdening the literature of such linguistic incongruities once the names are validly published and can claim priority.

Fossil plants are often named from their genuine or fancied resemblance to some living plant. Thus we have as examples of genera *Prepinus, Acerites, Ginkgophyllum, Cupressinoxylon,* and *Lycopodites,* and specific designations such as *Populus prefremontii, Pinus monticolensis,* and *Quercus pliopalmeri.* In the early days of paleobotany many fossil plants were given names in allusion to their resemblance to recent plants, which today we find rather awkward to use. *Prototaxites,* for example, is not a primitive member of the Taxaceae, but a giant organism more closely related to the Algae. Also *Drepanophycus* is a lycopodiaceous plant, and *Yuccites* is a gymnosperm. More recently a leaf fragment supposedly allied to the red oaks and named *Quercus precoccinea* was found to be a single leaflet of the digitately compound leaf of *Oreopanax.* Therefore, unless affinities can be established beyond all reasonable doubt, specific and generic names indicating relationships should be avoided. The rules contain no provision for setting aside a name simply because it later proves to be inappropriate. Inconsistencies also develop with the use of formational or period names. For example, *Populus eotremuloides* is slightly incongruous because the Payette formation from which it was originally secured is Miocene and not Eocene as once supposed. Likewise the unsuitability of the specific name *devonicus* for a Carboniferous plant is evident. It is impossible to avoid entirely situations such as those described above, but when there is the slightest chance of error in affinities or age, names of a more noncommital nature should be chosen. Names like *Asteroxylon* or *Lepidodendron* avoid this difficulty because they are purely descriptive. Places often furnish suitable roots for generic names, examples being *Caytonia, Rhynia, Thursophyton,* and *Hostimella.* Personal names are as often applicable for genera as for species. Some familiar examples are *Williamsonia, Baragwanathia, Zeilleria, Nathorstia,* and *Wielandiella.*

The Principle of Priority.—The principle of priority stipulates that the oldest published name of a plant is its valid name.[1] However, to avoid the necessity of searching through the literature of centuries in

[1] An exception is the list of *Nomina Conservanda.* Names which have been in general use for 50 years following their publication and which have been officially placed on this list must be retained as the valid names even though they cannot claim priority

search of the oldest name of a plant, dates have been arbitrarily set beyond which it is not necessary to go. Different dates have been set for different groups of plants. The starting point of the nomenclature of vascular plants is 1753, at which date Linnaeus's "Species Plantarum" was published. Since, however, very few fossil plants had been named by 1753, a later date, 1820, has been selected as the starting point for the naming of fossil plants of all groups. This date marks the initial appearance of the important works by Schlotheim and Sternberg in which many specific names in use today were proposed.

The temptation to set aside the principle of priority is sometimes strong, especially in the case of generic and specific names which were inappropriately chosen. Many authors, for example, insist on the name *Nematophyton* instead of *Prototaxites*, on grounds which are entirely reasonable but for the fact that the latter name has priority. A name can be set aside only if it was invalidly published.

To have a name set aside simply because the author who proposed it was mistaken concerning its affinities would lead only to confusion. Moreover, many names that have become established through long usage and are now accepted without the slightest objection by everyone were originally based upon erroneous interpretations of affinities. An example is the name *Calamites*, which was originally proposed because the conspicuously noded pith casts were thought to be fossil bamboo stems. It must be understood that investigators of fossil plants are not always able to place their material in its proper position in the plant kingdom. Preservation is not always sufficient to reveal the real affinities of a plant. Moreover, as our knowledge increases, our concepts of the limits of plant groupings and the constituents of these groups change accordingly.

There is as yet no list of *nomina conservanda* (conserved names) for fossil plants.

Generic Assignments.—Most of the problems peculiar to paleobotanical nomenclature, for which special provisions have been made in the rules, arise from the disassociated and fragmentary condition of the plant remains as they are usually found. For this reason it is necessary to assign separate generic names to the different plant parts. For example, the strobili of a well-known Paleozoic lycopod are known as *Lepidostrobus*, the detached sporophylls are called *Lepidostrobophyllum;* the sporangia, *Lepidocystis;* the spores, *Triletes;* the leaves, *Lepidophyllum;* the roots, *Stigmaria;* and the trunk, *Lepidodendron.* There are numerous instances where it is necessary to use a different name for similar parts that are differently preserved. For example, *Lagenospermum* is sometimes used for compressions of the seeds of *Lyginopteris*, which if petrified are called *Lagenostoma.* Different names are sometimes given to the surface and the interior of a trunk, for example, *Megaphyton* and *Psaronius*. Although to some botanists such taxonomic procedure may seem irksome or even positively ludicrous, it is nevertheless born of necessity. Without it concise designation of fossil plants would be impossible. In paleobotany, names are usually given to parts rather than to whole plants.

At the Sixth International Botanical Congress held in Amsterdam in 1935, a number of proposals were made which, although they have not been formally written into the rules, have been followed by some paleobotanists. Some of them will without doubt become standardized practice. One of these proposals is the creation of *organ, artificial*, and *combination* genera. An organ genus embraces detached parts belonging to the same morphological categories, such as cones, leaves, seeds, stems, roots, etc. If a number of parts from genetically unrelated plants are to be grouped, the assemblage is an *artificial* (or *form*) genus. Most of the genera of fossil plants are therefore organ or artificial genera. *Lepidostrobus* is an organ genus for the strobili belonging to the trunk *Lepidodendron*. *Trigonocarpus* is the seed belonging to *Medullosa*. In both of these examples the names apply to organs of related plants. *Carpolithus*, on the other hand, is an artificial genus because it is a name given to almost any fossil seed that cannot be definitely placed. It may include seeds belonging to several groups. The generic names coined by Brongniart and later workers for the foliage of fernlike plants are excellent examples of artificial genera. Some of them, it is true, do represent to a large extent natural groupings, but no consideration of affinity need be made in assigning species to them. All fronds, for example, that conform to the original description of *Alethopteris* probably belong to the Medullosaceae, but if a leaf of some group other than the Medullosaceae should conform to *Alethopteris* as it is defined, it could correctly be assigned to that generic category.

It frequently happens that a frond belonging to an artificial genus previously known only in the sterile condition may be found bearing fructifications having another generic name. Such specimens could be given a combination generic name, although the usual practice, especially when citing synonyms, is to quote one of the generic names followed by the other in parentheses, as for example, *Pecopteris* (*Dactylotheca*) *plumosa*.

In practice the artificial genus is applicable mainly to such objects as epidermal fragments, isolated spores and pollen grains, pieces of petrified secondary wood, stem compressions lacking diagnostic features, seed casts, and some foliage. It is seldom that a plant part, such as a complete stem, a seed or other fructification, or a well-preserved leaf that is sufficiently complete to permit a detailed description, cannot be assigned with some degree of certainty to one of the larger categories such as an order or class, and hence to an organ genus. It would seem advisable in future paleobotanical practice to restrict the multiplicity of artificial genus names as much as possible and to give a broad application to a few. *Carpolithus* for seeds, *Phyllites* and *Dicotylophyllum* for leaves, *Antholithus* for strobiloid inflorescences, *Pollenites* and *Sporites* for pollen and spores, *Dadoxylon* and *Araucarioxylon* for gymnospermous secondary wood, along with a few others, would take care of most of the requirements of material of unknown affinities.

When, as occasionally happens, organs which have been described separately are found to be parts of the same kind of plant, the question arises as to which name should be applied to the combination. The whole plant obviously is an organism of which only the separate parts have heretofore received generic designations. The usual practice under such circumstances has been to retain the oldest valid name that applies to any of the parts. For example, the stem *Lyginopteris oldhamia*, the leaf *Sphenopteris Hoeninghausi*, and the seed *Lagenostoma Lomaxi* were found to belong together, but further research finally showed that the cupules or husks within which the seeds were borne had long been known as *Calymmatotheca*. *Hoeninghausi*, however, is the oldest valid specific name for any of the parts (in this instance the frond), so *Calymmatotheca Hoeninghausi* is now accepted as the correct name for the plant. There are certain obvious disadvantages in using the oldest name for a combination of parts because it may be *Carpolithus* or *Dictyoxylon*, or some other name of very wide application, and to be compelled to replace an otherwise appropriate name with some outmoded or unsuitable designation is not only objectionable in itself but is often confusing. A proposal made at Amsterdam in 1935 provides that an author may at his discretion create a new genus, a *combination genus*, for two or more parts shown to be in organic union. A recently proposed combination genus is *Lebachia* for the gymnosperm that bore the foliage long known as *Walchia*. In this instance the plant receiving the combination genus name consists of the foliage *Walchia*, the seed cone *Walchiostrobus*, a seed which had been included in *Carpolithus*, and pollen referred to *Pollenites*. Another combination genus is the Mesozoic cycad *Bjuvia*, which is a combination of the leaf *Taeniopteris* and a megasporophyll known both as *Cycadospadix* and *Palaeocycas*.

For some plants such as *Lepidodendron* and *Sigillaria* combination generic names have never been proposed because the names used for the trunks have, from long usage, become accepted as applicable to the entire plants. *Rhynia* is an example of a generic name that from its first institution was applied to an entire plant. Detached parts would also be assigned to this genus because no other names have been proposed for them.   The name *Rhynia* has been applied to plants with the internal structure preserved, but if the question were to arise concerning the appropriate name for compressions resembling *Rhynia*, which could not be proved to belong to this genus, then a new form genus would need to be created for them if no suitable one were already in existence.

It should always be borne in mind that the purpose of rules is to secure uniformity and to prevent confusion.   The successful application often depends to a considerable extent upon the judgment of the author.

Valid Publication.—A new genus or species of fossil plants is not validly published unless the description is accompanied by illustrations or figures showing the essential characters.   Although the rules require a Latin diagnosis for Recent plants, such is not necessary for fossils. The description may be in any language although English, French, or German is preferable.   Russian authors usually add summaries in one of these languages.   The date at which the new genus or species is considered published, if after December 31, 1911, is the date of the simultaneous publication of the description and figure, or if these are published on different dates, the later of the two is the publication date.   New names, therefore, which are not accompanied by both a description and a figure are not validly published and have no claim to priority, and consequently may be disregarded by subsequent authors if they so prefer.   If the subsequent author adds a figure or description to a genus or species, which because of lack of it was previously invalid, the name then becomes valid and he is regarded as the legitimate coauthor.

A matter that authors of new genera and families sometimes neglect is to assign their new creations definitely to higher categories.   The rules explicitly state that every species belongs to a genus, every genus to a family, every family to an order, every order to a class, and every class to a division; but we occasionally encounter new genera which are not referred to families, and families which are orphaned from orders. These higher categories may need to be specially created to accomodate the newly proposed families or genera and they serve the essential purpose of completing the classification.   In case of artifical genera, however, it may be necessary to make special applications of this rule because the members of such genera are unrelated and may belong to different taxonomic groups.   One possibility is to group such genera into a sort of combined "ordinal family" bearing a special ending.   For example the fernlike leaf genera that are sometimes grouped into Archaeopterides, Pecopterides, Alethopterides, etc., are sometimes classified together under the common heading of Pteridophyllae.   Since the Pteridophyllae include both ferns and seed plants, they cannot be referred to any natural group smaller than the Pteropsida. Another common practice in class-ifying artificial genera is to place them under the category of *Incertae Sedis*. This should be avoided whenever possible, and the only advantage in even describing such questionable material is to provide it with a ready means of reference.

Special Problems.—In the naming of species many special problems often arise. It is often difficult to decide whether a specimen (or series of specimens) should be described as new, or whether it should be referred to existing species from which it may differ only in relatively insignificant points. Many species have been so briefly described and crudely illus-trated that it is impossible to compare new material satisfactorily with them. Under such circumstances, the types should be examined if possible. In naming leaf impressions of Upper Cretaceous and Tertiary age it is usually necessary to compare them with their nearest living equivalents, and in doing so attention should be given to the range of variation of leaves of the Recent species. Also the present geographic range and the natural associates of the Recent species should be heeded. One should be cautious about referring a fossil leaf to a genus of Recent plants if the fossil is associated with other genera, the living equivalents of which at present occupy very different habitats. On the other hand, the similarity of floral associations of the past and present is often unknown, and becomes increasingly obscure in older rocks.

The extent to which artificial genera should be used for Upper Cre-taceous and Tertiary leaf impressions of the flowering plants depends mostly upon the judgment of the author. Some have advocated the use of names of living genera only when the determinations can be supported by accompanying fructifications. Thus *Acer* would be applied to leaves associated with the characteristic winged fruits, but *Acerites* or *Acero-phyllum* to leaves not so accompanied. However, the ultimate advan-tages of such nomenclatorial quibbling are doubtful. The basic problem before the investigator is to determine as correctly as possible the affin-ities of the remains, and the naming of them is of secondary importance. The value of a name can be no greater than the accuracy with which the plant is determined.

Another problem often confronting the student of fossil plants is the relation of a species to stratigraphic boundaries. Should a form found in abundance in a certain formation be identified as the same species when it appears in another formation of a different age, or should it receive a new name even though it shows no distinctive morphological characteristics? No set rules of procedure can be dictated to cover such instances. It may depend upon how well the other species has been described, and its known range must also be taken into consideration. In general, less confusion will result when organisms from distinct formations are differently named, even at the risk of having one name later reduced to synonymy, than when too many variants are lumped into one species. A group soon becomes unwieldy when too many species are indiscriminately assigned to it. Also it is well to remember that a formation may represent a time sequence of several million years, often longer than the probable duration of a species. A species, however, is a taxonomic category, and only upon morphological characters can it be permanently established.

Nomenclatorial Types.—The International Rules specify that the application of names is determined by *nomenclatorial types*. A generic type is one of the species that was included within the genus when it was originally published, and a specific type is a specimen used in the original description of the species. The type specimen determines the kind of plant to which the name applies, and no other kind can validly carry that name. It is common practice to label the type specimen and to give whatever information may be necessary in locating it at future dates. If it is deposited in some herbarium or museum, the name of the institution is given, along with file or catalogue numbers if such are kept. For living plants the type is usually a specimen mounted upon a herbarium sheet. In the case of plants that cannot be mounted or satisfactorily preserved, such as aquatic microorganisms or fleshy fungi, a drawing or water-color sketch made directly from a specimen is retained as the type. With fossils, however, more than one specimen frequently enters into the specific description. In describing angiosperm leaf compressions, for example, it may be necessary to include in the description those specimens which are believed to represent natural variants within the species, and since no one leaf will show all the characters several have to be designated as types. It may be possible to select one specimen which represents an average form from which the variants will depart to a greater or less degree. This one may if desired be designated the *holotype,* and the normal variants as *paratypes.* It is not necessary to accompany a holotype with paratypes if the holotype adequately expresses the species. If the author does not feel inclined to select any particular specimen for a holotype, he may give them all equal rating as *syntypes.* When working with living plants, different type categories are seldom employed because the single specimen selected may show enough characters to express the species adequately, but with fossils one usually describes not complete individuals, but parts of individuals, and the assumption is that the individual is expressed in the sum total of the detached organs. In designating syntypes or paratypes only specimens from the same formation and locality should be included, and it is preferable that they all be in very intimate association.

A question may arise in connection with type designations when the descriptions are based upon thin sections of petrified plant parts. When studying stems or other organs in thin section, it is customary to prepare several sections along different planes, and as no one section will show all the characters necessary for the description, it is impossible to designate any individual section as the type. In such a case the fact must be recognized that the specimen from which the section was taken is the type, and the section should be merely designated as a section of the holotype or of the syntype, whichever it may be. Care should be taken, however, not to mix sections of different specimens under a single type designation.

The procedure in specifying types of fossil plants is obviously more complex than for living plants because of the disassociation of the parts. In many museums where the fossil plants are housed along with animal fossils it is the practice to follow prescribed type designations. Paleontologists employ an elaborate system of classification, which in simplified form may be used to advantage with plant fossils. There are many type categories, but the essential ones are as follows:

Generitype—the type species of the genus. The individual specimen that serves for the type of the generitype species may be called a *Diplotype*.

Holotype—a single specimen (or fragment) upon which a species is based.

Syntype—any specimen from the original collection upon which a species is based if no holotype is designated.

Paratype—any specimen other than the holotype from the collection upon which the species is based.

Lectotype—a specimen selected from among the syntypes and used for a revision of the species.

Neotype—a specimen from the original locality to replace the holotype if the latter is lost or destroyed.

All the above may be considered *primary types* in that they are concerned with the original definition of the genus or species. There are other types designed to supplement or embellish the original descriptions. A few of these are:

Hypotype—any described, figured, or listed specimen of a species.

Topotype—a specimen from the type locality of a species. If it becomes a neotype, it replaces the holotype.

Plastotype—a cast or exact replica of a type. Thus there may be a *plastoholotype, plastoparatype*, etc.

The importance of caring for type specimens cannot be overemphasized, and in any collection they should be carefully catalogued and stored in special places provided for them. The final status of a species is sometimes dependent upon the accessibility of the type for reexamination.

References

ARKELL, W.J., and J.A. MOY-THOMAS: Paleontology and the Taxonomic Problem. "The New Systematics" (Julian Huxley, editor), pp. 395–410, Oxford, 1940.

BAILEY, L.H.: "How Plants Get Their Names," New York, 1933.

FRIZZELL, D.L.: The terminology of types, *Am. Midland Naturalist*, **14,** 1933.

GRAHAM, R.: Suggestions regarding the taxonomy and nomenclature of Cretaceous and Tertiary plants, *Jour. Paleontology*, **13,** 1939.

HITCHCOCK, A.S.: "Methods of descriptive systematic botany," New York, 1925.

JONGMANS, W., T.G. HALLE, and W. GOTHAN: "Proposed Additions to the International Rules of Botanical Nomenclature Adopted by the Fifth International Botanical Congress, Cambridge, 1930," Heerlen, 1935.

RENDLE, A.B. (editor): International Rules of Botanical Nomenclature adopted by the Fifth International Botanical Congress, Cambridge, 1930, *Jour. Botany*, **72** (Supplement), 1934.

SCHENK, E.T., and J.H. McMASTERS, "Procedure in Taxonomy," Stanford University 1935.

INDEX[1]

---

[1] Boldface figures indicate that there is an illustration on the page referred to.